LONDON MATHEMATICAL SOCIETY LECTURE NOTE SERIES

Managing Editor: Professor J.W.S. Cassels, Department of Pure Mathematics and Mathematical Statistics, University of Cambridge, 16 Mill Lane, Cambridge CB2 1SB, England

The titles below are available from booksellers, or, in case of difficulty, from Cambridge University Press.

London Mathematical Society Lecture Note Series. 253

The *q*-Schur Algebra

S. Donkin
Queen Mary and Westfield College, London

CAMBRIDGE
UNIVERSITY PRESS

CAMBRIDGE UNIVERSITY PRESS
Cambridge, New York, Melbourne, Madrid, Cape Town, Singapore, São Paulo

Cambridge University Press
The Edinburgh Building, Cambridge CB2 2RU, UK

Published in the United States of America by Cambridge University Press, New York

www.cambridge.org
Information on this title: www.cambridge.org/9780521645584

First published 1998

A catalogue record for this publication is available from the British Library

ISBN-13 978-0-521-64558-4 paperback
ISBN-10 0-521-64558-1 paperback

Transferred to digital printing 2006

I dislike arguments of any kind.
They are always vulgar, and often convincing.

Oscar Wilde, *The Importance of Being Earnest*

Contents

Preface

These notes are concerned with the representation theories of the quantum general linear groups, the q-Schur algebras and the Hecke algebras of type A, and, most importantly, the relationships between these theories. Roughly speaking, we are presenting a q-analogue of the monograph by J. A. Green, [51], which treats the representation theory of general linear groups, Schur algebras and symmetric groups. The theory developed here is a generalization of the classical case (which one recovers by putting $q = 1$). But the main difference between [51] and this text is that whereas Green's approach is combinatorial ours is (for the most part) homological, informed (or made possible even) by Kempf's vanishing theorem (for quantum GL_n). This approach is a continuation of one developed in our earlier papers on Schur algebras and related algebras, [31], [32], [33], [34], [35].

The version of quantum GL_n that we shall use is the one introduced by R. Dipper and the author. The q-Schur algebras were introduced by Dipper and James, [21], as endomorphism algebras of certain modules over Hecke algebras. Mostly, we work over an arbitrary field k and q is an element of k, which is usually required to be non-zero. For a positive integer n and a non-negative integer r, we have the Schur algebra $S_q(n, r)$. In the approach taken here (modelled on the treatment of ordinary Schur algebras by Green, [51]) $S_q(n, r)$ is constructed as the dual algebra of a certain coalgebra. More precisely, the coordinate algebra $k[G_q(n)]$ of the quantum general linear group $G_q(n)$ of degree n is generated by "coefficient functions" c_{ij}, $1 \leq i, j \leq n$, and the inverse of the quantum determinant. Thus the (in general non-commutative) algebra $A_q(n)$ generated by the c_{ij} is a subbialgebra of $k[G_q(n)]$, and $A_q(n)$ has an algebra grading and coalgebra decomposition $A_q(n) = \bigoplus_{r=0}^{\infty} A_q(n, r)$ in which each c_{ij} has degree 1. The coalgebra $A_q(n, r)$ has dimension $\binom{n^2+r-1}{r}$ and its dual algebra is $S_q(n, r)$. A module for $S_q(n, r)$ is naturally an $A_q(n, r)$-comodule and hence a $k[G_q(n)]$-comodule, i.e. a module for the quantum group $G_q(n)$. Thus the category of $S_q(n, r)$-modules is naturally embedded as a full subcategory of the category of $G_q(n)$-modules, namely the category of $G_q(n)$-modules which are polynomial of degree r.

Suppose now that $r \leq n$. Then there is a distinguished idempotent $e \in S_q(n, r)$ such that $e S_q(n, r) e$ is the Hecke algebra $\mathrm{Hec}(r)$ defined by the symmetric group $\mathrm{Sym}(r)$ of degree r (regarded as a Coxeter group). Thus one has, as in the classical case $q = 1$ (see Green, [51, Chapter 6]), the Schur functor from the category of $S_q(n, r)$-modules to the category of $\mathrm{Hec}(r)$-modules, taking an $S_q(n, r)$-module V to the subspace eV, viewed as a module for $\mathrm{Hec}(r) = e S_q(n, r) e$. Our philosophy here is to proceed uniformly in this direction: that is, to first prove results about our quantum version of GL_n, then to use this knowledge to deduce results about the q-Schur algebras and finally, by a further "descent", to obtain results on the Hecke algebra. Our purpose then is twofold: not only to present new results but also to rederive known results by these descents. Perhaps an extreme example of the latter, given in Section 4.3, is the determination of the labelling of irreducible modules for Hecke algebras, first obtained by Dipper and James, [20], by decomposing $E^{\otimes r}$, the rth tensor power of the natural module for quantum GL_n, viewed as a tilting module. An exception to this philosophy is Section 2.2, where we use the representation theory of the Hecke algebra at $q = 0$ to describe explicitly the characters of the irreducible modules for the 0-Schur algebras.

These notes started life as part of a manuscript which dealt with the standard homological properties of $G_q(n)$ as well as some other topics. We published the homological parts separately, [36], and prepared as a companion paper the other topics (as detailed in the last paragraph of the introduction of [36]). However, as time went by, we kept adding to this manuscript, and a desire to keep the material together and the prospect of an opportunity to present the material from the uniform point of view described above led to the existence of the notes in their present form. The original intention of publication as a research article is responsible for the terse journal style of the main body of the text (and the fact that these notes are referred to in various places as "On Schur algebras and related algebras VI: The q-Schur algebra"). We have tried to compensate for this, and to make the notes reasonably self contained, by adding a long expository introductory chapter, which starts with the representation theory of algebraic groups and makes a gradual transition to the representation theory of quantum GL_n, and also by adding an appendix on quasihereditary algebras.

We defer a more detailed description of the contents of the notes until the end of the introductory chapter, so as to avail ourselves of the notation and definitions given there.

I am grateful to the School of Mathematical Sciences (especially to the Algebraic Lie Theory seminar) of Queen Mary and Westfield College, University of London, for the opportunity to present various parts of these notes at various times and also to the Institute for Experimental Mathematics, Essen, for the opportunity to lecture there (on the results given in Section 4.1 and Section 4.2(14), on the Ringel dual of the q-Schur algebras) in April 1994.

I am grateful to Anton Cox for his help in detecting numerous minor errors in earlier versions of these notes.

0. Introduction

0.1   In this chapter we endeavour to take the reader slowly from the familiar world of rational representations of algebraic groups to that of rational representations of general linear quantum groups, q-Schur algebras and Hecke algebras of type A. We do this partly to stress the close analogies between the theories, partly to establish some notation and list in a convenient form some results which are to be used in the sequel, and partly so that we will be able to describe in outline, towards the end of this chapter, the contents of the main part of these notes. Appropriate references for each section are given at the end of the chapter.

0.2   We fix an algebraically closed field K and a subfield k. We begin by recalling the basics of the theory of affine varieties and linear algebraic groups over K. Suppose given a set V and an algebra R of K-valued functions on V. For each point $x \in V$ we have the evaluation map $\varepsilon_x : R \to K$, defined by $\varepsilon_x(f) = f(x)$, for $f \in R$. The pair (V, R) is called an *affine variety* (over K) if the algebra R is finitely generated over K and the map $x \mapsto \varepsilon_x$, from V to the set $\mathrm{Hom}_{K-\mathrm{alg}}(R, K)$, of K-algebra homomorphisms from R to K, is bijective. An affine variety (V, R) is usually abbreviated to V. The algebra R, usually written $K[V]$, is called the coordinate algebra of V, or algebra of regular functions on V.

If (V, R) and (W, S) are affine varieties, a map $\phi : V \to W$ is a *morphism* of affine varieties if we have $g \circ \phi \in R$ for every $g \in S$. Such a map determines a K-algebra homomorphism $\phi^* : S \to R$, given by $\phi^*(g) = g \circ \phi$, and conversely, one checks that each K-algebra homomorphism $\theta : S \to R$ may be written $\theta = \phi^*$ for a unique morphism $\phi : V \to W$. For a subset Z of V we define $I_Z = \{f \in R \mid f(x) = 0 \text{ for all } x \in Z\}$. A subset Z is *closed* if there exist regular functions $f_1, \ldots, f_r$ on V such that $Z = \{x \in V \mid f_1(x) = \cdots = f_r(x) = 0\}$. The closed sets of V form the closed sets of the *Zariski topology* of V. If Z is closed in V then Z is naturally an affine variety with coordinate algebra $K[V]/I_Z$. (That is, the regular functions on Z are just the restriction to Z of regular functions on V.)

We write $\mathbf{A}^n$ for $K \times \cdots \times K$ (n times). For $1 \leq i \leq n$ we have the coordinate function $X_i : \mathbf{A}^n \to K$, defined by $X_i(x) = x_i$, for $x = (x_1, \ldots, x_n) \in \mathbf{A}^n$. The functions $X_1, \ldots, X_n$ are algebraically independent over K and one quickly verifies that $(\mathbf{A}^n, K[X_1, \ldots, X_n])$ is an affine algebraic variety. Moreover if V is an algebraic variety, with coordinate algebra A generated by $a_1, \ldots, a_r$, then the algebra map $\theta : K[X_1, \ldots, X_r] \to A$ given by $\theta(X_i) = a_i$, $1 \leq i \leq r$, corresponds to a morphism $\phi : V \to \mathbf{A}^r$ which identifies V with a closed subset of $\mathbf{A}^r$. (The image of ϕ is the set of $x \in \mathbf{A}^r$ such that $f(x) = 0$ for all f in the kernel of $\theta : K[X_1, \ldots, X_r] \to A$.) In this way we recover the usual description of affine K-varieties as closed subsets of affine r-space.

Given a finitely generated commutative K-algebra R without nilpotent elements (i.e. $f^m = 0$ implies $f = 0$, for $f \in R$ and $m > 0$) we may construct an affine variety with R as its coordinate algebra. For V we take $\mathrm{Hom}_{K-\mathrm{alg}}(R, K)$. For each $f \in R$ we have the function $\tilde{f} : V \to K$ defined by $\tilde{f}(x) = x(f)$, $x \in V$. We define $\tilde{R}$ to be the algebra of functions on V consisting of all $\tilde{f}$, $f \in R$. If $0 \neq f \in R$ we may choose a maximal ideal M of R not containing f. By the Nullstellensatz, inclusion $K \to R$ induces an isomorphism $K \to R/M$. Thus we have some K-algebra homomorphism $x : R \to K$ with kernel M and $x(f) \neq 0$. Hence the natural map $R \to \tilde{R}$ is injective and therefore an isomorphism. Identifying R with $\tilde{R}$ via this map we have that (V, R) is an affine variety. Thus the category of affine K-varieties is equivalent to the category of reduced (i.e. without nilpotent elements) finitely generated commutative K-algebras.

Let (V, R) and (W, S) be affine varieties. For $f \in R$, $g \in S$ we have the function $h_{f,g} : V \times W \to K$ defined by $h_{f,g}(x, y) = f(x)g(y)$, for $x \in V$, $y \in Y$. Let T be the algebra of K-valued functions on $V \times W$ generated by all such functions. The natural map $R \otimes_K S \to T$ (taking $f \otimes g$ to $h_{f,g}$) is an isomorphism and we thereby identify $R \otimes_K S$ with T. One checks that $(V \times W, T)$ is an affine K-variety and indeed this is the product of (V, R) and (W, S) in the category of affine varieties. (Alternatively, we could define $V \times W$ to be the affine variety whose coordinate algebra is $R \otimes_K S$, in view of the paragraph above.) Note that we have $\mathbf{A}^m \times \mathbf{A}^n \cong \mathbf{A}^{m+n}$.

0.3   By a *linear algebraic group* over K we mean a group G which is also an affine K-variety in such a way that the structure maps $m : G \times G \to G$ (multiplication) and $i : G \to G$ (inversion) are morphisms of varieties. Consider the general linear group $\mathrm{GL}_n(K)$ of invertible $n \times n$ matrices. We define $c_{ij} : \mathrm{GL}_n(K) \to K$ to be the (i,j) coordinate function, $1 \leq i, j \leq n$. We define $d : \mathrm{GL}_n(K) \to K$ to be the determinant function. The coordinate functions c_{ij}, $1 \leq i, j \leq n$, are algebraically independent over K. The coordinate algebra $K[\mathrm{GL}_n(K)]$ is, by definition, the algebra of K-valued functions on $\mathrm{GL}_n(K)$ generated by all coordinate functions c_{ij} together with d^{-1} (the function taking $x \in \mathrm{GL}_n(K)$ to the reciprocal of the determinant of x). Then $\mathrm{GL}_n(K)$ is a linear algebraic group (with coordinate algebra $K[\mathrm{GL}_n(K)]$).

0.4   Let G be a linear algebraic group. We say that a matrix representation $\rho : G \to \mathrm{GL}_n(K)$ is *rational* if ρ is a homomorphism of linear algebraic groups (i.e. a group homomorphism and a morphism of varieties). We say that a finite dimensional KG-module V is *rational* if it affords a rational matrix representation with respect to some (and hence every) basis. Let us be quite explicit. Choose a K-basis $v_1, \ldots, v_n$ of V and define coefficient functions $f_{ij} : G \to K$ by the equations

$$x v_i = \sum_{j=1}^{n} f_{ji}(x) v_j \qquad (x \in G, 1 \le i \le n).$$

The *coefficient space* of V is defined to be the space of K-valued functions on G spanned by the coefficient functions f_{ij}, $1 \le i,j \le n$. This space is independent of the choice of basis and we denote it by $\mathrm{cf}(V)$. Then the condition for V to be rational is that $\mathrm{cf}(V)$ should consist of regular functions, i.e. $\mathrm{cf}(V) \le K[G]$. Note that if V, W are finite dimensional KG-modules then we have $\mathrm{cf}(V \otimes_K W) = \mathrm{cf}(V).\mathrm{cf}(W)$, the K-span of all functions fg with $f \in \mathrm{cf}(V)$, $g \in \mathrm{cf}(W)$. In particular $V \otimes_K W$ is rational if V and W are rational. For a K-valued function f on G we write $\bar{f}$ for the K-valued function on G defined by the formula $\bar{f}(x) = f(x^{-1})$, $x \in G$. Note that $\bar{f} = i^*(f) \in K[G]$ if $f \in K[G]$. For a finite dimensional (left) KG-module V with coefficient space C we have that the coefficient space of the dual left module V^* is $\bar{C} = \{\bar{f} \mid f \in C\}$. It follows that if V is rational then so is V^*.

We say that a KG-module V of arbitrary dimension is rational if it is the union of finite dimensional rational submodules. If V is rational then so is every submodule and quotient module. A module is rational if it is generated by rational submodules.

0.5   Let G be a linear algebraic group over K. Then $K[G]$ is naturally a left KG-module for the left regular action, which we now describe. For $x \in G$, $f \in K[G]$ the function xf is given by the formula $(xf)(y) = f(yx)$, for $y \in G$. If $m^*(f) = \sum_{i=1}^{n} f_i \otimes f_i'$ then we have $xf = \sum_{i=1}^{n} f_i'(x) f_i$. In particular $xf \in K[G]$ and so $K[G]$ is naturally a left KG-module. We choose a basis $\{v_i \mid i \in I\}$ of $K[G]$. Then, for $i \in I$, we have $m^*(v_i) = \sum_{j \in I} v_j \otimes f_{ji}$ for elements f_{ij} of $K[G]$. We fix an $i \in I$. For $x \in G$ we have $xv_i = \sum_{j \in I} f_{ji}(x) v_j$. Since only finitely many of the f_{ji} are non-zero (for fixed i), for all $x \in G$ the element xv_i lies in the finite dimensional space spanned by the v_j for which f_{ji} is non-zero. Hence the KG-module V_i, say, generated by v_i, is finite dimensional. Let $W = V_i$ and let $w_1, \ldots, w_l$ be a basis of W. We have $m^*(w_r) = \sum_{s=1}^{l} w_s \otimes g_{sr}$, for elements $g_{rs} \in K[G]$ with $1 \le r,s \le l$. Hence $xw_r = \sum_{s=1}^{l} g_{sr}(x) w_s$ and the coefficient space of W is the K-span of the g_{rs}, in particular V_i is a rational module. But $K[G] = \sum_{i \in I} V_i$ and hence $K[G]$ is a rational G-module with respect to the left regular action.

0.6   An importance consequence of 0.5 is that every linear algebraic group is isomorphic to a closed subgroup of $\mathrm{GL}_n(K)$, for suitable n. Let G be a linear algebraic group. Let $a_1, \ldots, a_m$ be a set of algebra generators

of $K[G]$ and let W be a finite dimensional subspace of $K[G]$ which contains
these generators and is a submodule for the left regular G-module action.
Let $w_1, \ldots, w_n$ be a basis of W and let f_{ij} be the corresponding coefficient
functions. Thus we have $xw_i = \sum_{j=1}^n f_{ji}(x)w_j$, for $1 \leq i \leq n$. We define
$\rho : G \to \mathrm{GL}_n(K)$ by $\rho(x) = (f_{ij}(x))$, $x \in G$. Then ρ is a group homomorphism. We claim that ρ is a morphism of varieties, and hence a morphism of
algebraic groups. We have $c_{ij} \circ \rho = f_{ij}$, and hence $c \circ \rho \in K[G]$ for all c in the
subalgebra of $K[\mathrm{GL}_n(K)]$ generated by all c_{ij}. Moreover, since $K[\mathrm{GL}_n(K)]$ is
generated by all c_{ij} together with d^{-1}, it suffices to show that $d^{-1} \circ \rho \in K[G]$.
So let $f = d \circ \rho$ and $g = d^{-1} \circ \rho$. Then we have $fg = 1$ and hence f is a
regular function on G which is everywhere non-zero. However, it is a general
fact that if h is a regular function on an affine variety V which is everywhere
non-zero then $h^{-1} = 1/h$ is regular. (If not then the ideal of $K[V]$ generated
by h is a proper ideal and hence contained in a maximal ideal M, say. By
the Nullstellensatz, once more, M has codimension 1 in $K[V]$ and there is a
K-algebra homomorphism $\theta : K[V] \to K$ which vanishes on M. But we have
$\theta = \varepsilon_x$ for some $x \in V$ and $h(x) = \varepsilon_x(h) = \theta(h) = 0$, a contradiction.) Hence
$g = d^{-1} \circ \rho \in K[G]$. Thus $\rho : G \to \mathrm{GL}_n(K)$ is indeed a morphism of algebraic
groups. The image of ρ^* contains the generators $a_1, \ldots, a_m$ of $K[G]$. Thus
$\rho^* : K[\mathrm{GL}_n(K)] \to K[G]$ is surjective and it follows that the image of ρ is
closed in $\mathrm{GL}_n(K)$ and that ρ induces an isomorphism of algebraic groups
from G onto the image of ρ.

0.7   Let k be an arbitrary field. We take tensor products over k. A
coalgebra over k (or k-coalgebra) is a triple (C, δ, ε) consisting of a k-vector
space C and linear maps $\delta : C \to C \otimes C$ (comultiplication) and $\varepsilon : C \to k$
(the counit or augmentation map) satisfying the equations

$$(\delta \otimes \mathrm{id}) \circ \delta = (\mathrm{id} \otimes \delta) \circ \delta : C \to C \otimes C \otimes C$$

and

$$(\varepsilon \otimes \mathrm{id}) \circ \delta = (\mathrm{id} \otimes \varepsilon) \circ \delta = \mathrm{id} : C \to C$$

where id denotes the identity map on C. The first equation is called the
coassociativity condition and the second is called the counit condition.
    A *bialgebra* over k (or k-bialgebra) is a coalgebra (A, δ, ε) such that A is
a k-algebra and the structure maps $\delta : A \to A \otimes A$ and $\varepsilon : A \to k$ are algebra
homomorphisms. A bialgebra (A, δ, ε) is called a *Hopf algebra* if there exists
a linear map $\sigma : A \to A$ such that

$$\mu(\sigma \otimes \mathrm{id})\delta = \mu(1 \otimes \sigma)\delta = \tilde{\varepsilon}$$

where $\mu : A \otimes A \to A$ is the multiplication map and $\tilde{\varepsilon} : A \to A$ is defined by
$\tilde{\varepsilon}(a) = \varepsilon(a)1$, for $a \in A$. The above equation is called the antipode condition.

If such a linear map $\sigma : A \to A$ exists, it is uniquely determined by the antipode condition and called the *antipode* of (A, δ, ε). A bialgebra (or Hopf algebra) is said to be commutative if the algebra A is commutative. In general the antipode of a Hopf algebra (A, δ, ε) is an algebra anti-homomorphism and hence is an algebra homomorphism if A is commutative. If (A, δ, ε), $(A', \delta', \varepsilon')$ are coalgebras, a linear map $\phi : A \to A'$ is a morphism of coalgebras if $\delta' \circ \phi = (\phi \otimes \phi) \circ \delta$ and $\varepsilon' \circ \phi = \varepsilon$. A coalgebra, bialgebra or Hopf algebra (A, δ, ε) will often be abbreviated to A. If A, A' are bialgebras a map $\phi : A \to A'$ is a morphism of bialgebras if it is an algebra and coalgebra morphism. If A, A' are Hopf algebras with antipodes σ, σ' then a morphism of bialgebras $\phi : A \to A'$ automatically has the property that $\sigma' \circ \phi = \phi \circ \sigma$, and such a map ϕ is also called a morphism of Hopf algebras.

Let G be a linear algebraic group over K. Then it is easy to check that $(K[G], m^*, \varepsilon_1)$ is a commutative Hopf algebra over K with antipode i^*. Conversely, suppose that (A, δ, ε) is a commutative Hopf K-algebra with antipode σ and suppose that A is finitely generated and reduced (i.e. is without nilpotent elements). Then we associate with A the set $G = \operatorname{Hom}_{K-\mathrm{alg}}(A, K)$ and regard (G, A) as an affine variety, as in 0.2. We have $m : G \times G \to G$ given by $m(x, y) = (x \otimes y) \circ \delta$, $x, y \in G$, and $i : G \to G$ given by $i(x) = x \circ \sigma$, $x \in G$. Moreover G is a group with multiplication $m : G \times G \to G$, inversion $i : G \to G$ and identity ε. By construction we have $f \circ m = \delta(f) \in K[G] \otimes K[G]$ and $f \circ i = \sigma(f) \in K[G]$, for $f \in K[G]$. Thus G is an algebraic group and $m^* = \delta$, $i^* = \sigma$. In this way we obtain an equivalence of categories between linear algebraic groups over K and finitely generated, commutative, reduced Hopf algebras over K.

0.8   As well as preparing the way for our transition to quantum groups, the formalism of the previous paragraph provides a convenient language for discussing rationality properties of algebraic groups. So now let k be a subfield of our algebraically closed field K. For a K-vector space N, a *k-form* of N is a k-subspace M such that the natural map $K \otimes_k M \to N$ is an isomorphism. (This amounts to saying that some, and hence every, k-basis of M is a K-basis of N.) For a K-algebra S we say that R is an algebra k-form of S if R is a k-subalgebra of S and a (vector space) k-form of S. Let $(V, K[V])$ be an affine algebraic variety over K. By the expression "V is a *k-variety*" we indicate that we have in mind a fixed algebra k-form $k[V]$ of $K[V]$. For example, affine n-space $\mathbf{A}^n$ has coordinate algebra $K[X_1, \ldots, X_n]$ and is usually regarded as a k-variety by taking $k[\mathbf{A}^n] = k[X_1, \ldots, X_n]$. Let Z be a closed subset in the affine k-variety V. We say that Z is defined over k if the ideal I_Z is spanned, over K, by $k[V] \cap I_Z$. If this is the case then the natural map $K \otimes_k k[V] \to K[V]$ restricts to an isomorphism $K \otimes_k (k[V] \cap I_Z) \to I_Z$ and Z is viewed as a k-variety with $k[Z]$ consisting of the k-algebra of functions $f|_Z$, for $f \in k[V]$.

Let (C, δ, ε) be a coalgebra over K. We say that B is a coalgebra k-form of C if B is a vector space k-form of C which is closed under the structure maps, i.e. if $\delta(B) \leq B \otimes_k B$ and $\varepsilon(B) \leq k$. If B is a coalgebra k-form of C then B is naturally a k-coalgebra whose structure maps $B \to B \otimes_k B$ and $B \to k$ are the restrictions of δ and ε. If (C, δ, ε) is a bialgebra we say that B is a bialgebra k-form of C if B is both an algebra and coalgebra k-form of C. Finally, if (C, δ, ε) is a Hopf algebra with antipode σ then we say that B is a Hopf k-form of C if B is a bialgebra k-form and $\sigma(B) \leq B$. If B is a bialgebra k-form of the K-bialgebra C then B is naturally a bialgebra over k with comultiplication and counit as above. If B is a Hopf k-form of the Hopf K-algebra C, with antipode σ, then B is a Hopf algebra whose antipode is the restriction of σ.

By the expression "G is a *k-group*" or "G is a *linear algebraic group defined over k*" we indicate that G is a linear algebraic group over K and that we have in mind a Hopf k-form of $K[G]$, which we denote $k[G]$. We say that a closed subgroup H is defined over k if H is defined over k, as a closed set in G. In this case H has a natural k-group structure.

Suppose that G and H are k-groups. A map $\phi : G \to H$ is a morphism of k-groups if it is morphism of linear algebraic groups such that $\phi^*(k[H]) \leq k[G]$. Thus ϕ gives rise to a morphism of Hopf algebras $k[H] \to k[G]$ and, conversely, a morphism of Hopf algebras $k[H] \to k[G]$ corresponds to a unique morphism of k-groups $G \to H$.

We regard the linear algebraic group $GL_n(K)$ as a k-group via the Hopf form $k[GL_n(K)] = k[c_{11}, c_{12}, \ldots, c_{nn}, d^{-1}]$ of $K[GL_n(K)]$. From now on we shall also write GL_n for $GL_n(K)$, regarded as a k-group.

0.9   Let (C, δ, ε) be a k-coalgebra. By a right C-*comodule* we mean a pair (V, τ) consisting of a k-space V and a linear map $\tau : V \to V \otimes C$ such that

$$(\tau \otimes \mathrm{id}) \circ \tau = (\mathrm{id} \otimes \delta) \circ \tau : V \to V \otimes C \otimes C$$

and

$$(\mathrm{id} \otimes \varepsilon) \circ \tau = \mathrm{id} : V \to V$$

where id is the identity map on V. We often write simply V for the comodule (V, τ) and call τ the structure map of the comodule V. Let $\{v_i \mid i \in I\}$ be a basis of V. We have elements c_{ij}, $i, j \in I$, defined by the equations

$$\tau(v_i) = \sum_{j \in I} v_j \otimes c_{ji}$$

(for $i \in I$). The k-span of $\{c_{ij} \mid i, j \in I\}$ is called the coefficient space of V and denoted $\mathrm{cf}(V)$. It is independent of the choice of basis of V. Note that $\mathrm{cf}(V)$ is a subcoalgebra of C and that V is naturally a right $\mathrm{cf}(V)$-comodule.

Let (V, τ), (V', τ') be (right) comodules. A linear map $\phi : V \rightarrow V'$ is a morphism of comodules (or comodule homomorphism) if $\tau' \circ \phi = (\phi \otimes \mathrm{id}_C) \circ \tau$ (where id_C is the identity map on C). We write $\mathrm{Comod}(C)$ for the category of right C-comodules and $\mathrm{comod}(C)$ for the category of finite dimensional right C-comodules. A subspace U of a comodule V is a subcomodule if $\tau(U) \leq U \otimes C$, where τ is the structure map of V. A subcomodule U is naturally a comodule whose structure map $U \rightarrow U \otimes C$ is the restriction of $\tau : V \rightarrow V \otimes C$. The structure map τ also induces a map $V/U \rightarrow V/U \otimes C$, making V/U into a comodule. The inclusion map $U \rightarrow V$ and the quotient map $V \rightarrow V/U$ are homomorphisms of comodules. An important feature of the representation theory of coalgebras is the local finiteness property. If V is any right C-comodule then for every finite dimensional subspace T there is a finite dimensional subcomodule U of V such that $T \leq U$. The argument is similar to that of 0.5. Left comodules are defined similarly and have similar properties.

The dual space $C^* = \mathrm{Hom}_k(C, k)$ has the structure of an associative k-algebra with multiplication $\alpha\beta = (\alpha \otimes \beta)\delta$, for $\alpha, \beta \in C^*$, and identity $1_{C^*} = \varepsilon$. For an algebra S, we write $\mathrm{Mod}(S)$ for the category of left S-modules and $\mathrm{mod}(S)$ for the category of finite dimensional left S-modules. Let (V, τ) be a right C-comodule. We regard the k-space V as a left C^*-module via the product $\alpha v = (\mathrm{id} \otimes \alpha)\tau(v)$, for $\alpha \in C^*$, $v \in V$. For $X, Y \in \mathrm{Comod}(C)$, a linear map $\phi : X \rightarrow Y$ is a morphism of C-comodules if and only if it is a morphism of C^*-modules. In this way we obtain a full embedding of $\mathrm{Comod}(C)$ into $\mathrm{Mod}(C^*)$. If C is finite dimensional then we obtain, in this way, equivalences of categories $\mathrm{Comod}(C) \rightarrow \mathrm{Mod}(C^*)$ and $\mathrm{comod}(C) \rightarrow \mathrm{mod}(C^*)$.

0.10   We now return to our discussion of the representation theory of linear algebraic groups in general, and general linear groups in particular. Let G be a linear algebraic group over K. Let V be a (possibly infinite dimensional) rational representation and let $\{v_i \mid i \in I\}$ be a basis of V. We have coefficient functions $f_{ij} \in K[G]$ defined by

$$xv_i = \sum_{j \in I} f_{ji}(x)v_j$$

for $i \in I$. We define $\tau : V \rightarrow V \otimes K[G]$ by

$$\tau(v_i) = \sum_{j \in I} v_j \otimes f_{ji}$$

for $i \in I$. Then (V, τ) is a $K[G]$-comodule and we obtain, in this way, an equivalence of categories between rational left G-modules and right $K[G]$-comodules. Similar remarks apply to rational right G-modules and left $K[G]$-comodules.

We identify the categories of left rational G-modules and right $K[G]$-comodules via the above. We now want to discuss the representation theory of k-groups. Let G be a k-group. We shall simply define a left module for G to be a right $k[G]$-comodule and write $\mathrm{Mod}(G)$ (resp. $\mathrm{mod}(G)$) for $\mathrm{Comod}(k[G])$ (resp. $\mathrm{comod}(k[G])$). One could instead decree that a rational G-module over k is a rational G-module V together with a vector space k-form U of V such that for some (and hence every) basis $\{u_i \mid i \in I\}$ the corresponding coefficient functions f_{ij} all belong to $k[G]$. We leave it to the reader to check the equivalence of this with the comodule point of view.

0.11   Let G be a k-group. We write $G(k)$ for the set of $x \in G$ such that $\varepsilon_x(k[G]) \leq k$. It is easy to check that $G(k)$ is a subgroup of G (called the subgroup of k-*rational points*). If $G(k)$ is dense in the Zariski topology on G then we have available yet another description of the category of G-modules (defined over k). This is very close to our original formulation of the notion of a rational module for a linear algebraic group. For $f \in k[G]$ let $\tilde{f}$ be the k-valued function on $G(k)$ defined by $\tilde{f}(x) = \varepsilon_x(f)$, $x \in G(k)$. Let R be the algebra of k-valued functions on $G(k)$ consisting of all functions $\tilde{f}$, with $f \in k[G]$. Note that if $\tilde{f} = 0$ then the closed set $Z = \{x \in G \mid f(x) = 0\}$ of G contains $G(k)$, hence G, and therefore $f = 0$. Thus the map $k[G] \to R$, taking f to $\tilde{f}$, is injective, and we identify $k[G]$ with a k-algebra of functions on $G(k)$ by this map. We say that a finite dimensional $kG(k)$-module V is rational if for some (and hence every) basis $v_1, \ldots, v_n$, the corresponding k-valued functions f_{ij} on $G(k)$, defined by the equations

$$xv_i = \sum_{j=1}^{n} f_{ji}(x)v_j$$

for $1 \leq i \leq n$, all belong to $k[G]$. Given such a $kG(k)$-module we obtain a $k[G]$-comodule structure on V by defining $\tau : V \to V \otimes k[G]$ by $\tau(v_i) = \sum_{j=1}^{n} v_j \otimes f_{ji}$. We leave it to the reader to check that one obtains, in this way, an equivalence of categories between finite dimensional rational left $kG(k)$-modules and finite dimensional right $k[G]$-comodules, hence G-modules (defined over k).

0.12   We consider now the representation theory of a torus, i.e. a linear algebraic group of the form $\mathrm{GL}_1(K)^n = \mathrm{GL}_1(K) \times \cdots \times \mathrm{GL}_1(K)$ (n times). Thus $K[\mathrm{GL}_1(K)^n]$ is the Laurent polynomial algebra $K[t_1, t_1^{-1}, \ldots, t_n, t_n^{-1}]$, where $t_i(x_1, \ldots, x_n) = x_i$, $1 \leq i \leq n$, for $(x_1, \ldots, x_n) \in \mathrm{GL}_1(K)^n$. We consider $\mathrm{GL}_1(K)^n$ as a k-group, which we now write as GL_1^n, via the form $k[\mathrm{GL}_1^n] = k[t_1, t_1^{-1}, \ldots, t_n, t_n^{-1}]$. We put $X(n) = \mathbb{Z}^n$. For $\alpha = (\alpha_1, \ldots, \alpha_n) \in X(n)$ put $t^\alpha = t_1^{\alpha_1} \ldots t_n^{\alpha_n}$. We have the 1-dimensional GL_1^n-module k_α with structure map $k_\alpha \to k_\alpha \otimes k[\mathrm{GL}_1^n]$ taking $x \in k_\alpha$ to $x \otimes t^\alpha$. We have the comodule decomposition $k[\mathrm{GL}_1^n] = \bigoplus_{\alpha \in X(n)} kt^\alpha$. It follows that $\{k_\alpha \mid \alpha \in X(n)\}$ is a complete set of pairwise non-isomorphic irreducible GL_1^n-modules and that every GL_1^n-module is completely reducible.

We form the integral group ring $\mathbb{Z}X(n)$. This has $\mathbb{Z}$-basis $\{e(\alpha) \mid \alpha \in X(n)\}$ whose elements multiply according to the rule $e(\alpha)e(\beta) = e(\alpha + \beta)$. For a GL_1^n-module V and $\alpha \in X(n)$ we have the corresponding weight space $V^\alpha = \{v \in V \mid \tau(v) = v \otimes t^\alpha\}$, where $\tau : V \to V \otimes k[\mathrm{GL}_1^n]$ is the structure map. We say that $\alpha \in X(n)$ is a weight of V if $V^\alpha \neq 0$. The character ch V of a finite dimensional GL_1^n-module is defined by ch $V = \sum_{\alpha \in X(n)} (\dim V^\alpha) e(\alpha) \in \mathbb{Z}X(n)$.

Note that each $\alpha = (\alpha_1, \ldots, \alpha_m) \in X(n)$ gives rise to an algebraic group homomorphism $\mathrm{GL}_1^n \to \mathrm{GL}_1$, also denoted α and given by the formula $\alpha(x) = x_1^{\alpha_1} \ldots x_m^{\alpha_m}$, for $x = (x_1, \ldots, x_m) \in \mathrm{GL}_1^n$. If k is infinite then $\mathrm{GL}_1(k)^n$ is dense in $\mathrm{GL}_1(K)^n$ and, by the formalism of 0.11, for a rational GL_1^n-module V (over K) and $\alpha \in X(n)$ we have $V^\alpha = \{v \in V \mid xv = \alpha(x)v$ for all $x \in \mathrm{GL}_1^n\}$.

0.13 We now wish to discuss the rational representation theory of GL_n. Let k be a subfield of K. We view GL_n as a group over k, as in 0.8.

Let $\phi : B \to C$ be a morphism of k-coalgebras. If (V, τ) is a B-comodule then we may regard V as a C-comodule via the structure map $(\mathrm{id} \otimes \phi) \circ \tau : V \to V \otimes C$. We say that this C-comodule is obtained by ϕ-*inflation*, or just inflation if ϕ is inclusion. Suppose that B is a subcoalgebra of C. We say that a C-comodule V *belongs* to B if $\mathrm{cf}(V) \leq B$. The C-comodules belonging to B are the objects of a full subcategory of $\mathrm{Comod}(C)$ and inflation defines an equivalence of categories between B-comodules and C-comodules belonging to B.

Let A be a bialgebra. For right comodules $(V, \tau_V), (W, \tau_W)$ we have the tensor product right comodule $(V \otimes W, \tau_{V \otimes W})$. The structure map $\tau_{V \otimes W}$ is given by $\tau_{V \otimes W}(v \otimes w) = \sum_{i,j} v_i \otimes w_j \otimes f_i g_j$, for $v \in V$, $w \in W$ with $\tau_V(v) = \sum_i v_i \otimes f_i$ and $\tau_W(w) = \sum_j w_j \otimes g_j$. Moreover, if A is a Hopf algebra and (V, τ_V) is a finite dimensional right A-comodule then we have the dual right comodule (V^*, τ_{V^*}). The structure map $\tau^* : V^* \to V^* \otimes A$ may be described as follows. Let $v_1, \ldots, v_r$ be a basis of V and let $\alpha_1, \ldots, \alpha_r$ be the dual basis of V^*. Then we have $\tau^*(\alpha_j) = \sum_{i=1}^n \alpha_i \otimes \sigma(f_{ij})$, where $\tau(v_i) = \sum_{j=1}^r v_j \otimes f_{ji}$, for $1 \leq i \leq n$, and where σ is the antipode. In the case in which $A = K[G]$, for a linear algebraic group G over K, these constructions correspond, via the formalism of 0.10, to the usual group action on the tensor product of modules and on the dual of a finite dimensional module. Similar remarks apply to left comodules and right modules.

We set $A(n) = k[c_{ij} \mid 1 \leq i, j \leq n]$. Then $A(n)$ is a subbialgebra of $k[\mathrm{GL}_n]$ and we say that a GL_n-module V is polynomial if $\mathrm{cf}(V) \leq A(n)$, i.e.

if V belongs to $A(n)$. We identify the category of polynomial GL_n-modules with the category of $A(n)$-comodules, via the equivalence above. Thus a finite dimensional $K GL_n(K)$-module V is polynomial if and only if for some (and hence every) basis $\{v_i \mid i \in I\}$ the corresponding coefficient functions f_{ij}, determined by the equations

$$xv_i = \sum_{j \in I} f_{ji}(x)v_j$$

for $x \in GL_n(K)$ and $i \in I$, all lie in $K[c_{11}, \ldots, c_{nn}]$. More generally, if k is infinite then $GL_n(k)$ is dense in $GL_n(K)$ and, by the formalism of 0.11, we have the same description of polynomial GL_n-modules. That is, a finite dimensional polynomial GL_n-module is is a finite dimensional $kGL_n(k)$-module V such that for some (and hence every) basis $\{v_i \mid i \in I\}$ the corresponding coefficient functions f_{ij}, determined by the equations

$$xv_i = \sum_{j \in I} f_{ji}(x)v_j$$

for $x \in GL_n(k)$ and $i \in I$, all lie in $k[c_{11}, \ldots, c_{nn}]$. In [51], Green considers polynomial GL_n-modules (over an infinite field k) from this perspective.

We have the natural GL_n-module E with basis $e_1, \ldots, e_n$ and structure map $\tau : E \to E \otimes k[G]$ given by $\tau(e_i) = \sum_{j=1}^n e_j \otimes c_{ji}$, for $1 \le i \le n$. Note that E is a polynomial module. Moreover (since $A(n)$ is a bialgebra) $V \otimes W$ is a polynomial module if V and W are polynomial modules. Hence the rth tensor power $E^{\otimes r}$, the rth symmetric power $S^r E$ and the rth exterior power $\bigwedge^r E$ are polynomial modules, for r any non-negative integer. In particular the determinant module $D = \bigwedge^n E$ is polynomial.

From the point of view of algebraic group representation theory, one is interested in the category of rational modules. However, as we now describe, this differs only trivially from the category of polynomial modules, and the latter category has strong connections with the theory of representations of certain finite dimensional algebras. So let V be a finite dimensional GL_n-module. Then $cf(V)$ is a finite dimensional subspace of $k[G]$, which is the localization of $A(n)$ at the determinant d. Hence we have $cf(V) \le d^{-r}.N$, for some finite dimensional subspace N of $A(n)$, and so $cf(D^{\otimes r} \otimes V) = cf(D)^r.cf(V) = d^r.cf(V) \le A(n)$. Thus $D^{\otimes r} \otimes V$ is a polynomial module, for $r \gg 0$, and every finite dimensional rational module is isomorphic to one of the form $D^{\otimes -r} \otimes U$, for some $r \ge 0$ and polynomial module U (where $D^{\otimes -r}$ denotes the dual of $D^{\otimes r}$).

Thus all finite dimensional rational modules can be understood in terms of the polynomial ones. We regard $A(n)$ as a graded algebra by giving each c_{ij} degree 1. So $A(n)$ decomposes

$$A(n) = \bigoplus_{r=0}^{\infty} A(n, r) \tag{$*$}$$

into homogeneous components. The dimension of $A(n, r)$ is the number of monomials in n^2 variables of total degree r, and this is $\binom{n^2+1-r}{r}$. Since $\delta(c_{ij}) = \sum_{h=1}^{n} c_{ih} \otimes c_{hj}$, for $1 \leq i, j \leq n$, we have $\delta(A(n, 1)) \leq A(n, 1) \otimes A(n, 1)$, and then since δ is an algebra homomorphism we get $\delta(A(n, r)) \leq A(n, r) \otimes A(n, r)$. Hence $(*)$ is a coalgebra decomposition. We say that a polynomial module has *degree* r if it belongs to $A(n, r)$ and say that a polynomial module is *homogeneous* if it has degree r for some $r \geq 0$. Let V be a polynomial module (i.e. an $A(n)$-comodule) and for $r \geq 0$, let $V(r)$ be the largest subcomodule of V belonging to $A(n, r)$. It follows from $(*)$ that we have

$$V = \bigoplus_{r=0}^{\infty} V(r).$$

Hence every polynomial module is a direct sum of homogeneous modules. We define the Schur algebra $S(n, r) = A(n, r)^*$, the dual algebra. We identify right $A(n, r)$-comodules with left $S(n, r)$-modules, as in 0.9. To summarize: for any finite dimensional rational module U, we have that U is isomorphic to $D^{\otimes -r} \otimes V$, for some polynomial module V, each polynomial module decomposes into a direct sum of homogeneous modules and the category of polynomial modules of degree r is equivalent to the category of modules for the Schur algebra $S(n, r)$.

Thus from the representation theory of the the finite dimensional algebras $S(n, r)$, one can obtain all finite dimensional rational modules. However, the point of view relevant for this work is that one should use the representation theory of GL_n to illuminate the representation theory of the finite dimensional algebras $S(n, r)$ and from there, via the Schur functors, the representation theory of the symmetric groups. It is the q-analogue of this point of view (from quantum general linear groups to q-Schur algebras to Hecke algebras) that is explored in detail in the subsequent chapters.

0.14 Before going further, we give another version of the theory of weights for GL_n-modules. We now write $G(n)$ for GL_n. We write $T(n)$ for the group of diagonal matrices in $G(n)$. We identify $T(n)$ with GL_1^n via the map $\mathrm{GL}_1^n \to T(n)$ taking $x = (x_1, \ldots, x_n)$ to $\mathrm{diag}(x)$, the diagonal matrix with $(1,1)$-entry x_1, with $(2,2)$-entry x_2 and so on. In particular, $T(n)$ has the structure of a k-group, and we have the theory of weights for $T(n)$, as developed in 0.12. We define the weight spaces and character of a $G(n)$-module to be the weight spaces and character of its restriction to $T(n)$.

If V is a polynomial module of degree r we may also assign weight spaces in another way. We write $I(n, r)$ for the set of functions $i : [1, r] \to [1, n]$. We often represent $i \in I(n, r)$ as the r-tuple $(i_1, \ldots, i_r)$ (where $i_a = i(a)$, $1 \leq a \leq r$). For $i, j \in I(n, r)$ we put $c_{ij} = c_{i_1 j_1} \ldots c_{i_r j_r}$. Note that the symmetric group $\mathrm{Sym}(r)$ has a natural right action on $I(n, r)$. Moreover, we have $c_{ij} = c_{i'j'}$ if and only if the sequence of pairs $(i_1, j_1), \ldots, (i_r, j_r)$ is a permutation of the sequence of pairs $(i'_1, j'_1), \ldots, (i'_r, j'_r)$, i.e. if and only if $(i, j), (i', j') \in I(n, r) \times I(n, r)$ belong to the same $\text{Sym}(r)$-orbit. We write the dual basis ξ_{ij}, $i, j \in I(n, r)$, with the understanding that $\xi_{ij} = \xi_{i'j'}$ if (i, j) and (i', j') are $\text{Sym}(r)$-conjugate.

For a sequence $\alpha = (\alpha_1, \alpha_2, \ldots)$ of non-negative integers we set $|\alpha| = \alpha_1 + \alpha_2 + \cdots$. Let $\Lambda(n, r)$ denote the set of $\alpha \in \mathbb{N}_0^n$ such that $|\alpha| = r$. We define the *content* $\alpha = (\alpha_1, \ldots, \alpha_n) \in \Lambda(n, r)$ of $i \in I(n, r)$ by $\alpha_a = |i^{-1}(a)|$, $1 \leq a \leq n$. Thus we have $\xi_{ii} = \xi_{jj}$ if and only if i and j have the same content. For $\alpha \in \Lambda(n, r)$ we write ξ_α for ξ_{ii}, where $i \in I(n, r)$ has content α. Now

$$1 = \sum_{\alpha \in \Lambda(n,r)} \xi_\alpha$$

is a decomposition of 1 as an orthogonal sum of idempotents in $S(n, r)$. Moreover, for a polynomial module V of degree r and $\alpha \in \Lambda(n, r)$ we have $V^\alpha = \xi_\alpha V$.

0.15 We define the so-called *dominance order* on $X(n) = \mathbb{Z}^n$. For $\alpha = (\alpha_1, \ldots, \alpha_n)$, $\beta = (\beta_1, \ldots, \beta_n)$ we write $\alpha \leq \beta$ if $|\alpha| = |\beta|$ and $\alpha_1 + \cdots + \alpha_a \leq \beta_1 + \cdots + \beta_a$, for all $1 \leq a \leq n$. We say that $\lambda = (\lambda_1, \ldots, \lambda_n) \in X(n)$ is dominant if $\lambda_1 \geq \lambda_2$, $\lambda_2 \geq \lambda_3$, $\ldots$, $\lambda_{n-1} \geq \lambda_n$. We write $X^+(n)$ for the set of dominant weights. For each $\lambda \in X^+(n)$ there is an irreducible $G(n)$-module $L(\lambda)$ such that $\dim L(\lambda)^\lambda = 1$ and $\mu < \lambda$ for every weight μ of $L(\lambda)$ not equal to λ. Moreover, $\{L(\lambda) \mid \lambda \in X^+(n)\}$ is a complete set of irreducible, mutually non-isomorphic $G(n)$-modules. From this, and the fact that

$$\text{ch}\, L(\lambda) = e(\lambda) + \text{lower terms},$$

it follows that finite dimensional rational modules V, V' have the same composition factors (counted according to multiplicity) if and only if $\text{ch}\, V = \text{ch}\, V'$. Thus the character of a finite dimensional rational module plays the same role as the Brauer character of a finite dimensional module for a finite group.

0.16 We wish to give a sketch of how the parametrization of irreducible modules in 0.15 may be brought about. We shall need a theory of induction for coalgebras. Let C be a coalgebra. Let $V \in \text{Comod}(C)$ and let X be a k-vector space (which may come to us with some additional structure). We write $|X| \otimes V$ for the vector space $X \otimes V$ regarded as a right C-comodule via the structure map $\text{id} \otimes \tau : X \otimes V \to X \otimes V \otimes C$, where $\tau : V \to V \otimes C$ is the structure map of V. For any right comodule V, the structure map $\tau : V \to |V| \otimes C$ is a morphism of right comodules.

Let $\theta : C \to D$ be a coalgebra map. We have already seen that we have a functor $\text{Comod}(C) \to \text{Comod}(D)$, given by "$\theta$-inflation". We now write θ_0 for this functor. Now let $W \in \text{Comod}(D)$. We write $\theta^0(W)$ for the C-subcomodule of $|W| \otimes C$ consisting of the elements $f \in W \otimes C$ such that $(\tau \otimes \text{id}_C)(f) = (\text{id}_W \otimes (\theta \otimes \text{id}_C) \circ \delta)(f)$, where $\delta : C \to C \otimes C$ is comultiplication. If $\alpha : W \to W'$ is a morphism of D-comodules, then $(\alpha \otimes \text{id}_C) : |W| \otimes C \to |W'| \otimes C$ restricts to a C-comodule map $\theta^0(W) \to \theta^0(W')$, which we denote $\theta^0(\alpha)$. In this way we get a left exact functor $\theta^0 : \text{Comod}(D) \to \text{Comod}(C)$. There is a natural map $\nu : \theta_0(\theta^0(W)) \to W$, namely the restriction of $(\text{id} \otimes \varepsilon) : W \otimes C \to W$, where $\varepsilon : C \to k$ is the counit. Moreover, the functors θ_0 and θ^0 are adjoint in that, for any $V \in \text{Comod}(C)$ and $W \in \text{Comod}(D)$, the map $\text{Hom}_C(V, \theta^0(W)) \to \text{Hom}_D(\theta_0(V), W)$, taking $\alpha : V \to \theta^0(W)$ to $\nu \circ \alpha : \theta_0(V) \to W$, is a linear isomorphism.

Now let $\phi : G_1 \to G_2$ be a morphism of k-groups. Then we have the associated comorphism $\theta : k[G_2] \to k[G_1]$. We write $\text{Res}(\phi)$ for $\theta_0 :$ $\text{Mod}(G_2) \to \text{Mod}(G_1)$ and $\text{Ind}(\phi)$ for $\theta^0 : \text{Mod}(G_1) \to \text{Mod}(G_2)$. We call $\text{Res}(\phi) : \text{Mod}(G_2) \to \text{Mod}(G_1)$ the ϕ-*restriction* functor and $\text{Ind}(\phi) :$ $\text{Mod}(G_2) \to \text{Mod}(G_1)$ the ϕ-*induction* functor. If $\phi : H \to G$ is inclusion then we also write $\text{Res}(\phi)(U)$ as $\text{Res}_H^G U$, or just $U|_H$, and call it the restriction of U to H, for $U \in \text{Mod}(G)$; we also write $\text{Ind}(\phi)(V)$ as $\text{Ind}_H^G V$, and call this the G-module induced from V, for $V \in \text{Mod}(H)$.

We now consider the general linear k-group $G(n)$. We take $B(n)$ to be the Borel subgroup consisting of the lower triangular matrices. We have a homomorphism $B(n) \to T(n)$ taking a lower triangular matrix to the corresponding diagonal matrix. For each $\lambda \in X(n)$ we have the 1-dimensional $T(n)$-module k_λ, and we regard this as a $B(n)$-module via inflation. The modules $\text{Ind}_{B(n)}^{G(n)} k_\lambda$ are finite dimensional and we have $\text{Ind}_{B(n)}^{G(n)} k_\lambda \neq 0$ if and only if λ is dominant. For $\lambda \in X^+(n)$ we put $\nabla(\lambda) = \text{Ind}_{B(n)}^{G(n)} k_\lambda$. Then one proves that $\nabla(\lambda)$ has a simple socle $L(\lambda)$, say, and that $\{L(\lambda) \mid \lambda \in X^+(n)\}$ is a complete set of pairwise non-isomorphic $G(n)$-modules.

In fact it is not difficult to produce, for each dominant weight λ, an irreducible module $L(\lambda)$ such that $\dim L(\lambda)^\lambda = 1$ and $\mu < \lambda$ for every weight μ of $L(\lambda)$ not equal to λ. Suppose first that all entries of $\lambda = (\lambda_1, \ldots, \lambda_n)$ are non-negative. For any finite sequence $\alpha = (\alpha_1, \alpha_2, \ldots)$ with entries in $[0, n]$, we define the polynomial module $\bigwedge^\alpha E = \bigwedge^{\alpha_1} E \otimes \bigwedge^{\alpha_2} E \otimes \cdots$. One may check that if α is the transpose of the partition λ then $\bigwedge^\alpha E$ has unique highest weight λ and that λ occurs with multiplicity 1. Hence there is a unique composition factor, with these properties, which we may take to be $L(\lambda)$. For an arbitrary dominant λ, we may choose $r \gg 0$ such that all entries of $\mu = \lambda + r(1, 1, \ldots, 1)$ are non-negative and take for $L(\lambda)$ the module $D^{\otimes -r} \otimes L(\mu)$.

Let $\Lambda^+(n, r) = \Lambda(n, r) \cap X^+(n)$. The character of $\nabla(\lambda)$, $\lambda \in X^+(n)$, is given by Weyl's character formula and, for $\lambda \in \Lambda^+(n, r)$, may be viewed as the corresponding Schur symmetric function. Moreover the characters of the $\nabla(\lambda)$ and $L(\lambda)$, $\lambda \in \Lambda^+(n, r)$, are connected by a unitriangular matrix.

One of the main problems of this area is to obtain an explicit formula for ch $L(\lambda)$, for $\lambda \in \Lambda^+(n,r)$, for example as an explicit $\mathbb{Z}$-linear combination of the ch $\nabla(\mu)$, $\mu \in \Lambda^+(n,r)$. A solution to this problem would also give an explicit formula for the Brauer characters of all irreducible modules for the symmetric groups.

0.17   It follows from 0.14 that the weights of any $G(n)$-module which is polynomial of degree r belong to $\Lambda(n,r)$. The modules $L(\lambda)$, $\lambda \in \Lambda^+(n,r)$, are all polynomial of degree r and we therefore obtain that $\{L(\lambda) \mid \lambda \in \Lambda^+(n,r)\}$ is a complete set of pairwise non-isomorphic irreducible $S(n,r)$-modules. Moreover, the modules $\nabla(\lambda)$, $\lambda \in \Lambda^+(n,r)$, are all polynomial. Thus the main problem (of finding the characters of the irreducible $G(n)$-modules) may be formulated entirely within the representation theory of the Schur algebras $S(n,r)$. We have the $S(n,r)$-modules $\nabla(\lambda), L(\lambda)$, for $\lambda \in \Lambda^+(n,r)$, and a knowledge of the composition multiplicities $[\nabla(\lambda) : L(\mu)]$, for $\lambda, \mu \in \Lambda^+(n,r)$, would give an explicit formula for the character of the irreducible modules $L(\lambda)$, $\lambda \in \Lambda^+(n,r)$.

Not only is there a strong connection between simple modules for the k-group $G(n)$, and the finite dimensional algebra $S(n,r)$, but also between the homological algebra of these objects. More precisely, let U, V be homogeneous $G(n)$-modules which have the same degree r. Then we have a natural isomorphism

$$\mathrm{Ext}^i_{S(n,r)}(U,V) \cong \mathrm{Ext}^i_{G(n)}(U,V)$$

for each $i \geq 0$. The observation that this was the case (for a class of groups and algebras including the general linear groups and Schur algebras) was the starting point of our investigations into Schur algebras and related algebras (see [31],[32],[34],[35],[40]) of which the present work is a continuation.

0.18   The above section describes the first "process of descent" and we now come to the second, from Schur algebras to group algebras of symmetric groups. We assume that $r \leq n$ and let $\omega = (1,\ldots,1) \in \Lambda(n,r)$. We put $S = S(n,r)$ and $e = \xi_\omega \in S$. We call eSe a subalgebra of S even though the identity element, e, of eSe is not the identity of $S(n,r)$ (unless $r = 1$). Let $u = (1,2,\ldots,r) \in I(n,r)$. Then the subalgebra eSe has basis $\xi_{u\pi,u}$, $\pi \in \mathrm{Sym}(r)$, and these elements multiply according to the rule

$$\xi_{u\pi,u}\xi_{u\pi',u} = \xi_{u\pi\pi',u}.$$

Thus eSe is a realization of the group algebra over k of the symmetric group of degree r. We have the exact functor $f : \mathrm{mod}(S) \to \mathrm{mod}(eSe)$, known as the Schur functor, given by $fV = eV = V^\omega$ on objects and given by restriction of mappings on morphisms. If $L \in \mathrm{mod}(S)$ is simple then eL is either simple or zero. Moreover the set of modules $eL(\lambda)$, as λ runs over those $\lambda \in \Lambda^+(n,r)$ for which $eL(\lambda)$ is not zero, form a complete set of pairwise non-isomorphic simple eSe-modules.

Let p be a positive integer. We say that $\lambda = (\lambda_1, \ldots, \lambda_n) \in \Lambda^+(n,r)$ is *column p-regular* if $\lambda_i - \lambda_{i+1} < p$ for all $1 \le i < n$. We say that $\lambda \in \Lambda^+(n,r)$ is *row p-regular* if the transpose partition λ' is column p-regular. Thus $\lambda = (\lambda_1, \ldots, \lambda_n)$ is row p-regular if there is no i with $0 < i \le n - p$ such that $\lambda_{i+1} = \cdots = \lambda_{i+p} > 0$. All elements of $\Lambda^+(n,r)$ are declared to be column 0-regular and row 0-regular. Now let p be the characteristic of k. We write $\Lambda^+(n,r)_{\text{col}}$ (resp. $\Lambda^+(n,r)_{\text{row}}$) for the set of column p-regular (resp. row p-regular) elements in $\Lambda^+(n,r)$.

The set of $\lambda \in \Lambda^+(n,r)$ such that $eL(\lambda) \neq 0$ is exactly $\Lambda^+(n,r)_{\text{col}}$ and thus we have the parametrization $\{fL(\lambda) \mid \lambda \in \Lambda^+(n,r)_{\text{col}}\}$ of irreducible modules of the symmetric group of degree r. Applying the Schur functor to other naturally occurring modules for $S(n,r)$-modules (i.e. polynomial $G(n)$-modules of degree r) produces other modules familiar from the representation theory of the symmetric group. For example applying the Schur functor to $\nabla(\lambda)$ yields the Specht module and applying the Schur functor to the injective envelope, as an $S(n,r)$-module, of $L(\lambda)$ yields the Young module, labelled by λ.

0.19   The story so far has been both a review of the classical theory and a dry run for the q-analogue, which we now describe. One has a well known q-analogue of the group algebra of the symmetric group $\text{Sym}(r)$, namely the Hecke algebra $\text{Hec}(r)$. We fix a field k and an element $q \in k$. We denote the length of $w \in \text{Sym}(r)$ by $l(w)$. Then $\text{Hec}(r)$ has k-basis T_w, $w \in \text{Sym}(r)$, and multiplication satisfying

$$T_w T_{w'} = T_{ww'}, \qquad \text{if } l(ww') = l(w) + l(w')$$
$$(T_s + 1)(T_s - q) = 0$$

for $w, w' \in \text{Sym}(r)$, and a basic transposition $s \in W$.

The idea is to study the representation theory of the Hecke algebra $\text{Hec}(r)$, as in the classical case, by a sequence of two descents: from a quantization (i.e. q-analogue) of $G(n)$ to a q-Schur algebra and from there to the Hecke algebra via a Schur functor. The point of view taken in these notes is that the information should flow uniformly in this direction. (Other points of view are possible, see the work of Dipper and James, e.g. [21], and there is an exception to this principle in Section 2.2, where we use the representation theory of Hecke algebras to determine the characters of the irreducible modules for the q-Schur algebras at $q = 0$.)

0.20   We view $G(n)$ as a group over k. Recall that a $G(n)$-module is, by definition, a right comodule for the Hopf algebra $k[G(n)]$. We shall introduce a deformation of this Hopf algebra. Thus our quantum general linear group will, in reality, be a Hopf k-algebra. However, we do not wish to abandon, at this point, the carefully developed intuition and notation of group representation theory and therefore stretch the analogy between groups and Hopf algebras somewhat beyond its legal limits. That is, we shall regard the category of quantum groups as the opposite category of the category of Hopf algebras. Thus, when we say "G is a quantum group over k", we mean that we have in mind a Hopf algebra which is denoted $k[G]$, and when we say that $\phi : G_1 \to G_2$ is a morphism of quantum groups we mean that we have in mind a homomorphism of Hopf algebras $\hat{\phi} : k[G_2] \to k[G_1]$. We call $\hat{\phi}$ the *comorphism* of ϕ. We use the expression "H is a quantum subgroup of G" (or just "H is a subgroup of G") to indicate that we have in mind a Hopf ideal I_H of $k[G]$ (i.e. I_H is an ideal of $k[G]$ such that $\delta(I_H) \leq k[G] \otimes I_H + I_H \otimes k[G]$, $\varepsilon(I_H) = 0$ and $\sigma(I_H) \leq I_H$) and that $k[H] = k[G]/I_H$. In this case we have the morphism $\phi : H \to G$, called inclusion, whose comorphism $\hat{\phi}$ is the natural map $k[G] \to k[G]/I_H$.

Let G be a quantum group. A left G-module is, by definition, a right $k[G]$-comodule. We write $\mathrm{Mod}(G)$ for $\mathrm{Comod}(k[G])$ and write $\mathrm{mod}(G)$ for $\mathrm{comod}(k[G])$. Let $\phi : G_1 \to G_2$ be a morphism of quantum groups and let $\theta = \hat{\phi} : k[G_2] \to k[G_1]$. As in 0.16, we write $\mathrm{Res}(\phi)$ for $\theta_0 : \mathrm{Mod}(G_2) \to \mathrm{Mod}(G_1)$ and $\mathrm{Ind}(\phi)$ for $\theta^0 : \mathrm{Mod}(G_1) \to \mathrm{Mod}(G_2)$. We call $\mathrm{Res}(\phi) : \mathrm{Mod}(G_2) \to \mathrm{Mod}(G_1)$ the ϕ-*restriction* functor and $\mathrm{Ind}(\phi) : \mathrm{Mod}(G_1) \to \mathrm{Mod}(G_2)$ the ϕ-*induction* functor. If $\phi : H \to G$ is inclusion then we also write $\mathrm{Res}(\phi)(U)$ as $\mathrm{Res}_H^G U$, or just $U|_H$, and call it the restriction of U to H, for $U \in \mathrm{Mod}(G)$; we also write $\mathrm{Ind}(\phi)(V)$ as $\mathrm{Ind}_H^G V$, and call this the G-module induced from V, for $V \in \mathrm{Mod}(H)$.

To construct our quantization of the general linear group $G(n)$, we start with the associative algebra F freely generated by X_{ij}, $1 \leq i, j \leq n$. Define $\delta : F \to F \otimes F$ to be the algebra map such that $\delta(X_{ij}) = \sum_{r=1}^{n} X_{ir} \otimes X_{rj}$ and $\varepsilon : F \to k$ to be the algebra map such that $\varepsilon(X_{ij}) = \delta_{ij}$ (the Kronecker delta), for $1 \leq i, j \leq n$. Then (F, δ, ε) is a bialgebra over k. We view F as a graded k-algebra, by giving X_{ij} degree 1, for $1 \leq i, j \leq n$. Let I be the ideal of F generated by the elements

$$
\begin{array}{ll}
X_{ir}X_{is} - X_{is}X_{ir} & \text{for all } i, r, s \\
X_{jr}X_{is} - qX_{is}X_{jr} & \text{for all } i < j \text{ and } r \leq s \\
X_{js}X_{ir} - X_{ir}X_{js} - (q-1)X_{is}X_{jr} & \text{for all } i < j \text{ and } r < s
\end{array}
$$

(for $1 \leq i, j, r, s \leq n$). Then I is a biideal, i.e. we have $\delta(I) \leq F \otimes I + I \otimes F$ and $\varepsilon(I) = 0$. The quotient algebra $A_q(n) = F/I$ is therefore a bialgebra with structure maps, also denoted δ, ε, induced from those of F. Thus $A_q(n)$ is the k-algebra generated by c_{ij}, $1 \leq i, j \leq n$, subject to the relations

$$
\begin{array}{ll}
c_{ir}c_{is} = c_{is}c_{ir} & \text{for all } i, r, s \\
c_{jr}c_{is} = qc_{is}c_{jr} & \text{for all } i < j, \text{ and } r \leq s \\
c_{js}c_{ir} = c_{ir}c_{js} + (q-1)c_{is}c_{jr} & \text{for all } i < j, \text{ and } r < s
\end{array}
$$

(for $1 \leq i, j, r, s \leq n$). The bialgebra structure on $A_q(n)$ is given by

$$\delta(c_{ij}) = \sum_{r=1}^{n} c_{ir} \otimes c_{rj} \quad \text{and} \quad \varepsilon(c_{ij}) = \delta_{ij}$$

for $1 \leq i, j \leq n$. Moreover, the generators of I above are homogeneous so that I is a graded ideal and $A_q(n)$ inherits the structure of a graded algebra with c_{ij} having degree 1, for $1 \leq i, j \leq n$. Note that putting $q = 1$ gives the bialgebra $A(n)$ discussed above.

For $a = (a_{11}, a_{12}, \ldots, a_{nn}) \in \mathbb{N}_0^{n^2}$ we put $c^a = c_{11}^{a_{11}} c_{12}^{a_{12}} \ldots c_{nn}^{a_{nn}}$. The elements c^a, $a \in \mathbb{N}_0^{n^2}$, form a k-basis of $A_q(n)$. From the grading on $A_q(n)$ we get a coalgebra decomposition $A_q(n) = \bigoplus_{r=0}^{\infty} A_q(n, r)$. Hence we get an algebra $S_q(n, r) = A_q(n, r)^*$ of dimension $\binom{n^2 + r - 1}{r}$, for $r \geq 0$. The algebras $S_q(n, r)$ are called the q-Schur algebras, and were originally constructed (by other means) by Dipper and James.

We have now constructed a q-analogue of the algebra $A(n)$ and of the Schur algebras $S(n, r)$. To make possible our investigation of the representation of $\mathrm{Hec}(r)$ by means of the sequence of descents outlined above, we need to construct a Hopf algebra $k[G_q(n)]$, to take the place of $k[G(n)]$ in the case $q = 1$. Recall that $k[G(n)]$ is the localization of $A(n)$ at the determinant. We define the quantum determinant

$$d_q = \sum_{\pi \in \mathrm{Sym}(n)} \mathrm{sgn}(\pi) c_{1,1\pi} c_{2,2\pi} \ldots c_{n,n\pi} \in A_q(n).$$

Then we have $\varepsilon(d_q) = 1$ and $\delta(d_q) = d_q \otimes d_q$, i.e. d_q is a group-like element of $A_q(n)$. Assume now that $q \neq 0$. We have $c_{ij} d_q = q^{i-j} d_q c_{ij}$, for $1 \leq i, j \leq n$. It follows that we can localize the bialgebra $A_q(n)$ at d_q. It turns out that the localization $A_q(n)_{d_q}$ is a Hopf algebra and we define the quantum general linear group $G_q(n)$ to be the quantum group whose coordinate algebra is $A_q(n)_{d_q}$.

0.21 We now consider the representation theory of $G_q(n)$. We do this, as in the classical case ($q = 1$), by means of subgroups $T_q(n)$ and $B_q(n)$. The defining ideal of $T_q(n)$ is the Hopf ideal of $k[G_q(n)]$ generated by all c_{ij}, with $i \neq j$. We have an isomorphism of Hopf algebras $k[T(n)] \rightarrow k[T_q(n)]$ taking t_i to $c_{ii} + I_{T_q(n)}$, for $1 \leq i \leq n$. In particular, $T_q(n)$ is independent of q, and we identify it with $T(n)$ via the above isomorphism. Thus we have a theory of weights for $G_q(n)$ and, as before, for $V \in \mathrm{mod}(G_q(n))$ we define the character $\mathrm{ch}\, V \in \mathbb{Z}X(n)$ to be the character of $V|_{T(n)}$. The defining ideal of $B_q(n)$ is the Hopf ideal of $k[G]$ generated by all c_{ij} with $i < j$. We have $T(n) \leq B_q(n) \leq G_q(n)$. We also have a homomorphism $\phi : B_q(n) \rightarrow T(n)$, whose comorphism $\hat{\phi} : k[T(n)] \rightarrow k[G_q(n)]$ takes t_i to

$\bar{c}_{ii} = c_{ii} + I_{B_q(n)}$, for $1 \leq i \leq n$. Thus each irreducible $T(n)$-module k_α, for $\alpha = (\alpha_1, \ldots, \alpha_n) \in X(n)$, may be regarded as a $B_q(n)$-module, by ϕ-restriction. Precisely, the structure map $\tau : k_\alpha \to k_\alpha \otimes k[B_q(n)]$ is given by $\tau(x) = x \otimes \bar{c}^\alpha$, for $x \in k_\alpha$, where $\bar{c}^\alpha = \bar{c}_{11}^{\alpha_1} \ldots \bar{c}_{nn}^{\alpha_n}$. Then $\{k_\alpha \mid \alpha \in X(n)\}$ is a full set of simple $B_q(n)$-modules. We have the following fundamental properties of induction from $B_q(n)$ to $G_q(n)$ and its derived functors. The second property is known as Grothendieck vanishing.

(1)   The derived functor $R^i\mathrm{Ind}_{B_q(n)}^{G_q(n)}$ takes finite dimensional modules to finite dimensional modules, for $i \geq 0$, and $R^i\mathrm{Ind}_{B_q(n)}^{G_q(n)} = 0$ for $i > \binom{n}{2}$.

Of central importance to our approach is Kempf's vanishing theorem, which is the second assertion of (2) below.

(2)   For $\lambda \in X(n)$, we have $\mathrm{Ind}_{B_q(n)}^{G_q(n)} k_\lambda \neq 0$ if and only if $\lambda \in X^+(n)$ and furthermore we have $R^i\mathrm{Ind}_{B_q(n)}^{G_q(n)} k_\lambda = 0$, for $\lambda \in X^+(n)$ and $i > 0$.

We define $\nabla_q(\lambda) = \mathrm{Ind}_{B_q(n)}^{G_q(n)} k_\lambda$, for $\lambda \in X^+(n)$. We define $L_q(\lambda)$ to be the socle of $\nabla_q(\lambda)$, for $\lambda \in X^+(n)$.

(3)   For $\lambda \in X^+(n)$, the module $\nabla_q(\lambda)$ has character given by Weyl's character formula. Moreover $L_q(\lambda)$ is a simple module of highest weight λ and furthermore $\{L_q(\lambda) \mid \lambda \in X^+(n)\}$ is a complete set of pairwise non-isomorphic simple $G_q(n)$-modules.

Thus we are in the same situation as in the non-quantized case. Two finite dimensional $G_q(n)$-modules have the same composition factors (counting multiplicities) if and only if they have the same character. The main problem is to find the characters of the irreducible modules. We have a supply of modules whose characters are well understood, i.e. the $\nabla_q(\lambda)$'s, and the characters of the simple modules $L_q(\lambda)$ are related to those of the $\nabla_q(\lambda)$'s by a unitriangular matrix. The main problem is thus equivalent to determining the decomposition matrix $([\lambda : \mu])$, where, for $\lambda, \mu \in X^+(n)$, we are writing $[\lambda : \mu]$ for the multiplicity of $L_q(\mu)$ as a composition factor of $\nabla_q(\lambda)$.

0.22   We now turn to the q-Schur algebra $S_q(n, r)$. For $\lambda \in \Lambda^+(n, r)$, the $G_q(n)$-module $\nabla_q(\lambda)$ is polynomial of degree r, and thus naturally an $S_q(n, r)$-module. Furthermore $\{L_q(\lambda) \mid \lambda \in \Lambda^+(n, r)\}$ is a complete set of pairwise non-isomorphic simple $S_q(n, r)$-modules. Moreover we have, as in 0.17, natural isomorphisms

$$\mathrm{Ext}^i_{S_q(n,r)}(U, V) \cong \mathrm{Ext}^i_{G_q(n)}(U, V)$$

for $i \geq 0$ and $U, V \in \mathrm{mod}(S_q(n,r))$.

One also has a q-version of the theory of weights for $S(n,r)$ described in 0.14. We first construct a suitable basis of $A_q(n,r)$. For $i = (i_1, \ldots, i_r), j = (j_1, \ldots, j_r) \in I(n,r)$ we put $c_{ij} = c_{i_1 j_1} \ldots c_{i_r j_r}$. For $\alpha = (\alpha_1, \ldots, \alpha_n) \in \Lambda(n,r)$, we write $I^-(\alpha)$ for the set of all $j = (j_1, \ldots, j_r) \in I(n,r)$ such that $j_1 \geq \cdots \geq j_{\alpha_1}, j_{\alpha_1+1} \geq \cdots \geq j_{\alpha_1+\alpha_2}$, and so on. We write U for the set of all pairs $(i,j) \in I(n,r) \times I(n,r)$ such that i is weakly increasing of content α, say, and $j \in I^-(\alpha)$. Then $\{c_{ij} \mid (i,j) \in U\}$ is a basis of $A(n,r)$. Thus we get a dual basis $\{\xi_{ij} \mid (i,j) \in U\}$ of $S_q(n,r)$. For $\alpha = (\alpha_1, \ldots, \alpha_n) \in \Lambda(n,r)$ we put $i^\alpha = (1, \ldots, 1, 2, \ldots, 2, 3, \ldots)$ (where 1 occurs α_1 times, 2 occurs α_2 times, and so on). We put $\xi_\alpha = \xi_{i^\alpha, i^\alpha}$. Then each ξ_α is an idempotent and we have the orthogonal decomposition

$$1 = \sum_{\alpha \in \Lambda(n,r)} \xi_\alpha$$

as before. Moreover, if $q \neq 0$ and $V \in \mathrm{mod}(G_q(n))$, we have $V^\alpha = \xi_\alpha V$, for $\alpha \in \Lambda(n,r)$.

0.23 We now assume $r \leq n$. As in the classical case, we single out the idempotent $e = \xi_\omega$ for special treatment, where $\omega = (1, \ldots, 1) \in \Lambda(n,r)$. Then we have:

(1) $eS_q(n,r)e$ has basis $\{\xi_{u,u\pi} \mid \pi \in \mathrm{Sym}(r)\}$, where $u = (1, 2, \ldots, r) \in I(n,r)$.

We define $T_w = \xi_{u,uw^{-1}}$, for $w \in \mathrm{Sym}(r)$. For $1 \leq a < r$ we write s_a for the basic transposition $(a, a+1)$. Then we have the following.

(2) Let $1 \leq a < r$ and $w \in \mathrm{Sym}(r)$. We have

$$T_{s_a} T_w = \begin{cases} T_{s_a w}, & \text{if } l(s_a w) = l(w) + 1; \\ qT_{s_a w} + (q-1)T_w, & \text{if } l(s_a w) = l(w) - 1. \end{cases}$$

These are defining relations for the Hecke algebra $\mathrm{Hec}(r)$. Thus we get a surjection $\mathrm{Hec}(r) \to eS_q(n,r)e$. However, both $\mathrm{Hec}(r)$ and $eS_q(n,r)e$ have dimension $r!$, so that $\mathrm{Hec}(r) \to eS_q(n,r)e$ is an isomorphism, and $eS_q(n,r)e$ is the Hecke algebra of type A_{r-1} on generators T_{s_a}.

Thus we have, as in the classical case, the Schur functor

$$f_q : \mathrm{mod}(S_q(n,r)) \to \mathrm{Hec}(r),$$

taking an object $V \in \mathrm{mod}(S_q(n,r))$ to $eV \in \mathrm{mod}(eS_q(n,r)e)$. The non-zero modules among the $f_q L_q(\lambda)$ parametrize the simple modules for $\mathrm{Hec}(r)$. To describe this set we need some more notation. If q is not a root of unity then we set $\Lambda^+(n,r)_{col} = \Lambda^+(n,r)$. If $q = 1$ and k has characteristic 0, we again set $\Lambda^+(n,r)_{col} = \Lambda^+(n,r)$. In the remaining cases we define a positive integer s to be the (non-zero) characteristic of k if $q = 1$ and to be l if q is a primitive lth root of unity (and $l > 1$). We write $\Lambda^+(n,r)_{col}$ for the set of column s-regular elements of $\Lambda^+(n,r)$.

(3)   For $\lambda \in \Lambda^+(n,r)$, the Hec(r)-module $f_q L_q(\lambda)$ is non-zero if and only if $\lambda \in \Lambda^+(n,r)_{col}$. Moreover $\{f_q L_q(\lambda) \mid \lambda \in \Lambda^+(n,r)_{col}\}$ is a complete set of pairwise non-isomorphic simple Hec(r)-modules.

0.24   Having described the general framework, we now briefly describe the contents of these notes. There is a well-known basis of bideterminants of the polynomial algebra $A(n)$, in the n^2 commuting variables c_{ij}. The bideterminants are products of minors of the matrix (c_{ij}). This basis gives rise to a certain filtration of $A(n)$ as a $(G(n), G(n))$-bimodule. In Chapter 1, we give a q-analogue of this. The bideterminants are realized as coefficient functions of exterior powers of the natural $G_q(n)$-module. In the second chapter, we describe the formalism of the q-Schur functor and use it to describe the character of the irreducible $S_q(n,r)$-modules when $q = 0$.

The third chapter is concerned with the infinitesimal theory of $G_q(n)$, assuming that q is a primitive lth root of 1, with $l > 1$; this proceeds by analogy with Jantzen's theory of G_1T modules, for a reductive group G with maximal torus T. We have the finite quantum subgroup $\hat{G}_q(n)_1$, whose defining ideal $I_{G_q(n)_1}$ is generated by all c_{ij}^l, with $i \neq j$. We discuss the representation theory of $\hat{G}_q(n)_1$, in particular the irreducible modules, Steinberg's tensor product theorem, the principal indecomposable modules, we describe the theory of tilting modules for quantum GL_n and use the infinitesimal theory to completely determine the tilting modules for quantum GL_2. This has an application to decomposition numbers for Hecke algebras, which is given in 4.5.

The final chapter is concerned with the connections between the Hecke algebra and the q-Schur algebra, as well as various other topics. Among the other topics is the determination of the global dimension of $S_q(n,r)$, for $r \leq n$, given in 4.8. This is strongly influenced by the recent work of Totaro dealing with the classical case $q = 1$.

We conclude this exposition with a warning that the index q, added in the above to the notation given in the classical case, will be dropped in the main text. This is unlikely to lead to confusion, since we do not often need to compare with the classical case, and avoids some unnecessary notational complications. Thus we shall simply write $G(n)$ for $G_q(n)$, write $S(n,r)$ for $S_q(n,r)$, write $L(\lambda)$ for $L_q(\lambda)$ and so on.

0.25   The approach to linear algebraic groups in 0.2 to 0.6 is taken from

Chapter 1 of Steinberg, [**74**]. For more on k-groups and their relationship with Hopf k-algebras (Sections 0.7, 0.8 and 0.10, 0.11, 0.12) see Chapter AG of Borel, [**6**]. For further reading relevant to 0.9 see the thorough account of the representation theory of coalgebras given by Green in [**50**]. For the aspects of the representation theory of GL_1^n and GL_n featured in 0.12, 0.13, 0.14 and 0.15 see Green, [**51**], especially Chapter 3 and Sections 2.1, 2.2, 2.3. For the approach to the induced modules $\nabla(\lambda)$ in 0.16, see Jantzen, [**61**], Part II, Chapter 5. For the isomorphism $\mathrm{Ext}^*_{S(n,r)}(U,V) \to \mathrm{Ext}^*_{G(n)}(U,V)$ of 0.17 see [**31**]. For the material of 0.18 as formulated here see Green, [**51**], Chapter 6. Section 0.19 needs no reference. The construction of quantum GL_n given in 0.20 is taken from [**18**]. The material on quantum GL_n in 0.21 is taken from [**36**]. The q-Schur algebras were introduced by Dipper and James, [**21**]. For the isomorphism $\mathrm{Ext}^*_{S_q(n,r)}(U,V) \to \mathrm{Ext}^*_{G_q(n)}(U,V)$ of 0.22 see [**36**]. The rest of 0.22 and the remaining sections concern topics to be described in more detail in the main body of the text. The parametrization of the irreducible modules for Hecke algebras of type A, given in 0.22, is due to Dipper and James, [**20**].

1. Exterior Algebra

The purpose of this chapter is to give a basis of $A(n)$ consisting of bidetermi-nants. This basis is a q-analogue of the basis given by several authors in the classical case: see Mead, [67]; Doubilet, Rota and Stein, [43]; De Concini, Eisenbud and Procesi, [17]; and Green, [52]. Our treatment differs from these in that we view the bideterminants throughout as coefficient functions on modules for quantum GL_n which are tensor products of exterior powers of the natural module.

We shall later give a basis of the induced module $\nabla(\lambda)$ (for $\lambda \in \Lambda^+(n, r)$) consisting of bideterminants, Proposition 4.5.2, in the spirit of Green's treat-ment in the classical case, [51; (4.5a)].

We shall also see later that the tensor products of exterior powers of the natural module are distinguished by being tilting modules for GL_n and that every indecomposable polynomial tilting module is a direct summand of such a module. However, these features are not used in this chapter, which is entirely combinatorial.

In the first section we introduce the notion of a comodule pairing and list some elementary properties for future use. In the second section we describe an algebra which is a quotient of the tensor algebra on the exterior algebra of the natural module E for the quantum general linear group. In the last section we define (two kinds of) bideterminants as certain coefficient functions and prove the basis theorem.

1.1 Preliminaries

We begin by establishing some notation and listing, without proof, some elementary properties of representations of comodules for future use. For a set X we denote by id_X (or simply id) the identity map on X. We fix a field k. Let (A, δ, ε) be a k-coalgebra. We often denote by τ_E (resp. τ_V) the structure map $E \to E \otimes A$ (resp. $V \to A \otimes V$) of a right (resp. left) A-comodule E (resp. V). Suppose that $\phi : Y \to Z$ is a morphism of comodules. Recall that we have $\mathrm{cf}(Y) \leq \mathrm{cf}(Z)$ if ϕ is injective and that $\mathrm{cf}(Z) \leq \mathrm{cf}(Y)$ if ϕ is surjective, [50; (1.2c)].

For a k-space X we write $|X| \otimes E$ for the k-space $X \otimes E$, regarded as a comodule via the structure map $\mathrm{id} \otimes \tau_E : X \otimes E \to X \otimes E \otimes A$. Similarly we define a right comodule $E \otimes |X|$. We define left comodules $V \otimes |X|$ and $|X| \otimes V$ in the same way. Note that the structure maps $E \to |E| \otimes A$ and $V \to A \otimes |V|$ are morphisms of comodules.

We shall identify a bilinear map $b : X \times Y \to k$ with its k-linear extension $X \otimes Y \to k$. We say that a bilinear map $b : V \times E \to k$ is a *pairing* if it is non-singular and $(\mathrm{id} \otimes b) \circ (\tau_V \otimes \mathrm{id}) = (b \otimes \mathrm{id}) \circ (\mathrm{id} \otimes \tau_E)$. If b is a pairing we write Φ_b for the map $(\mathrm{id} \otimes b) \circ (\tau_V \otimes \mathrm{id}) = (b \otimes \mathrm{id}) \circ (\mathrm{id} \otimes \tau_E) : V \otimes E \to A$.

By an (A, A)-bicomodule we mean a triple (U, λ, ρ) consisting of a k-space U and linear maps $\lambda : U \to A \otimes U$, $\rho : U \to U \otimes A$ such that (U, λ) is a left A-comodule, (U, ρ) is a right A-comodule and we have $(\mathrm{id} \otimes \rho) \circ \lambda = (\lambda \otimes \mathrm{id}) \circ \rho : U \to A \otimes U \otimes A$. Note that (A, δ, δ) is an (A, A)-bicomodule. Note also that $V \otimes E$ is naturally an (A, A)-comodule with structure maps $\tau_V \otimes \mathrm{id} : V \otimes E \to A \otimes V \otimes E$ and $\mathrm{id} \otimes \tau_E : V \otimes E \to V \otimes E \otimes A$. Let T and U be A-bicomodules. A map $\phi : T \to U$ is a morphism of bicomodules if it is a morphism of left comodules and a morphism of right comodules. The following is an easy consequence of the definitions.

Lemma 1.1.1 *Suppose that $b : V \times E \to k$ is a pairing.*
(i) $\Phi_b : V \otimes E \to A$ is a morphism of (A, A)-bicomodules.
(ii) $\mathrm{cf}(V) = \mathrm{cf}(E)$.

For left (resp. right) A-comodules X, Y we denote by $\mathrm{Hom}_A(X, Y)$ the space of comodule homomorphisms from X to Y. For a k-space Z we denote the dual space $\mathrm{Hom}_k(Z, K)$ by Z^*. For a linear map $\alpha : Z_1 \to Z_2$ we write α^* for the dual map $Z_2^* \to Z_1^*$. For a comodule X we have a natural isomorphism $\mathrm{Hom}_A(X, A) \to X^*$, taking $\theta \in \mathrm{Hom}_A(X, A)$ to $\varepsilon \circ \theta \in X^*$. In particular the functor $\mathrm{Hom}_A(-, A)$, from A-comodules to k-spaces, is exact and so A is injective as a comodule over itself. (A detailed account of the representation theory of coalgebras is to be found in [50].) We now suppose that E is finite dimensional and let α_i, $i \in I$, be the basis dual to e_i, $i \in I$. We make E^* into a right module with structure map τ_{E^*} satisfying $\tau_{E^*}(\alpha_i) = \sum_{j \in I} c_{ij} \otimes \alpha_j$. The construction does not depend on the choice of basis. Similarly the dual V^* of the left A-comodule V is naturally a right A-comodule. We leave it to the reader to check the following.

Lemma 1.1.2 *(i) Let X_1 and X_2 be either both finite dimensional left A-comodules or both finite dimensional right A-comodules. Then we have the natural isomorphism*

$$\mathrm{Hom}_A(X_1, X_2) \to \mathrm{Hom}_A(X_2^*, X_1^*)$$

taking $\alpha \in \mathrm{Hom}_A(X_1, X_2)$ to $\alpha^* \in \mathrm{Hom}_A(X_2^*, X_1^*)$.
(ii) If V (resp. E) is injective then V^* (resp. E^*) is projective and if V (resp. E) is projective then V^* (resp. E^*) is injective.
(iii) Suppose we have a pairing $V \times E \to k$. Then we have $V^* \cong E$ and $E^* \cong V$, as A-comodules. If V (resp. E) is injective then E (resp. V) is projective. If V (resp. E) is projective then E (resp. V) is injective.
(iv) Suppose we have pairings $b_1 : V_1 \times E_1 \to k$, $b_2 : V_2 \times E_2 \to k$ of finite dimensional A-comodules. For each $\phi \in \mathrm{Hom}_k(V_1, V_2)$ we have the adjoint map $\mathrm{ad}(\phi) \in \mathrm{Hom}_k(E_2, E_1)$, defined by $b_1(v_1, \mathrm{ad}(\phi)e_2) = b_2(\phi(v_1), e_2)$, for $v_1 \in V_1$, $e_2 \in E_2$. Furthermore ϕ is an A-comodule map if and only if $\mathrm{ad}(\phi)$ is an A-comodule map.

1.2 Exterior algebra

For integers $a \le b$ we write $[a, b]$ for the set of integers m such that $a \le m \le b$.
For a set X we write $\mathrm{Sym}(X)$ for the group of permutations of X. Thus
$\mathrm{Sym}(r) = \mathrm{Sym}[1, r]$, for $r \ge 1$.

We fix a positive integer n. Let R be a commutative ring and $q \in R$.
Let $A(n) = A_{R,q}(n)$ be the R-algebra generated by c_{ij}, $1 \le i, j \le n$, subject
to the relations

AI	$c_{ir} c_{is} = c_{is} c_{ir}$ for all i, r, s;
AII	$c_{jr} c_{is} = q c_{is} c_{jr}$ for all $i < j$ and $r \le s$;
AIII	$c_{js} c_{ir} = c_{ir} c_{js} + (q - 1) c_{is} c_{jr}$ for all $i < j$ and $r < s$

(with $i, j, r, s \in [1, n]$). Note that AIII, in the presence of AI, AII, is equivalent to

AIII' $c_{ir} c_{js} + c_{jr} c_{is} = c_{js} c_{ir} + c_{is} c_{jr}$, for all $1 \le i \le j \le n$, $1 \le r < s \le n$.

For elements a, b, c, d of a ring X we write $\begin{vmatrix} a & b \\ c & d \end{vmatrix}$ for $ad - bc \in X$. Thus
AIII' may be expressed

AIII'' $\begin{vmatrix} c_{ir} & c_{is} \\ c_{jr} & c_{js} \end{vmatrix} = - \begin{vmatrix} c_{jr} & c_{js} \\ c_{ir} & c_{is} \end{vmatrix}$

for all $1 \le i \le j \le n$, $1 \le r < s \le n$.

Note that $A(n)$ has a natural R-algebra grading $A(n) = \bigoplus_{r=0}^{\infty} A(n, r)$,
where each c_{ij} has degree 1. From [18; 1.1.8 Theorem] we have that the construction commutes with base change, i.e. the natural map $R \otimes_{\mathbf{Z}[t]} A_{\mathbf{Z}[t],t}(n) \to$
$A_{R,q}(n)$ (where t is an indeterminant) is an R-algebra isomorphism, inducing
an R-module isomorphism $R \otimes_{\mathbf{Z}[t]} A_{\mathbf{Z}[t],t}(n, r) \to A_{R,q}(n, r)$ in each degree.

Suppose $R = k$, a field. We denote by $M = M(n)$ the quantum monoid
with coordinate algebra $k[M] = A(n)$, where k is a field. Here, by analogy
with 0.20, we use the expression "H is a quantum monoid" simply to indicate
that we have in mind a k-bialgebra denoted $k[H]$ and called the coordinate
algebra of M. By definition a left (resp. right) H-module is a right (resp.
left) $k[H]$ comodule and a homomorphism of H-modules is a $k[H]$-comodule
map.

Now assume that $q \ne 0$. Then, as in 0.20, the quantum group $G = G(n)$
with coordinate algebra $k[G] = A(n)_d$ is obtained by localizing $A(n)$ at the
(quantum) determinant d. If $0 \ne q$ then any left (resp. right) M-module
V, with structure map $\tau : V \to V \otimes k[M]$ (resp. $\tau : V \to k[M] \otimes V$), is
naturally a G-module with structure map $\tau' = \iota \circ \tau$, where ι is inclusion
$V \otimes k[M] \to V \otimes k[G]$ (resp. $k[M] \otimes V \to k[G] \otimes V$). A G-module V may be
obtained from an M-module in this way if and only if the coefficient space
$\mathrm{cf}(V)$ lies in $k[M]$ (i.e. V is a polynomial G-module).

We now take E to be the natural left M-module and V to be the natural
right M-module. Thus E has k-basis $e_1, \ldots, e_n$ and structure map taking e_i

to $\sum_{j=1}^{n} e_j \otimes c_{ji}$ and V has k-basis $v_1, \ldots, v_n$ and structure map taking v_i to $\sum_{j=1}^{n} c_{ij} \otimes v_j$ (for $1 \leq i \leq n$). We recall from [18; 2.1] the construction of the exterior powers of the natural modules E and V. Let $T(E) = \bigoplus_{r=0}^{\infty} E^{\otimes r}$ be the tensor algebra on E. We define J to be the ideal of $T(E)$ generated by the elements e_h^2, with $1 \leq h \leq n$, and $e_i e_j + q e_j e_i$, with $1 \leq i < j \leq n$. Then J is an M-submodule of $T(E)$. We define $\bigwedge(E) = T(E)/J$ and write $\wedge$ for the multiplication in $\bigwedge(E)$. Since J is homogeneous we have an induced grading and M-module decomposition $\bigwedge(E) = \bigoplus_{r=0}^{\infty} \bigwedge^r E$. The rth exterior power $\bigwedge^r E$ has k-basis consisting of the elements $e_{i_1} \wedge \cdots \wedge e_{i_r}$, with $n \geq i_1 > \cdots > i_r \geq 1$. Moreover, it follows from the invariance of the character of $\bigwedge^r E$ under the action of $\mathrm{Sym}(r)$ that $\bigwedge^r E$ is an irreducible left M-module, for $1 \leq r \leq n$, see [36; Remark 3.7]. Similarly we define K to be the ideal of $T(V)$ generated by the elements v_h^2, with $1 \leq h \leq n$, and $v_i v_j + q v_j v_i$, with $1 \leq i < j \leq n$. Then K is an M-submodule of $T(V)$. We define $\bigwedge(V) = T(V)/K$ and write $\wedge$ for the multiplication in $\bigwedge(V)$. Since K is homogeneous we have an induced grading and M-module decomposition $\bigwedge(V) = \bigoplus_{r=0}^{\infty} \bigwedge^r V$. The rth exterior power $\bigwedge^r V$ has k-basis consisting of the elements $v_{i_1} \wedge \cdots \wedge v_{i_r}$, with $1 \leq i_1 < \cdots < i_r \leq n$. Moreover, $\bigwedge^r V$ is an irreducible right M-module, for $1 \leq r \leq n$.

Recall, from 0.14, that $I(n,r)$ denotes the set of functions $i : [1,r] \to [1,n]$. We shall often write $i \in I(n,r)$ as the sequence $(i_1, \ldots, i_r)$ (where $i_a = i(a)$, $1 \leq a \leq r$). Let $i = (i_1, \ldots, i_r) \in I(n,r)$. We define $e_i = e_{i_1} \otimes \cdots \otimes e_{i_r} \in E^{\otimes r}$ and $v_i = v_{i_1} \otimes \cdots \otimes v_{i_r} \in V^{\otimes r}$. We define $\hat{e}_i = e_{i_1} \wedge \cdots \wedge e_{i_r} \in \bigwedge^r E$, the image of e_i under the natural map $E^{\otimes r} \to \bigwedge^r E$ and define $\hat{v}_i = v_{i_1} \wedge \cdots \wedge v_{i_r} \in \bigwedge^r V$, the image of v_i under the natural map $V^{\otimes r} \to \bigwedge^r V$.

We write $P = P(n)$ for a certain set of sequences of non-negative integers. We define P to be 0 together with sequences $\alpha = (\alpha_1, \ldots, \alpha_m)$, for some positive integer m, with entries $\alpha_1, \ldots, \alpha_m \in [1,n]$. For $\alpha, \beta \in P$ we define $(\alpha|\beta)$ to be α if $\beta = 0$, to be β if $\alpha = 0$ and to be the concatenation $(\alpha_1, \ldots, \alpha_l, \beta_1, \ldots, \beta_m)$ if $\alpha = (\alpha_1, \ldots, \alpha_l)$, $\beta = (\beta_1, \ldots, \beta_m)$. We make P into a monoid with binary operation $(\alpha|\beta)$. For $\alpha \in P$ we set $\bigwedge^\alpha E = k$ if $\alpha = 0$ and $\bigwedge^\alpha E = \bigwedge^{\alpha_1} E \otimes \cdots \otimes \bigwedge^{\alpha_m} E$ if $\alpha = (\alpha_1, \ldots, \alpha_m)$. In the same way, we set $\bigwedge^\alpha V = k$ if $\alpha = 0$ and $\bigwedge^\alpha V = \bigwedge^{\alpha_1} V \otimes \cdots \otimes \bigwedge^{\alpha_m} V$ if $\alpha = (\alpha_1, \ldots, \alpha_m)$.

We set $P(n,0) = \{0\}$. For $r > 0$ we write $P(n,r)$ for the set of $\alpha = (\alpha_1, \ldots, \alpha_m) \in P(n)$ such that $\sum_{i=1}^{m} \alpha_i = r$. We write $P^+(n,r)$ for the set of partitions in $P(n,r)$ (for $r \geq 0$). For $\alpha \in P$ we denote by $\bar{\alpha}$ the partition associated with α, i.e. the partition obtained by arranging the terms in α in descending order. We shall use the natural (dominance) partial order on partitions. Thus, for partitions $\lambda = (\lambda_1, \lambda_2, \ldots)$ and $\mu = (\mu_1, \mu_2, \ldots)$ we write $\lambda \geq \mu$ if $|\lambda| = |\mu|$ and $\lambda_1 + \cdots + \lambda_a \geq \mu_1 + \cdots + \mu_a$ for all $a \geq 1$ (see 0.15). For a partition λ we denote by λ' the conjugate partition. For partitions λ, μ we have the partition $\lambda \bigcup \mu$, whose parts are those of λ and μ

arranged in descending order (see [63; I,1]). Note that for $\alpha, \beta, \gamma \in P$ with $\bar{\alpha} > \bar{\beta}$ we have $\bar{\alpha} \bigcup \bar{\gamma} > \bar{\alpha} \bigcup \bar{\gamma}$, i.e. $(\alpha|\gamma) > (\beta|\gamma)$, for any $\gamma \in P$.

Let R denote the tensor algebra on $\bigoplus_{i=1}^{n} \bigwedge^{i} E$. Thus we have an M-module decomposition $R = \bigoplus_{\alpha \in P} R^{\alpha}$, where $R^{\alpha} = \bigwedge^{\alpha} E$, $\alpha \in P$.

For $\alpha \in P$ we write J^{α} for the M-submodule of R^{α} generated by the images of all M-module homomorphisms $R^{\beta} \to R^{\alpha}$ for $\beta \in P$ with $\bar{\beta} \not\leq \bar{\alpha}$. We set $J = \bigoplus_{\alpha \in P} J^{\alpha}$.

Proposition 1.2.1 *(i) J is an ideal of R.*
(ii) For $\alpha \in P$, the module R^{α}/J^{α} has unique highest weight $\bar{\alpha}'$ and

$$\dim (R^{\alpha}/J^{\alpha})^{\bar{\alpha}'} = 1.$$

Proof   (i) Let $\alpha, \beta \in P$. Clearly $J^{\alpha} R^{\beta}, R^{\alpha} J^{\beta} \subset J$ if α or β is 0. We suppose $\alpha, \beta \neq 0$ and show that $R^{\alpha} J^{\beta} \subset J$ (the other case is similar). It suffices to show that $R^{\alpha} \otimes \mathrm{Im}(\theta) \subset J$ for any M-map $\theta : R^{\gamma} \to R^{\beta}$ with $\bar{\gamma} \not\leq \bar{\beta}$. However, we have $R^{\alpha} \otimes \mathrm{Im}(\theta) = \mathrm{Im}(\phi)$, where $\phi = (\mathrm{id} \otimes \theta) : R^{(\alpha|\gamma)} \to R^{(\alpha|\beta)}$. Thus $R^{\alpha} \otimes \mathrm{Im}(\theta)$ is contained in J unless $\overline{(\alpha|\gamma)} \leq \overline{(\alpha|\beta)}$. However, that would give $\bar{\alpha} \bigcup \bar{\gamma} \leq \bar{\alpha} \bigcup \bar{\beta}$ and hence, by [63,I,(1.8)], $\bar{\alpha}' + \bar{\gamma}' \geq \bar{\alpha}' + \bar{\beta}'$, and so $\bar{\gamma}' \geq \bar{\beta}'$ and $\bar{\gamma} \leq \bar{\beta}$, a contradiction.
(ii) Suppose $0 \neq \alpha = (\alpha_1, \ldots, \alpha_m) \in P$. If α is a partition then it is easy to see that $e_1 \wedge \cdots \wedge e_{\alpha_1} \otimes e_1 \wedge \cdots \wedge e_{\alpha_2} \otimes \cdots$ spans the α' weight space of $\bigwedge^{\alpha} E$ and α' is the unique highest weight of $R^{\alpha} = \bigwedge^{\alpha} E$. For general α the module R^{α} has the same character, hence the same weight multiplicities, as $R^{\bar{\alpha}}$. Hence R^{α} has unique highest weight $\bar{\alpha}'$ and this occurs with multiplicity 1. Note this holds also in the case $\alpha = 0$. To get the required assertion, we need to know that $\bar{\alpha}'$ is not a weight of the image of a homomorphism $R^{\beta} \to R^{\alpha}$ with $\bar{\beta} \not\leq \bar{\alpha}$. But this would give that $\bar{\alpha}'$ is a weight of R^{β} and so $\bar{\alpha}' \leq \bar{\beta}'$ and hence $\bar{\beta} \leq \bar{\alpha}$, a contradiction.

We now introduce some additional notation. Let r be a positive integer. We write $I_0(n,r)$ for the set of $i \in I(n,r)$ with distinct entries and define $I_1(n,r)$ to be the set of strictly decreasing sequences in $I(n,r)$. For a partition $\lambda = (\lambda, \ldots, \lambda_n)$ we write $\mathrm{Tab}(\lambda)$ for the set of λ-tableaux with entries in $[1,n]$. We call a λ-tableau *standard* if the entries are strictly increasing along rows and weakly increasing down columns. (Note that we have reversed rows and columns in the notion of standard used in [51].) We call a λ-tableau *antistandard* if the entries are strictly decreasing along rows and weakly decreasing down columns. We write $\mathrm{AStan}(\lambda)$ for the set of antistandard λ-tableaux. We write $\mathrm{Tab}_0(\lambda)$ for the set of λ-tableaux S such that the entries in each row of S are distinct and write $\mathrm{Tab}_1(\lambda)$ for the set of λ-tableaux such that the entries in each row of S are strictly decreasing. For $S \in \mathrm{Tab}_0(\lambda)$ we write $\bar{S}$ for the element of $\mathrm{Tab}_1(\lambda)$ obtained by arranging the elements in each row of S in descending order. For a partition $\lambda \in P$ and $S \in \text{Tab}(\lambda)$ we put

$$\hat{e}_S = e_{S(1,1)} \wedge \cdots \wedge e_{S(1,\lambda_1)} \otimes e_{S(2,1)} \wedge \cdots \wedge e_{S(2,\lambda_2)} \otimes \cdots .$$

We note that $\hat{e}_S = 0$ if $S \notin \text{Tab}_0(\lambda)$. We note also that for $S \in \text{Tab}_0(\lambda)$ we have $\hat{e}_S = (-q)^l \hat{e}_{\bar{S}}$ for some $l \geq 0$ (from the description of $\bigwedge(E)$ given above) and that the elements $\hat{e}_S$, $S \in \text{Tab}_1(\lambda)$, form a basis of $\bigwedge^\lambda E$. The *content* of a λ-tableau S is $c = (c_1, \ldots, c_n) \in \mathbb{Z}^n$, where c_r is the number positions in the tableau in which the entry r appears, for $1 \leq r \leq n$.

We order r-tuples of non-negative integers lexicographically. We order elements of $\text{Tab}_1(\lambda)$ of given content lexicographically by row. Thus if S and T are in $\text{Tab}_1(\lambda)$ with rows $S^1, \ldots, S^m$ and $T^1, \ldots, T^m$ respectively (where $\lambda = (\lambda_1, \ldots, \lambda_m)$) then $S < T$ if S and T have the same content and, for some $1 \leq a \leq m$, we have $S^i = T^i$ for $i < a$ and $S^a < T^a$.

Lemma 1.2.2 For $\alpha \in P(n)$ we have an M-pairing $b : \bigwedge^\alpha V \times \bigwedge^\alpha E \to k$ such that $(\hat{v}_S, \hat{e}_T) = \delta_{S,T}$, for $S, T \in \text{Tab}_1(\alpha)$.

Proof We have a pairing $b_1 : V^{\otimes r} \times E^{\otimes r} \to k$ such that $b_1(v_i, e_j) = \delta_{ij}$, for $i, j \in I(n, r)$, see [36], the discussion preceding Lemma 3.3. We have the natural map $\psi : \bigwedge^r E \to E^{\otimes r}$, $\psi(\hat{e}_j) = \sum_{\pi \in \text{Sym}(r)} \text{sgn}(\pi) e_{j\pi}$, for $j \in I_1(n, r)$, and hence a map $b_2 : V^{\otimes r} \times \bigwedge^r E \to k$ given by $b_2(x, y) = b_1(x, \psi(y))$ (for $x \in V^{\otimes r}$, $y \in \bigwedge^r E$). Specifically, we have $b_2(v_i, \hat{e}_j) = \sum_{\pi \in \text{Sym}(r)} \text{sgn}(\pi) \delta_{i,j\pi}$, for $i \in I(n, r)$ and $j \in I_1(n, r)$. The kernel N, say, of the natural map $V^{\otimes r} \to \bigwedge^r V$ is spanned by elements v_i such that i has a repeated entry, together with the elements $v_i - \text{sgn}(\sigma) v_{i\sigma}$, $\sigma \in \text{Sym}(r)$. However, we have $b_2(v, \hat{e}_j) = 0$, for $j \in I_1(n, r)$, if v is one of these elements. Hence $b_2(v, e) = 0$ for all $v \in N$, $e \in \bigwedge^r E$ and b_2 induces a map $b_3 : \bigwedge^r V \otimes \bigwedge^r E \to k$. For $i, j \in I_1(n, r)$ we have $b_3(\hat{v}_i, \hat{e}_j) = b_2(v_i, \hat{e}_j) = b_1(v_i, e_j) = \delta_{ij}$. Thus, for each i, we have a pairing $\bigwedge^{\alpha_i} V \times \bigwedge^{\alpha_i} E \to k$ as above and for $b : \bigwedge^\alpha V \times \bigwedge^\alpha E \to k$ we take the product form (see [36; Section 3]).

We write $X(r, s)$ for the set of $\sigma \in \text{Sym}(r+s)$ such that $\sigma(a) < \sigma(b)$ whenever $a < b$ and $a, b \in [1, r]$ or $a, b \in [r+1, r+s]$.

Lemma 1.2.3 The linear map $\psi : \bigwedge^{r+s} E \to \bigwedge^r E \otimes \bigwedge^s E$ given by $\psi(\hat{e}_i) = \sum_{\sigma \in X(r,s)} \text{sgn}(\sigma) \bar{e}_{i\sigma}$, for $i \in I_1(n, r+s)$, where $\bar{e}_{i\sigma}$ denotes the image of $e_{i\sigma}$ under the natural map $E^{\otimes(r+s)} \to E^{\otimes r} \otimes E^{\otimes s}$, is an M-module map.

Proof We have the natural pairings $\bigwedge^{r+s} V \times \bigwedge^{r+s} E \to k$ and $\bigwedge^r V \otimes \bigwedge^s V \times \bigwedge^r E \otimes \bigwedge^s E \to k$. We leave it to the reader to check that the map ψ is the adjoint of multiplication $\bigwedge^r V \otimes \bigwedge^s V \to \bigwedge^{r+s} V$ (cf. [18; Section 2.1]) and so is an M-module map, by Lemma 1.1.2(iv).

Theorem 1.2.4 *Let λ be a partition in P. For each $S \in \mathrm{Tab}(\lambda)$ there exist elements c_T of k, for $T \in \mathrm{AStan}(\lambda)$, such that $c_T = 0$ unless $T \geq \bar{S}$ and $\hat{e}_S \equiv \sum_{T \in \mathrm{AStan}(\lambda)} c_T \hat{e}_T \pmod{J^\lambda}$. In particular, the elements $\hat{e}_S + J^\lambda$, $S \in \mathrm{AStan}(\lambda)$, form a spanning set of R^λ / J^λ.*

Proof If $\lambda = 0$ there is nothing to prove. If $\lambda = (\lambda_1)$ (a 1-part partition) then $\bigwedge^\lambda E = \bigwedge^{\lambda_1} E$ is simple, $J^\lambda = 0$ and again the result is clear.

Now suppose that λ has 2 parts, $\lambda = (l, m)$, say, with $l \geq m > 0$. If $S \notin \mathrm{Tab}_0(\lambda)$, then $\hat{e}_S = 0$. We assume inductively that $S \in \mathrm{Tab}_0(\lambda)$ and that the result holds for all $U \in \mathrm{Tab}_0(\lambda)$ with $\bar{U} > \bar{S}$. We have $\hat{e}_S = (-q)^l \hat{e}_{\bar{S}}$, for some $r \geq 0$, so we may assume that $S \in \mathrm{Tab}_1(\lambda)$. Let $S(1,1) = i_1, \ldots,$ $S(1,l) = i_l$ and $S(2,1) = j_1, \ldots, S(2,m) = j_m$. Thus we have $i_1 > \cdots > i_l$, $j_1 > \cdots > j_m$. If S is antistandard there is nothing to prove. So we assume $S \notin \mathrm{AStan}(\lambda)$ and let a be as small as possible such that $i_a < j_a$. Thus we have $j_1 > \cdots > j_a > i_a > \cdots > i_l$. We have the map $\psi : \bigwedge^{l+1} E \to \bigwedge^{l+1-a} E \otimes \bigwedge^a E$ of Lemma 1.2.3 and the multiplication maps $\theta : \bigwedge^{a-1} E \otimes \bigwedge^{l+1-a} E \to \bigwedge^l E$ and $\eta : \bigwedge^a E \otimes \bigwedge^{m-a} E \to \bigwedge^m E$. Hence we have the map $\Phi = (\theta \otimes \eta) \circ (\mathrm{id} \otimes \psi \otimes \mathrm{id}) : \bigwedge^{a-1} E \otimes \bigwedge^{l+1} E \otimes \bigwedge^{m-a} E \to \bigwedge^l E \otimes \bigwedge^m E$. Let $\mu = (a-1, l+1, m-a)$. The first part of $\bar{\mu}$ is $l+1$ so that $\bar{\mu} \not\leq \lambda$ and therefore $\mathrm{Im}(\Phi) \leq J^\lambda$. Let $h = (h_1, \ldots, h_{l+1}) = (j_1, \ldots, j_a, i_a, \ldots, i_l)$, the result of arranging the sequence $(i_a, \ldots, i_l, j_1, \ldots, j_a)$ in descending order. We have

$$\Phi(e_{i_1} \wedge \cdots \wedge e_{i_{a-1}} \otimes \hat{e}_h \otimes e_{j_{a+1}} \wedge \cdots \wedge e_{j_m})$$
$$= \sum_{\sigma \in X} \mathrm{sgn}(\sigma)(\theta \otimes \eta)(e_{i_1} \wedge \cdots \wedge e_{i_{a-1}} \otimes \bar{e}_{h\sigma} \otimes e_{j_{a+1}} \wedge \cdots \wedge e_{j_m})$$

where $X = X(l+1-a, a)$. For $\sigma \in X$ we write $f_\sigma = (h_{\sigma(1)}, \ldots, h_{\sigma(l+1-a)})$ and $g_\sigma = (h_{\sigma(l+2-a)}, \ldots, h_{\sigma(l+1)})$. Thus we have

$$\Phi(e_{i_1} \wedge \cdots \wedge e_{i_{a-1}} \otimes \hat{e}_h \otimes e_{j_{a+1}} \wedge \cdots \wedge e_{j_m})$$
$$= \sum_{\sigma \in X} \mathrm{sgn}(\sigma) e_{i_1} \wedge \cdots \wedge e_{i_{a-1}} \wedge \hat{e}_{f_\sigma} \otimes \hat{e}_{g_\sigma} \wedge e_{j_{a+1}} \wedge \cdots \wedge e_{j_m}$$

and hence

$$\sum_{\sigma \in X} \mathrm{sgn}(\sigma) e_{i_1} \wedge \cdots \wedge e_{i_{a-1}} \wedge \hat{e}_{f_\sigma} \otimes \hat{e}_{g_\sigma} \wedge e_{j_{a+1}} \wedge \cdots \wedge e_{j_m} \in J^\lambda.$$

This may also be expressed

$$\sum_{\sigma \in X} \mathrm{sgn}(\sigma) \hat{e}_{S_\sigma} \in J^\lambda \qquad (*)$$

where S_σ is the λ-tableau whose first row is $(i_1, \ldots, i_{a-1}, h_{\sigma(1)}, \ldots, h_{\sigma(l+1-a)})$ and second row is $(h_{\sigma(l+2-a)}, \ldots, h_{\sigma(l+1)}, j_{a+1}, \ldots, j_m)$. Note that, for $\sigma \in$

X, the first $a - 1$ entries of the first row of the tableau S_σ are $i_1, \ldots, i_{a-1}$ and the remaining entries belong to the set $\{i_a, \ldots, i_l, j_1, \ldots, j_a\}$. There is a unique permutation $\sigma_0 \in X$ such that

$$(h_{\sigma_0(1)}, \ldots, h_{\sigma_0(l+1)}) = (i_a, \ldots, i_l, j_1, \ldots, j_a).$$

Thus we have $S_{\sigma_0} = S$. For any $\sigma_0 \neq \sigma \in X$, some j_s, with $1 \leq s \leq a$, occurs in the first $l - a + 1$ entries of $(h_{\sigma(1)}, \ldots, h_{\sigma(l+1)})$, so that j_s occurs in the first row of S_σ, in position a or later. Since $j_s \geq j_a > i_a$, we have $\bar{S}_\sigma > S$, for $\sigma \neq \sigma_0$. From $(*)$ we get

$$\hat{e}_S \equiv \sum_{\sigma \in X, \sigma \neq \sigma_0} \text{sgn}(\sigma \sigma_0) \hat{e}_{S_\sigma} \pmod{J^\lambda}$$

and therefore

$$\hat{e}_S \equiv \sum_T a_T \hat{e}_T \pmod{J^\lambda}$$

for scalars a_T with $a_T = 0$ for $\bar{T} \not> S$. By the inductive hypothesis each $\hat{e}_T$, with $T > S$, has an expression (mod J^λ) of the required form and therefore so does $\hat{e}_S$.

Now let λ be arbitrary. We assume inductively that for each $U \in \text{Tab}(\lambda)$ with $\bar{U} > \bar{S}$ the element $\hat{e}_U$ has an expression of the required form. Again we may suppose that $S \in \text{Tab}_1(\lambda)$ and that S is not antistandard. Thus there is some h such that the tableau obtained by deleting all rows except the hth and $(h+1)$st is not antistandard. We define $\alpha = (\lambda_1, \ldots, \lambda_{h-1})$, $\beta = (\lambda_h, \lambda_{h+1})$, $\gamma = (\lambda_{h+2}, \ldots)$. Thus we have $R^\lambda = R^\alpha \otimes R^\beta \otimes R^\gamma$. Let A denote the tableau formed by the first $h - 1$ rows of S, let B denote the tableau formed by rows $h, h+1$ of S and let C denote the tableau formed by rows $h+2$ etc. of S. Now by the case considered already we have $\hat{e}_B - \sum_{B'} c_{B'} \hat{e}_{B'} \in J^\beta$, for $B' \in \text{Tab}(\beta)$ with $c_{B'} = 0$ unless $\bar{B}' > B$. By Proposition 1.2.1(i) we have that $R^\alpha \otimes J^\beta \otimes J^\gamma \leq R^\lambda$ and $\hat{e}_S = \hat{e}_A \otimes \hat{e}_B \otimes \hat{e}_C$ so we get

$$\hat{e}_S - \sum_{B'} c_{B'} \hat{e}_A \otimes \hat{e}_{B'} \otimes \hat{e}_C \in J^\lambda.$$

Now, for $c_{B'} \neq 0$, we have $\bar{B}' > B$ and hence $\hat{e}_A \otimes \hat{e}_{B'} \otimes \hat{e}_C = \hat{e}_{S'}$ with $\bar{S}' > S$. The result follows now from the inductive hypothesis on S.

Remark The above development works also with the natural right M-module V in place of E. Thus, for $r, s \geq 0$, one has an M-module map $\phi : \bigwedge^{r+s} V \to \bigwedge^r V \otimes \bigwedge^s V$ given by $\phi(\hat{v}_i) = \sum_{\sigma \in X} (-q)^{l(\sigma)} \bar{v}_{i\sigma}$, for i a decreasing sequence in $I(n, r + s)$. We define, for $\alpha \in P$, the module $Q^\alpha = \bigwedge^\alpha V$ to be k if $\alpha = 0$ and $\bigwedge^{\alpha_1} V \otimes \cdots \otimes \bigwedge^{\alpha_m} V$ if $\alpha = (\alpha_1, \ldots, \alpha_m)$. We form the algebra $Q = \bigoplus_{\alpha \in P} Q^\alpha$, graded as an algebra and M-module. For $\alpha \in P$ we define $I^\alpha \leq Q^\alpha$ to be the submodule generated by the image of all M-module homomorphisms $Q^\beta \to Q^\alpha$ with $\bar\beta > \bar\alpha$. Then $I = \bigoplus_{\alpha \in P} I^\alpha$ is an ideal and there exists a "straightening formula", $v_T \equiv \sum_{T' \in \mathrm{AStan}(\alpha)} c_{T'} v_{T'} \pmod{I^\alpha}$, for $T \in \mathrm{Tab}(\alpha)$.

1.3 Bideterminants

We work over an arbitrary commutative ring R.

Definitions For $i = (i_1, \ldots, i_r)$, $j = (j_1, \ldots, j_r)$ with entries in $[1, n]$ we define the *determinant* $(i : j) = \sum_{\pi \in \mathrm{Sym}(r)} \mathrm{sgn}(\pi) c_{i, j\pi}$. Let S, T be tableaux of shape λ with entries in $[1, n]$. We define the *bideterminant* $(S : T) = (S^1 : T^1)(S^2 : T^2) \ldots (S^m : T^m)$, where $\lambda = (\lambda_1, \ldots, \lambda_m)$ and S^j (resp. T^j) is the jth row of S (resp. T), for $1 \le j \le m$.

In fact there are two contenders for the title determinant and for the title bideterminant. Let i, j, S, T, λ be as above. We define

$$\langle i : j \rangle = \sum_{\pi \in \mathrm{Sym}(r)} (-q)^{l(\pi)} c_{i\pi, j}$$

and $\langle S : T \rangle = \langle S^1 : T^1 \rangle \ldots \langle S^m : T^m \rangle$.

For a tableau S, of shape $\lambda = (\lambda_1, \ldots, \lambda_m)$ and with entries in $[1, n]$, we define (by analogy with the definition of $\hat{e}_S$ in Section 1.2)

$$\hat{v}_S = v_{S(1,1)} \wedge \cdots \wedge v_{S(1,\lambda_1)} \otimes v_{S(2,1)} \wedge \cdots \wedge v_{S(2,\lambda_2)} \otimes \cdots.$$

Lemma 1.3.1 *Suppose $R = k$, a field. Let $\lambda \in P(n)$ and let $\tau_1 : \bigwedge^\lambda V \to k[M] \otimes \bigwedge^\lambda V$ and $\tau_2 : \bigwedge^\lambda E \to \bigwedge^\lambda E \otimes k[M]$ be the structure maps.*
(i) For $S \in \mathrm{Tab}_1(\lambda)$ we have

$$\tau_1(\hat{v}_S) = \sum_{T \in \mathrm{Tab}_1(\lambda)} (S : T) \otimes \hat{v}_T$$

and

$$\tau_2(\hat{e}_S) = \sum_{T \in \mathrm{Tab}_1(\lambda)} \hat{e}_T \otimes \langle T : S \rangle.$$

(ii) For $S, T \in \mathrm{Tab}_1(\lambda)$ we have $(S : T) = \langle S : T \rangle$.

Proof (i) Let $1 \le r \le m$ and put $a = \lambda_r$, $S^r = i = (i_1, \ldots, i_a)$, $I_0 = I_0(n, a)$, $I_1 = I_1(n, a)$. Then we have

$$\tau_1(\hat{v}_{S^r}) = \tau_1(\hat{v}_i) = \tau_1(v_{i_1} \wedge \cdots \wedge v_{i_a})$$

$$= \sum_{j_1, \ldots, j_a} c_{i_1 j_1} \ldots c_{i_a j_a} \otimes v_{j_1} \wedge \cdots \wedge v_{j_a} = \sum_{j \in I_0} c_{ij} \otimes \hat{v}_j$$

$$= \sum_{j \in I_1, \pi \in \mathrm{Sym}(r)} c_{i, j\pi} \otimes \hat{v}_{j\pi} = \sum_{j \in J_1, \pi \in \mathrm{Sym}(r)} \mathrm{sgn}(\pi) c_{i, j\pi} \otimes \hat{v}_j$$

$$= \sum_{j \in I_1} (i : j) \otimes \hat{v}_j.$$

Thus we get

$$\tau_1(\hat{v}_S) = \tau(\hat{v}_{S^1} \otimes \ldots \otimes \hat{v}_{S^m})$$

$$= \sum_{T^r \in I_1(n,\lambda_r), r=1,\ldots,m} (S^1 : T^1)\ldots(S^m : T^m) \otimes \hat{v}_{T^1} \otimes \cdots \otimes \hat{v}_{T^m}$$

$$= \sum_{T \in \mathrm{Tab}_1(\lambda)} (S : T) \otimes \hat{v}_T.$$

The proof of the second assertion is similar.
(ii) We use the pairing $b : \bigwedge^\lambda V \times \bigwedge^\lambda E \to k$ of Lemma 1.2.2. The result follows by applying $(\mathrm{id} \otimes b) \circ (\tau_1 \otimes \mathrm{id}) = (b \otimes \mathrm{id}) \circ (\mathrm{id} \otimes \tau_2)$ to $\hat{v}_S \otimes \hat{e}_T$.

Lemma 1.3.2 *Let $\alpha = (\alpha_1, \ldots, \alpha_m) \in P(n)$ and let S, T be α-tableaux.*
(i) If S or T has a repeated entry in some row then $(S : T) = 0$.
(ii) If S' is obtained by permuting the entries in a row of S according to the permutation π and T' is obtained by permuting the entries in a row of T according to the permutation ρ then we have $(S' : T') = \mathrm{sgn}(\pi\rho)(S : T)$.

Proof The assertions follow by base change from the case $R = \mathbb{Z}[t]$, $q = t$ (an indeterminant). This case follows from the case $R = \mathbb{Q}(t)$, $q = t$. In particular we may assume that R is a field. Since we have $(S : T) = (S^1 : T^1)\ldots(S^m : T^m)$ both assertions immediately reduce to the case of a single-rowed partition. Thus we must prove that for $i, j \in I(n, r)$ we have $(i : j) = 0$ if i or j has a repeated entry and $(i\pi : j\rho) = \mathrm{sgn}(\pi\rho)(i : j)$, for $\pi, \rho \in \mathrm{Sym}(r)$. We have $(h : j\rho) = \mathrm{sgn}(\rho)(h : j)$, for any $h \in I(n, r)$, directly from the definition. Thus it suffices to show (in the generic case) that $(i : j) = 0$ if i has a repeated entry and that $(i\sigma : j) = \mathrm{sgn}(\sigma)(i : j)$, for i and j strictly decreasing. Let $\tau : \bigwedge^r V \to k[M] \otimes \bigwedge^r V$ denote the structure map. We have

$$\tau(\hat{v}_i) = \tau(\mathrm{sgn}(\pi)\hat{v}_{i\pi}) = \mathrm{sgn}(\pi)\tau(v_{i_{\pi(1)}} \wedge \cdots \wedge v_{i_{\pi(r)}})$$

$$= \mathrm{sgn}(\pi) \sum_{j_1,\ldots,j_r} c_{i_{\pi(1)}j_1} \cdots c_{i_{\pi(r)}j_r} \otimes v_{j_1} \wedge \cdots \wedge v_{j_r}$$

$$= \mathrm{sgn}(\pi) \sum_{j \in I_1(n,r), \rho \in \mathrm{Sym}(r)} \mathrm{sgn}(\rho) c_{i\pi, j\rho} \otimes \hat{v}_j$$

$$= \mathrm{sgn}(\pi) \sum_{j \in I_1(n,r)} (i\pi : j) \otimes \hat{v}_j.$$

But we also have $\tau(\hat{v}_i) = \sum_{j \in I_1(n,r)} (i : j) \otimes \hat{v}_j$ and comparing terms in $\hat{v}_j$ with the above gives that $(i\pi : j) = \mathrm{sgn}(\pi)(i : j)$.

Now suppose that i has a repeated entry. Then the natural map $\phi : V^{\otimes r} \to \bigwedge^r V$ takes v_i to 0. Since ϕ is an M-module map, we have $\tau \circ \phi(v_i) = (\mathrm{id} \otimes \phi) \circ \tau(v_i)$. Thus we have

$$0 = \sum_{j \in I(n,r)} c_{ij} \otimes \hat{v}_j = \sum_{j \in I_1(n,r), \pi \in \mathrm{Sym}(r)} \mathrm{sgn}(\pi) c_{i,j\pi} \otimes \hat{v}_j.$$

Equating terms in $\hat{v}_j$ gives $(i:j) = 0$ for $j \in I_1(n,r)$.

In order to prove our main result we shall need to know that, up to M-module isomorphism, $\bigwedge^\alpha E$ does not depend on the order of the parts of $\alpha \in P(n)$. This is of course obvious in the classical case $q = 1$ but requires some argument in general. However, it is a consequence of the theory of tilting modules which we discuss in Chapter 4. (Tilting modules with the same character are isomorphic: this is clear from 3.3(1).) To make this chapter self contained, we give below a sketch proof from first principles. In the following lemma we shall take R to be a field. Let Hec(m) denote the Hecke algebra of Sym(m). Let $\{T_w \mid w \in \text{Sym}(m)\}$ be the usual generators of Hec(m) (see 0.23). Recall we have an action of Hec(m) on $E^{\otimes m}$, as M-module endomorphisms, as described in [18; Section 3.1]. For $s = (a, a+1)$, a basic transposition (with $1 \le a < m$), and $i \in I(n,m)$ we have

$$T_s e_i = \begin{cases} q e_{is}, & \text{if } i_a \le i_{a+1} \\ e_{is} + (q-1)e_i, & \text{if } i_a > i_{a+1}. \end{cases}$$

For $w \in \text{Sym}(m)$ denote by $N(w)$ the set of pairs (a,b) such that $1 \le a < b \le m$ and $w(a) > w(b)$. (Thus we have $l(w) = |N(w)|$, for $w \in \text{Sym}(m)$.) It follows by an easy induction on $l(w)$ that $T_w e_i = q^{l(w)} e_{iw^{-1}}$ provided that $i_a \le i_b$ for all $(a,b) \in N(w)$ (for $w \in \text{Sym}(m)$, $i \in I(n,m)$). In particular we have $T_w e_i = q^{l(w)} e_{iw^{-1}}$ if $i_1 \le \cdots \le i_m$.

For $1 \le a < m$ we write s_a for the basic transposition $(a, a+1)$.

Lemma 1.3.3 Let $R = k$, a field, and suppose $q \ne 0$.
(i) Let $r, s \ge 1$. There is an M-module isomorphism $\phi : \bigwedge^r E \otimes \bigwedge^s E \to \bigwedge^s E \otimes \bigwedge^r E$ such that $\phi(\eta(d)) = \zeta(T_\sigma d)$, for all $d \in E^{\otimes(r+s)}$, where $\eta : E^{\otimes(r+s)} \to \bigwedge^r E \otimes \bigwedge^s E$, $\zeta : E^{\otimes(r+s)} \to \bigwedge^s E \otimes \bigwedge^r E$ are the natural maps and $\sigma \in \text{Sym}(r+s)$ is defined by $\sigma(i) = i+s$, $1 \le i \le r$, and $\sigma(r+j) = j$, $1 \le j \le s$. Furthermore we have $\phi(\hat{e}_i \otimes \hat{e}_j) = q^{l(\sigma)} \hat{e}_j \otimes \hat{e}_i$ for all $i = (i_1, \ldots, i_r) \in I(n,r)$, $j = (j_1, \ldots, j_s) \in I(n,s)$ with $i_1 < \cdots < i_r < j_1 < \cdots < j_s$.
(ii) If $\alpha, \beta \in P(n)$ and $\bar{\alpha} = \bar{\beta}$ then we have $\bigwedge^\alpha V \cong \bigwedge^\beta V$, as right M-modules, and $\bigwedge^\alpha E \cong \bigwedge^\beta E$, as left M-modules.

Proof (i) For $X \subseteq \text{Sym}(r+s)$ define $\varepsilon_X = \sum_{w \in X} (-q)^{-l(w)} T_w \in \text{Hec}(r+s)$ and define $\tilde{\varepsilon}_X : E^{\otimes(r+s)} \to E^{\otimes(r+s)}$ by $\tilde{\varepsilon}_X h = \varepsilon_X h$, for $h \in E^{\otimes(r+s)}$. We now put $X = \text{Sym}[1,r] \times \text{Sym}[r+1, r+s]$, $Y = \{\pi \in \text{Sym}(r+s) \mid \pi[1,r] = [1+s, r+s]$ and $\pi[r+1, r+s] = [1,s]\}$ and $Z = \text{Sym}[1,s] \times \text{Sym}[1+s, r+s]$. We have $Y = \sigma X = Z\sigma$, where $\sigma \in Y$ is defined by $\sigma(i) = i+s$, $1 \le i \le r$. Furthermore, as one may easily check, we have $l(\sigma\pi) = l(\sigma) + l(\pi)$, for $\pi \in X$, and $l(\pi\sigma) = l(\pi) + l(\sigma)$, for $\pi \in Z$. Hence we have

$$\varepsilon_Y = (-q)^{-l(\sigma)} T_\sigma \varepsilon_X = (-q)^{-l(\sigma)} \varepsilon_Z T_\sigma. \tag{\dagger}$$

We now show that $\tilde{\varepsilon}_Y$ factors through the natural map $E^{\otimes(r+s)} \to \bigwedge^r E \otimes \bigwedge^s E$. Let N be the kernel of this map. Let $a \in [1, r-1] \bigcup [r+1, r+s-1]$. Then we have $e_i + qe_{is_a} \in N$ if $i_a < i_{a+1}$ and also $e_j \in N$ if $j_a = j_{a+1}$ (for $i, j \in I(n, r+s)$). Moreover it follows from the definition of the exterior powers that N is spanned by all such elements. So let a, i, j be as above. We have a factorization $\varepsilon_X = \varepsilon_{X'}(1 - q^{-1}T_{s_a})$, where $X' \subseteq X$ is the set of left coset representatives of $\text{Sym}[a, a+1]$ of shortest length. It follows that $\tilde{\varepsilon}_X h = 0$, for $h = e_i + qe_{is_a}$ and $h = e_j$. Thus, by (†), $\tilde{\varepsilon}_Y$ factors through the natural map $E^{\otimes(r+s)} \to \bigwedge^r E \otimes \bigwedge^s E$, giving an induced map $\bar{\varepsilon}_Y : \bigwedge^r E \otimes \bigwedge^s E \to \text{Im}(\tilde{\varepsilon}_Y) = K$, say.

We claim that $\bar{\varepsilon}_Y$ is an isomorphism. It suffices to prove that the image has dimension at least $\dim \bigwedge^r E \otimes \bigwedge^s E = \binom{n}{r}\binom{n}{s}$. By (†), it suffices to prove that the dimension of the image of $\tilde{\varepsilon}_X$ is at least $\binom{n}{r}\binom{n}{s}$. Define $\theta : E^{\otimes(r+s)} \to E^{\otimes(r+s)}$ to be the k-linear map taking $e_{i_1} \otimes \cdots \otimes e_{i_r} \otimes e_{j_1} \otimes \cdots \otimes e_{j_s}$ to itself if $i_1 < \ldots < i_r$, $j_1 < \ldots < j_s$, and to 0 otherwise. Then the image of the composite map $\theta \circ \tilde{\varepsilon}_X$ contains all tensors of the form $e_{i_1} \otimes \cdots \otimes e_{i_r} \otimes e_{j_1} \otimes \cdots \otimes e_{j_s}$ with $i_1 < \ldots < i_r$, $j_1 < \ldots < j_s$. Thus the image of $\theta \circ \tilde{\varepsilon}_X$ has dimension at least $\binom{n}{r}\binom{n}{s}$, and hence the image of $\tilde{\varepsilon}_X$ has at least this dimension, as required.

Similarly, $\tilde{\varepsilon}_Z : E^{\otimes(r+s)} \to E^{\otimes(r+s)}$ factors through $E^{\otimes(r+s)} \to \bigwedge^s E \otimes \bigwedge^r E$, giving an injective map $\bar{\varepsilon}_Z : \bigwedge^s E \otimes \bigwedge^r E \to \text{Im}(\tilde{\varepsilon}_Z)$, and by (†) we have $\text{Im}(\tilde{\varepsilon}_Z) = K$. Thus we have an isomorphism $\phi = (-q)^{l(\sigma)}\bar{\varepsilon}_Z^{-1} \circ \bar{\varepsilon}_Y : \bigwedge^r E \otimes \bigwedge^s E \to \bigwedge^s E \otimes \bigwedge^r E$.

Let $\eta : E^{\otimes(r+s)} \to \bigwedge^r E \otimes \bigwedge^s E$ and $\zeta : E^{\otimes(r+s)} \to \bigwedge^s E \otimes \bigwedge^r E$ be the natural maps. For $d \in E^{\otimes(r+s)}$ we have $(-q)^{-l(\sigma)}\bar{\varepsilon}_Z(\phi(\eta(d))) = \bar{\varepsilon}_Y(\eta(d)$ and hence $(-q)^{-l(\sigma)}\bar{\varepsilon}_Z(\phi(\eta(d))) = \tilde{\varepsilon}_Y(d)$ so that $\bar{\varepsilon}_Z(\zeta(T_\sigma d)) = \bar{\varepsilon}_Z(\phi(\eta(d)))$, by (†), and therefore

$$\phi(\eta(d)) = \zeta(T_\sigma d). \qquad (\ddagger)$$

Now suppose that $h = (h_1, \ldots, h_{r+s}) \in I(n, r+s)$ is strictly increasing. Then we have $T_\sigma e_h = q^{l(\sigma)}e_{h\sigma^{-1}}$. Now if $i = (i_1, \ldots, i_r) \in I(n, r)$, $j = (j_1, \ldots, j_s) \in I(n, s)$ satisfy $i_1 < \cdots < i_r < j_1 < \cdots < j_s$ then taking $h = (i_1, \ldots, i_r, j_1, \ldots, j_s)$ and $d = e_i \otimes e_j$ in ($\ddagger$) we get $\phi(\hat{e}_i \otimes \hat{e}_j) = q^{l(\sigma)}(\hat{e}_j \otimes \hat{e}_i)$, as required.

(ii) In view of Lemma 1.1.2(iv) and the pairing of Lemma 1.2.2 it suffices to prove that $\bigwedge^\alpha E \cong \bigwedge^\beta E$. This follows from (i) and the fact that any permutation is a product of basic transpositions.

We define a filtration on $A(n, r) = A_{R,q}(n, r)$. For $\lambda \in P^+(n, r)$ we define $C(\lambda)$ to be the R-span of all bideterminants $(S : T)$, with $S, T \in \text{Tab}(\lambda)$. Notice that $C(\lambda)$ is spanned by all bideterminants $(S : T)$ with $S, T \in \text{Tab}_1(\lambda)$, by Lemma 1.3.2, and that if R is a field then $C(\lambda)$ is the coefficient space of $\bigwedge^\lambda V$ and of $\bigwedge^\lambda E$, by Lemma 1.3.1. We define $F(\lambda) = F_{R,q}(\lambda)$ to be the sum of all $C(\mu)$'s with $\mu \in P^+(n, r)$ and $\mu \not\triangleleft \lambda$. We define

$F'(\lambda) = F'_{R,q}(\lambda)$ to be the sum of all $C(\mu)$'s with $\mu \in P^+(n,r)$ and $\mu \not\leq \lambda$. We are now ready to prove the main result of this chapter.

Theorem 1.3.4   *For any commutative ring R, any unit $q \in R$ and any $\lambda \in P^+(n,r)$, the bideterminants $(S:T)$, with S,T antistandard of shape $\mu \not< \lambda$, form an R-basis of $F(\lambda)$. In particular, $A_{R,q}(n,r)$ has R-basis $(S:T)$, with S,T antistandard of shape $\mu \in P^+(n,r)$.*

Proof   We first suppose that $R = k$, a field. For $\lambda \in P^+(n,r)$ let d_λ denote the number of antistandard tableaux of shape λ. If $R = \mathbb{C}$ and $q = 1$ then d_λ is the dimension of the irreducible M-module $L(\lambda')$, whose highest weight is the transpose λ', see [51; (5.3b),(2.6e)]. Moreover, by [51; (3.5a)(iii)], the modules $L(\lambda')$, $\lambda \in P^+(n,r)$, form a complete set of pairwise non-isomorphic irreducible modules for the dual algebra $S(n,r)$ of the coalgebra $A(n,r)$. By Wedderburn's theorem, and absolute irreducibility [51; (3.5a)], we have $S(n,r) \cong \bigoplus_{\lambda \in P^+(n,r)} \mathrm{End}_{\mathbb{C}}(L(\lambda'))$ and hence $\dim S(n,r) = \sum_{\lambda \in P^+(n,r)}(\dim L(\lambda'))^2 = \sum_{\lambda \in P^+(n,r)} d_\lambda^2$. However, the dimension of $A(n,r)$ is independent of the field of definition and of the choice of parameter q (see 0.20). Hence we get

$$\dim A_{k,q}(n,r) = \sum_{\lambda \in P^+(n,r)} d_\lambda^2 \qquad\qquad (\S).$$

Fix $\lambda \in P^+(n,r)$ and let $\tau_1 : \bigwedge^\lambda V \to k[M] \otimes \bigwedge^\lambda V$ and $\tau_2 : \bigwedge^\lambda E \to \bigwedge^\lambda E \otimes k[M]$ be the structure maps. We have the natural map $\Phi = (\mathrm{id} \otimes b) \circ (\tau_1 \otimes \mathrm{id}) = (b \otimes \mathrm{id}) \circ (\mathrm{id} \otimes \tau_2) : \bigwedge^\lambda V \otimes \bigwedge^\lambda E \to k[M]$ (where $b : \bigwedge^\lambda V \times \bigwedge^\lambda E \to k$ is the natural pairing of Lemma 1.2.2). Now Φ is an (M,M)-bimodule map and the image is the coefficient space of $\bigwedge^\lambda V$ and $\bigwedge^\lambda E$. Thus Φ has image in $F(\lambda)$ and induces an (M,M)-bimodule map $\Psi : \bigwedge^\lambda V \otimes \bigwedge^\lambda E \to F(\lambda)/F'(\lambda)$. We claim that Φ maps $I^\lambda \otimes \bigwedge^\lambda E + \bigwedge^\lambda V \otimes J^\lambda$ into $F'(\lambda)$. We shall prove $\Phi(I^\lambda \otimes \bigwedge^\lambda E) = 0$; the proof that $\Phi(\bigwedge^\lambda V \otimes J^\lambda) = 0$ is similar. It suffices to show that if U is the image of an M-homomorphism $\theta : \bigwedge^\mu V \to \bigwedge^\lambda V$ with $\bar{\mu} \not\leq \lambda$ then $\Psi(U \otimes \bigwedge^\lambda E) \leq F'(\lambda)$. Moreover, by Lemma 1.3.3 we can assume $\mu \in P^+(n,r)$. Now $\Phi \circ (\theta \otimes \mathrm{id}) : \bigwedge^\mu V \otimes \bigwedge^\lambda E \to k[M]$ is a morphism of left M-modules and hence has image in the coefficient space of $\bigwedge^\mu V \otimes |\bigwedge^\lambda E|$ (see [50; (1.2c)(i),(1.2f)]) which is $C(\mu)$. Hence Ψ maps $I^\lambda \otimes \bigwedge^\lambda E + \bigwedge^\lambda \otimes J^\lambda$ into $F'(\lambda)$ and therefore induces a surjective map $\Omega : \bar{\bigwedge}^\lambda V \otimes \bar{\bigwedge}^\lambda E \to F(\lambda)/F'(\lambda)$, where $\bar{\bigwedge}^\lambda V = \bigwedge^\lambda V/I^\lambda$ and $\bar{\bigwedge}^\lambda E = \bigwedge^\lambda E/J^\lambda$. By Theorem 1.2.4 we have that $F(\lambda)$ is spanned, modulo $F'(\lambda)$, by the elements $\Omega(\hat{v}_S \otimes \hat{e}_T + (I^\lambda \otimes \bigwedge^\lambda E + \bigwedge^\lambda V \otimes J^\lambda))$, with S,T antistandard of shape λ. Thus the elements $\Psi(\hat{v}_S \otimes \hat{e}_T)$, with $S,T \in \mathrm{AStan}(\lambda)$, span $F(\lambda)$ modulo $F'(\lambda)$. Thus $F(\lambda)$ is spanned by $F'(\lambda)$ and the bideterminants $(S:T) = \Phi(\hat{v}_S \otimes \hat{e}_T)$, with $S,T \in \mathrm{AStan}(\lambda)$.

Now, for $\lambda \in P^+(n, r)$, we define $a(\lambda)$ to be the cardinality of $\{\mu \in P^+(n, r) \mid \mu \not< \lambda\}$. We claim that $F(\lambda)$ is spanned by all bideterminants $(S : T)$, with S, T antistandard of shape $\mu \not< \lambda$, and we prove this by induction on $a(\lambda)$. We assume that the claim holds for $\mu \in P^+(n, r)$ with $a(\mu) < a(\lambda)$. By the paragraph above, $F(\lambda)$ is spanned by all $(S : T)$, with S, T antistandard of shape λ, together with $F'(\lambda) = \sum_{\mu \not< \lambda} F(\mu)$. Moreover, for $\mu \not\leq \lambda$, we have $a(\mu) < a(\lambda)$ so that, by the inducitve hypothesis, $F(\mu)$ is spanned by all $(S : T)$, with S, T antistandard of shape $\tau \not< \mu$. Hence $F(\lambda)$ is spanned by all $(S : T)$ with S, T antistandard of shape $\mu \not< \lambda$, as required.

Taking $\lambda = (1^n)$ we get $F(\lambda) = A(n, r)$ and hence $A(n, r)$ is spanned by all bideterminants $(S : T)$ with S, T antistandard of shape $\mu \in P^+(n, r)$. By (§) the number of these is $\dim A(n, r)$ and hence the elements $(S : T)$, with S, T antistandard of the same shape, are linearly independent. Hence we have for arbitrary $\lambda \in P^+(n, r)$ that the bideterminants $(S : T)$, with S, T antistandard of shape $\mu \not< \lambda$, form a k-basis of $F(\lambda)$.

We now deal with the case of an arbitrary coefficient ring R and unit $q \in R$. Clearly this follows, by base change via the natural isomorphism $R \otimes_{\mathbf{Z}[t, t^{-1}]} A_{\mathbf{Z}[t, t^{-1}], t}(n, r) \to A_R(n, r)$, from the case $R = \mathbf{Z}[t, t^{-1}]$, $q = t$ (an indeterminant). Certainly the elements $(S : T)$ (with S, T antistandard of the same shape) are independent over $\mathbf{Z}[t, t^{-1}]$ since they are over $\mathbf{Q}(t)$ by the above (and we have the natural isomorphism $\mathbf{Q}(t) \otimes_{\mathbf{Z}[t, t^{-1}]} A_{\mathbf{Z}[t, t^{-1}], t}(n, r) \to A_{\mathbf{Q}(t), t}(n, r)$). It therefore suffices to show that the antistandard bideterminants span $F(\lambda)$ (for $\lambda \in P^+(n, r)$) over an arbitrary commutative ring R. We let $F_1(\lambda)$ denote the R-span of the bideterminants $(S : T)$, with S, T antistandard of shape $\mu \geq \lambda$. Then, for every field k, inclusion $F_1(\lambda) \to F(\lambda)$ induces an isomorphism $k \otimes_R F_1(\lambda) \to k \otimes_R F(\lambda)$ for every field homomorphism $R \to k$, by the above. However, if X is a submodule of a finitely generated module Y over a commutative ring R such that $k \otimes_R X \to k \otimes_R Y$ is onto, for every homomorphism from R into a field k, then $X = Y$. Hence we have $F_1(\lambda) = F(\lambda)$, as required.

Remark This does not work if $q = 0$. For example consider the case $n = 2$, $r = 3$ and $\lambda = (2, 1)$. Then $F(\lambda)$ is the coefficient space of $\bigwedge^2 E \otimes E$ and is spanned by the elements dc_{ij}, with $1 \leq i, j \leq 2$, where d is the determinant $c_{11}c_{12} - c_{12}c_{21}$. By direct calculation using AIII or by [**18**; 4.1.9 Theorem], we have $dc_{12} = 0$. Hence $F(\lambda)$ has dimension at most (and in fact exactly) 3. But there are 2 antistandard λ tableaux and hence 4 bideterminants $(S : T)$ with S, T antistandard of shape λ.

It is easy to check that $\mathrm{cf}(E \otimes \bigwedge^2 E)$ has k-basis $c_{11}d, c_{12}d, c_{22}d$ and that $\mathrm{cf}(E \otimes \bigwedge^2 E) \neq \mathrm{cf}(\bigwedge^2 E \otimes E)$; hence $\bigwedge^2 E \otimes E \not\cong E \otimes \bigwedge^2 E$ and Lemma 1.3.3 also fails if $q = 0$.

2. The Schur Functor and a Character Formula

In the first part of this chapter we shall make some remarks on the Schur functor, from modules for the Schur algebra to modules for the Hecke algebra. In the latter part we concentrate on the case $q = 0$ and obtain a formula for the character of the simple modules. The first part proceeds by analogy with the classical case $q = 1$, [51], and there is some overlap here with S. Martin, [66]. A further study of the Schur functor and the relationship between the representation theories of Schur algebra and Hecke algebra is made in Chapter 4 (see especially Section 4.4). The second part of this chapter uses work of Norton, [69], on 0-Hecke algebras. I am grateful to Richard Dipper for bringing the work of Norton to my attention.

2.1 The Schur functor

We begin by recalling from [18; 2.1] the construction of the symmetric powers of the natural modules E and V. We define J to be the ideal of the tensor algebra $T(E)$ generated by the elements $e_i e_j - e_j e_i$, with $1 \leq i, j \leq n$. Then J is an M-submodule of $T(E)$. We define $S(E) = T(E)/J$. Since J is homogeneous we have an induced grading and M-module decomposition $S(E) = \bigoplus_{r=0}^{\infty} S^r E$. The rth symmetric power $S^r E$ has k-basis consisting of the elements $e_{i_1} \ldots e_{i_r}$, with $n \geq i_1 \geq \cdots \geq i_r \geq 1$. Similarly we define K to be the ideal of the tensor algebra $T(V)$ generated by the elements $q v_i v_j - v_j v_i$, for $1 \leq i < j \leq n$. Then K is an M-submodule of $T(V)$. We define $S(V) = T(V)/K$. Since K is homogeneous we have an induced grading and M-module decomposition $S(V) = \bigoplus_{r=0}^{\infty} S^r V$. The rth symmetric power $S^r V$ has k-basis consisting of the elements $v_{i_1} \ldots v_{i_r}$, with $1 \leq i_1 \leq \cdots \leq i_r \leq n$. For $i \in I(n, r)$, we define $\bar{e}_i \in S^r E$ to be the image of e_i under the natural map $E^{\otimes r} \to S^r E$ and define $\bar{v}_i \in S^r V$ to be the image of v_i under the natural map $V^{\otimes r} \to S^r V$. For $\alpha = (\alpha_1, \ldots, \alpha_n) \in \Lambda(n, r)$ we define the left M-module $S^\alpha E = S^{\alpha_1} E \otimes \cdots \otimes S^{\alpha_n} E$ and the right M-module $S^\alpha V = S^{\alpha_1} \otimes \cdots \otimes S^{\alpha_n} V$.

We briefly discuss multiplication in the Schur algebra $S(n, r) = A(n, r)^*$. For $i \in I(n, r)$ and $\alpha \in \Lambda(n, r)$ we write $i \in \alpha$ to indicate that i has content α. We write $i \sim j$, for $i, j \in I(n, r)$, to indicate that i and j have the same content. We define elements $i^\alpha, j^\alpha \in \alpha$ by $i^\alpha = (1, \ldots, 1, 2, \ldots, 2, 3, \ldots)$ and $j^\alpha = (\ldots, 3, 2, \ldots, 2, 1, \ldots, 1)$ (where 1 occurs α_1 times, 2 occurs α_2 times and so on). We define $I^+(\alpha)$ to be the set of $i = (i_1, \ldots, i_r) \in I(n, r)$ such that $i_1 \leq \cdots \leq i_{\alpha_1}, i_{\alpha_1+1} \leq \cdots \leq i_{\alpha_1+\alpha_2}$ and so on and define $I^-(\alpha)$ to be the set of $i = (i_1, \ldots, i_r) \in I(n, r)$ such that $i_1 \geq \cdots \geq i_{\alpha_1}, i_{\alpha_1+1} \geq \cdots \geq i_{\alpha_1+\alpha_2}$ and so on. We define ${}^\alpha A(n, r)$ to be the k-span of the elements c_{ij} with $i \in \alpha$ and define $A(n, r)^\alpha$ to be the k-span of the elements c_{ij} with $j \in \alpha$. Then ${}^\alpha A(n, r)$ is a left M-submodule of $A(n, r)$ and $A(n, r)^\alpha$ is a right M-submodule of $A(n, r)$. It is easy to check, from the defining relations, that $A(n)$ is spanned by the elements $c_{i^\alpha j}$, with $\alpha \in \Lambda(n, r)$ for some r and

$j \in I^-(\alpha)$, and that $A(n)$ is spanned by the elements $c_{ij\alpha}$, with $\alpha \in \Lambda(n,r)$ for some r and $i \in I^+(\alpha)$. It follows that for $\alpha \in \Lambda(n,r)$, the elements $c_{i\alpha j}$ with $j \in I^-(\alpha)$ span $^\alpha A(n,r)$ and the elements $c_{ij\alpha}$, with $i \in I^+(\alpha)$ span $A(n,r)^\alpha$. For $\alpha \in \Lambda(n,r)$ the linear map $E^{\otimes r} \to A(n,r)$ taking e_j to $c_{i\alpha j}$ (for $j \in I(n,r)$) induces an M-module epimorphism $\phi^\alpha : S^\alpha E \to {}^\alpha A(n,r)$. Thus we have a surjective homomorphism $\phi : \bigoplus_{\alpha \in \Lambda(n,r)} S^\alpha E \to A(n,r)$. In the special case $q = 1$ this map is an isomorphism, by [51; Section 4.8], but the dimension of both domain and codomain of ϕ is independent of q and hence ϕ is an isomorphism in general. Similar remarks apply to the corresponding right modules and so we have the following.

(1) (i) *For each $\alpha \in \Lambda(n,r)$ we have an isomorphism of left M-modules $\phi^\alpha : S^\alpha E \to {}^\alpha A(n,r)$ such that $\phi^\alpha(\bar{e}_j) = c_{i\alpha j}$, for $j \in I(n,r)$, and an isomorphism of right M-modules $\psi^\alpha : S^\alpha V \to A(n,r)^\alpha$, such that $\psi^\alpha(\bar{v}_i) = c_{ij\alpha}$, for $i \in I(n,r)$.*
(ii) *We have an isomorphism $\phi : \bigoplus_{\alpha \in \Lambda(n,r)} S^\alpha E \to A(n,r)$ of left M-modules and an isomorphism $\psi : \bigoplus_{\alpha \in \Lambda(n,r)} S^\alpha V \to A(n,r)$ of right M-modules.*

In particular, for $\alpha \in \Lambda(n,r)$, the elements $c_{i\alpha j}$, $j \in I^-(\alpha)$, form a basis of $^\alpha A(n,r)$ and the elements $c_{ij\alpha}$, $i \in I^+(\alpha)$, form a basis of $A(n,r)^\alpha$. For $\beta \in \Lambda(n,r)$ we write $^\alpha A(n,r)^\beta$ for the span of the elements c_{ij} with $i \in \alpha$, $j \in \beta$. Now by [18; 2.2.1] we have $A(n,r) = \bigoplus_{\alpha,\beta \in \Lambda(n,r)} {}^\alpha A(n,r)^\beta$ from which we get the following.

(2) $^\alpha A(n,r)^\beta$ *has a basis consisting of the elements $c_{i\alpha j}$ with $j \in I^-(\alpha)$ of content β and has a basis consisting of the elements $c_{ij\beta}$ with $i \in I^+(\beta)$ of content α.*

Let U denote the set of pairs (i,j) of elements of $I(n,r)$ such that i is weakly increasing of content α, say, and $j \in I^-(\alpha)$. We define ξ_{ij}, $(i,j) \in U$, to be the basis of $S(n,r)$ dual to the basis c_{ij}, $(i,j) \in U$, of $A(n,r)$. Let U' denote the set of pairs (i,j) of elements of $I(n,r)$ such that j is weakly decreasing of content β, say, and $i \in I^+(\beta)$. We define ξ'_{ij}, $(i,j) \in U$, to be the basis of $S(n,r)$ dual to the basis c_{ij}, $(i,j) \in U'$, of $A(n,r)$.

From (2) we note that if $(i,j) \in U$ and $s,t \in I(n,r)$ then $\xi_{ij}(c_{st}) = \xi'_{ij}(c_{st}) = 0$ unless $i \sim s$ and $j \sim t$. We put $\xi_\alpha = \xi_{i\alpha,i\alpha}$ and note that $\xi_\alpha(f) = \varepsilon(f)$ for $f \in {}^\alpha A(n,r)^\alpha$ and $\xi_\alpha(f) = 0$ for $f \in {}^\beta A(n,r)^\gamma$ with $(\beta,\gamma) \neq (\alpha,\alpha)$. Similar remarks apply to ξ'_α and so we get:

(3) $\xi_\alpha = \xi'_\alpha$, *for $\alpha \in \Lambda(n,r)$.*

Let $(i,j),(s,t) \in U$ and $a,b \in I(n,r)$. We have

$$(\xi_{ij} * \xi_{st})(c_{ab}) = \sum_{h \in I(n,r)} \xi_{ij}(c_{ah})\xi_{st}(c_{hb}).$$

From this and earlier remarks one sees:

(4)   (i) $\xi_{ij} * \xi_{st} = 0$ unless $j \sim s$.
(ii) $1 = \sum_{\alpha \in \Lambda(n,r)} \xi_\alpha$ is an orthogonal decomposition.

Fix $\alpha, \beta \in \Lambda(n,r)$. Let $\bar{B}$ denote the span of the set B of elements $\xi_{i\alpha j}$, where j has content β, and let $\bar{B}'$ denote the span of the set B' of elements $\xi_{ij\beta}$, where i has content α. Then $\bar{B} = \bar{B}'$ is exactly the set of elements which vanish on $^{\alpha_0}A(n,r)^{\beta_0}$, for all $(\alpha_0, \beta_0) \neq (\alpha, \beta)$. It is easy to deduce the following.

(5)   The elements $\xi_{i\alpha j}$ (resp. $\xi_{ij\beta}$) with $j \in I^-(\alpha)$ of content β (resp. $i \in I^+(\alpha)$ of content α) form a basis of $\xi_\alpha S \xi_\beta$.

We now suppose $r \leq n$ and put $\omega = (1,1,\ldots,1) \in \Lambda(n,r)$, $u = (1,2,\ldots,r)$ and $v = (r,\ldots,2,1)$. Putting $e = \xi_\omega$ we have the following.

(6) eSe has a basis $\{\xi_{u,u\pi} \mid \pi \in \mathrm{Sym}(r)\}$ and a basis $\{\xi'_{v\pi,v} \mid \pi \in \mathrm{Sym}(r)\}$.

We shall also need the following.

(7)   We have a left S-module isomorphism $\theta : Se \to E^{\otimes r}$ satisfying $\theta(e) = e_v$ and a right S-module isomorphism $\eta : eS \to V^{\otimes r}$ satisfying $\eta(e) = v_u$. In particular $E^{\otimes r}$ is a projective left S-module and $V^{\otimes r}$ is a projective right S-module.

We prove only the left module version. Let $\tilde{\theta} : S \to E^{\otimes r}$ be the S-module homomorphism defined by $\tilde{\theta}(s) = se_v$, $s \in S$. We have $\tilde{\theta}(\xi'_{iv}) = \xi'_{iv} e_v = \sum_{j \in I(n,r)} \xi'_{iv}(c_{jv})e_j = e_i$. This shows that $\tilde{\theta}$ is surjective and that $\tilde{\theta}(e) = e_v$. Thus $S(1-e) \leq \mathrm{Ker}(\tilde{\theta})$ and so the restriction θ of $\tilde{\theta}$ to Se is surjective. However, it follows from (5) that Se has basis ξ_{iv}, $i \in I(n,r)$. By dimensions we obtain that $\theta : Se \to E^{\otimes r}$ is an isomorphism.

For a left $S(n,r)$-module X we write X^α for $\xi_\alpha X$, for $\alpha \in \Lambda(n,r)$. Regarding X as a right $A(n,r)$-comodule, hence a $k[G(n)]$-comodule, i.e. a $G(n)$-module, we have that X^α is the α weight space of X, see 0.22. Moreover, we have $X^\alpha = \{x \in X \mid \tau(x) \in X \otimes A(n,r)^\alpha\}$, by [18; 2.2] (where $\tau : X \to X \otimes A(n,r)$ is the structure map of X, regarded as a right $A(n,r)$-comodule). Similar remarks apply to right $S(n,r)$-modules. In particular $A(n,r)$ is naturally an $(S(n,r), S(n,r))$-bimodule. The left and right actions are given by $\xi \cdot c = \sum_i \xi(c'_i)c_i$ and $c \cdot \xi = \sum_i \xi(c_i)c'_i$, for $\xi \in S(n,r)$ and $c \in A(n,r)$ with $\delta(c) = \sum_i c_i \otimes c'_i$. It is easy to check that $^\alpha A(n,r) = A(n,r)\xi_\alpha$ and $A(n,r)^\alpha = \xi_\alpha A(n,r)$. We have the canonical linear isomorphism $\Phi : X^* \to \mathrm{Hom}_{S(n,r)}(X, A(n,r))$, given by $\Phi(\eta)(x) = (\eta \otimes \mathrm{id})\tau(x)$, for $x \in X$. Moreover Φ is an isomorphism of right $S(n,r)$-modules and induces an isomorphism $X^* \xi_\alpha \rightarrow \mathrm{Hom}_{S(n,r)}(X, A(n,r))\xi_\alpha$, i.e. $(X\xi_\alpha)^* \rightarrow \mathrm{Hom}_{S(n,r)}(X, A(n,r)\xi_\alpha)$. Thus we have that X^α and $\mathrm{Hom}_{S(n,r)}(X, {}^\alpha A(n,r))$ are isomorphic k-spaces. Combining this with (1)(i) and the corresponding results for right modules we obtain the following. (See also the proof of [**32**; (2.4)] for the classical case.)

(8) Let $\alpha \in \Lambda(n,r)$. For any finite dimensional left (resp. right) S-module X we have $\mathrm{Hom}_S(X, S^\alpha E) \cong X^\alpha$ (resp. $\mathrm{Hom}_S(X, S^\alpha V) \cong X^\alpha$). In particular $S^\alpha E$ (resp. $S^\alpha V$) is an injective left (resp. right) S-module.

We shall now discuss the Schur functor. We are interested in the algebra $eS(n,r)e$. For $\sigma \in \mathrm{Sym}(r)$ we put $b_\sigma = \xi_{u,u\sigma}$, $b'_\sigma = \xi'_{v\sigma,v}$. It is easy to check the following from the defining relations.

(9) Let $1 \le a < r$ and $\sigma \in \mathrm{Sym}(r)$. We have

$$c_{us_a,u\sigma} = \begin{cases} qc_{u,u\sigma s_a}, & \text{if } \sigma(a) < \sigma(a+1); \\ c_{u,u\sigma s_a} + (q-1)c_{u,u\sigma}, & \text{if } \sigma(a) > \sigma(a+1) \end{cases}$$

and

$$c_{v\sigma,vs_a} = \begin{cases} qc_{v\sigma s_a,v}, & \text{if } \sigma(a) < \sigma(a+1); \\ v_{v\sigma s_a,v} - (q-1)c_{v\sigma,v}, & \text{if } \sigma(a) > \sigma(a+1). \end{cases}$$

From (9) we obtain the following.

(10) Let $1 \le a < r$ and $\sigma \in \mathrm{Sym}(r)$. We have

$$b_{s_a}b_\sigma = \begin{cases} b_{\sigma s_a}, & \text{if } l(\sigma s_a) = l(\sigma) + 1; \\ qb_{\sigma s_a} + (q-1)b_\sigma, & \text{if } l(\sigma s_a) = l(\sigma) - 1 \end{cases}$$

and

$$b'_\sigma b'_{s_a} = \begin{cases} b'_{\sigma s_a}, & \text{if } l(\sigma s_a) = l(\sigma) + 1; \\ qb'_{\sigma s_a} - (q-1)b'_\sigma, & \text{if } l(\sigma s_a) = l(\sigma) - 1. \end{cases}$$

We write T_σ for $b_{\sigma^{-1}}$, $\sigma \in \mathrm{Sym}(r)$. The first part of (10) may thus be reformulated as follows.

(10)$'$ Let $1 \le a < r$ and $\sigma \in \mathrm{Sym}(r)$. We have

$$T_{s_a}T_\sigma = \begin{cases} T_{s_a\sigma}, & \text{if } l(s_a\sigma) = l(\sigma) + 1; \\ qT_{s_a\sigma} + (q-1)T_\sigma, & \text{if } l(s_a\sigma) = l(\sigma) - 1. \end{cases}$$

Thus $eS(n,r)e$ is the Hecke algebra of $\mathrm{Sym}(r)$ with basic generators

$$T_{s_1}, \ldots, T_{s_{r-1}}.$$

We often write $eS(n,r)e$ as $\mathrm{Hec}(r)$, or just $H(r)$ or even simply H.

As in [**51**; Chapter 6], we have the Schur functor $f : \mathrm{mod}(S(n,r)) \to \mathrm{mod}(\mathrm{Hec}(r))$ defined as follows. For $X \in \mathrm{mod}(S)$, $f(X)$ is the k-space eX viewed as an H-module by restricting the action. For a morphism $\eta : X \to Y$, of S-modules, $f(\eta) : eX \to eY$ is the restriction of η.

Remarks (i) For $q = 1$ and $\sigma \in \mathrm{Sym}(r)$ we have $T_\sigma = \xi_{u,u\sigma^{-1}} = \xi_{u\sigma,u}$ so that, in the classical case, $\{T_\sigma \mid \sigma \in \mathrm{Sym}(r)\}$ is the basis which Green uses to identify eSe with the group algebra $k\mathrm{Sym}(r)$, [**51**; (6.1d)].
(ii) We have the natural functor $F : \mathrm{Hom}_S(Se, -) : \mathrm{mod}(S) \to \mathrm{mod}(B^{\mathrm{op}})$, where $B = \mathrm{End}_S(Se) = eSe$ and B^{op} is the opposite algebra. We also have an algebra isomorphism $\phi : H \to (eSe)^{\mathrm{op}}$ given by $\phi(T_\sigma) = b_{\sigma^{-1}}$, for $\sigma \in \mathrm{Sym}(r)$ (by (6) and (10)). This gives rise to an equivalence of categories $\tilde{\phi} : \mathrm{mod}((eSe)^{\mathrm{op}}) \to \mathrm{mod}(H)$. It is clear from the construction that $f = \tilde{\phi} \circ F$. Note that we have also observed that $\mathrm{Hec}(r)$ is naturally isomorphic to $\mathrm{End}_S(Se) = \mathrm{End}_M(E^{\otimes r})$.
(iii) Suppose $q \neq 0$. Suppose further that q is not a root of unity or that k has characteristic 0 and $q = 1$. Then $S(n,r)$ is semisimple, by [**36**; Section 4,(8)], and $\{\nabla(\alpha) \mid \alpha \in \Lambda^+(n,r)\}$ is a complete set of inequivalent simple modules. Let $n \geq r$. We get that $\mathrm{Hec}(r) \cong \mathrm{End}_{S(n,r)}(E^{\otimes r})$ is semisimple. Moreover, we have $f\nabla(\alpha) = \nabla(\alpha)^\omega \neq 0$, for each $\alpha \in \Lambda^+(n,r)$, so that $\{f\nabla(\alpha) \mid \alpha \in \Lambda^+(n,r)\}$ is a complete set of inequivalent irreducible $\mathrm{Hec}(r)$-modules, by [**51**; (6.2g) Theorem], and it follows that $f : \mathrm{mod}(S(n,r)) \to \mathrm{mod}(H(r))$ is an equivalence of categories.

We now define a non-degenerate associative bilinear form on H. Recall that we have the pairing $V^{\otimes r} \times E^{\otimes r} \to k$, satisfying $(v_i, e_j) = \delta_{ij}$, for $i, j \in I(n,r)$. In general if C is a coalgebra and $(\, ,\,) : X \otimes Y \to k$ is a pairing of C-comodules then it follows directly from the defining property of a pairing that $(xs, y) = (x, sy)$, for all $s \in C^*$ (regarding X as a right and Y as a left S-module in the natural way.) Regarding $V^{\otimes r}$ as a right S-module and $E^{\otimes r}$ as a left S-module we thus have $(xs, y) = (x, sy)$, for all $x \in V^{\otimes r}$, $y \in E^{\otimes r}$ and $s \in S$. We introduce a form on S by defining $(x, y) = (v_u x, y e_v)$ (for $x, y \in S$). For $x, y, z \in S$ we have $(xy, z) = (v_u xy, z e_v) = (v_u x, yz e_v) = (x, yz)$, i.e. $(\, ,\,)$ is associative. Now for $\sigma, \pi \in \mathrm{Sym}(r)$ we have

$$v_u b_\sigma = \sum_j b_\sigma(c_{uj}) v_j = v_{u\sigma}$$

and

$$b'_\pi e_v = \sum_j b'_\pi(c_{jv}) e_j = e_{v\pi}$$

so that

$$(b_\sigma, b'_\pi) = (v_{u\sigma}, e_{v\pi}) = \begin{cases} 1, & \text{if } u\sigma = v\pi; \\ 0, & \text{otherwise.} \end{cases}$$

Putting $w_0 = (1, r)(2, r - 1) \ldots$ we obtain:

(11) $(b_\sigma, b'_{w_0\pi}) = \delta_{\sigma\pi}$, *in particular* (,) *is non-singular on* H, *and hence* H *is a Frobenius algebra.*

We note that $(b_\sigma, 1) = (1, b'_\sigma)$ is 1 if $\sigma = w_0$ and 0 otherwise. Hence we have, for $x, y \in H$:

(12) (x, y) *is equal to the coefficient of* b_{w_0} *when* xy *is expressed as a linear combination of the elements* b_σ, $\sigma \in \mathrm{Sym}(r)$, *and is equal to the coefficient of* b'_{w_0} *when* xy *is expressed as a linear combination of the elements* b'_π, $\pi \in \mathrm{Sym}(r)$.

To make further progress it is convenient to make use of some of the homological properties of G and $S(n, r)$ established in [36]. Suppose that $q \neq 0$. We say that a G-module filtration $0 = X_0 \leq X_1 \leq \cdots \leq X_r \leq \cdots$ of $X \in \mathrm{mod}(G)$ is *good* if $X = \bigcup_{i=1}^\infty X_i$ and for each $i \geq 1$, we have that X_i/X_{i-1} is either 0 or isomorphic to $\nabla(\lambda_i)$, for some $\lambda_i \in X^+(n)$. We write $X \in \mathcal{F}(\nabla)$ to indicate that $X \in \mathrm{mod}(G)$ admits a good filtration. We may also call a good filtration a ∇-filtration.

Let w_0 be the longest element of $\mathrm{Sym}(n)$. For $\lambda \in X(n)$ we set $\lambda^* = -w_0\lambda$. For $\lambda \in X^+(n)$ we write $\Delta(\lambda)$ for the dual module $\nabla(\lambda^*)^*$.

(13) *(i) For dominant weights* λ, μ *and* $i \geq 1$ *we have* $\mathrm{Ext}_G^i(\nabla(\lambda), \nabla(\mu)) = 0$ *unless* $\lambda > \mu$.
(ii) For a G-module X *the following are equivalent:*
(a) $X \in \mathcal{F}(\nabla)$;
(b) $\mathrm{Ext}_G^1(\Delta(\lambda), X) = 0$ *for all dominant weights* λ;
(c) $\mathrm{Ext}_G^i(\Delta(\lambda), X) = 0$ *for all* $i \geq 1$ *and all dominant weights* λ.
(iii) Let $0 \to X' \to X \to X'' \to 0$ *be a short exact sequence of* G-modules. *If* $X', X \in \mathcal{F}(\nabla)$ *then* $X'' \in \mathcal{F}(\nabla)$ *and if* $X', X'' \in \mathcal{F}(\nabla)$ *then* $X \in \mathcal{F}(\nabla)$.

The proofs of (i) and (ii) are as in the case of semisimple algebraic groups: for (i) see [29; Lemma 3.2.1] (a consequence of a result of Cline, Parshall, Scott and van der Kallen, [13; (3.2) Corollary]); and for (ii) see [49] (or [26; Corollary 1.3] for the finite dimensional case) and [36; Section 4(2)]. Part (iii) follows from (ii). Alternatively, one may remark that $S(n, r)$ is a quasihereditary algebra with standard modules $\{\Delta(\lambda) \mid \lambda \in \Lambda^+(n, r)\}$ and costandard modules $\{\nabla(\lambda) \mid \lambda \in \Lambda^+(n, r)\}$, with respect to the natural partial order on $\Lambda^+(n, r)$ (see e.g. [36; Section 4]). Moreover, one has a natural isomorphism $\mathrm{Ext}_{S(n,r)}^i(X, Y) \to \mathrm{Ext}_G^i(X, Y)$, for each $i \geq 0$, by 0.17. Thus one may deduce the above (for X finite dimensional) from standard properties of modules over a quasihereditary algebra, see Appendix, A2.2 Proposition.

Definition   We shall call an epimorphism of G-modules $\phi : X \to Y$ *good* if X, Y and the kernel of ϕ belong to $\mathcal{F}(\nabla)$. We shall call a monomorphism of G-modules $\phi : X \to Y$ *good* if X and Y belong to $\mathcal{F}(\nabla)$ (and in this case the cokernel of ϕ also belongs to $\mathcal{F}(\nabla)$, by (13)(iii)).

(14)   If $\phi_i : X_i \to Y_i$ *is a good epimorphism (resp. monomorphism), for* $1 \leq i \leq m$, *then* $\phi_1 \otimes \cdots \otimes \phi_m : X_1 \otimes \cdots \otimes X_m \to Y_1 \otimes \cdots \otimes Y_m$ *is a good epimorphism (resp. monomorphism).*

We give the argument for epimorphisms. It suffices to consider the case $m = 2$. Let $\psi = \phi_1 \otimes \phi_2$. Note that $X_1 \otimes X_2, Y_1 \otimes Y_2 \in \mathcal{F}(\nabla)$, by [**36**; Section 4(3)(i)]. Thus it suffices to show that $\mathrm{Ker}(\psi) \in \mathcal{F}(\nabla)$. Let $K_i = \mathrm{Ker}(\phi_i)$, $i = 1, 2$. Then ψ has kernel $K_1 \otimes X_2 + X_1 \otimes K_2$. However, we have a short exact sequence $0 \to K_1 \otimes K_2 \to K_1 \otimes X_2 \bigoplus X_1 \otimes K_2 \to K_1 \otimes X_2 + X_1 \otimes K_2 \to 0$ and $K_1 \otimes K_2$ and $K_1 \otimes X_2 \bigoplus X_1 \otimes K_2$ belong to $\mathcal{F}(\nabla)$, by [**36**; Section 4, (3)(i)] (and (13)(ii) above). Hence $\mathrm{Ker}(\psi) \in \mathcal{F}(\nabla)$, by (13)(iii), as required.

Remark   These definitions and arguments (and results) also apply to rational modules for reductive groups.

(15)   *Suppose that* $q \neq 0$.
(i) (a) *For* $\lambda \in \Lambda^+(n, r)$ *the space* $\mathrm{Hom}_G(\bigwedge^{\lambda'} E, \nabla(\lambda))$ *is 1-dimensional and any non-zero element is a good epimorphism.*
(b) *For* $\lambda \in \Lambda^+(n, r)$ *the space* $\mathrm{Hom}_G(\nabla(\lambda), S^\lambda E)$ *is 1-dimensional and any non-zero element is a good monomorphism.*
(ii)(a) *The space* $\mathrm{Hom}_G(E^{\otimes r}, S^r E)$ *(resp.* $\mathrm{Hom}_G(\bigwedge^r E, E^{\otimes r})$*) is 1-dimensional and any non-zero element is a good epimorphism (resp good monomorphism).*
(b) *For any* $\alpha \in \Lambda(n, r)$, *the natural map* $E^{\otimes r} \to S^\alpha E$ *(resp.* $\bigwedge^\alpha E \to E^{\otimes r}$*) is a good epimorphism (resp. good monomorphism).*

The module $\bigwedge^{\lambda'} E$ has a good filtration, [**36**; Section 4(3)(i)], and [**36**; Remark 3.7], and has unique highest weight λ, which occurs with multiplicity 1. From these facts (i)(a) follows via a standard argument, see e.g. [**29**; proof of (11.4.1) Theorem]. Part (i)(b) follows in the same way using (8). Noting that $S^r E \cong \nabla(r, 0, \ldots, 0)$, [**36**; Remark 3.7], we get (ii)(a) in the same way. Part (ii)(b) follows from (ii)(a) and (14).

(16)   *Suppose that* $q \neq 0$.
(i) *If* $X, Y \in \mathcal{F}(\nabla)$ *are polynomial modules of degree* r *then restriction* $\mathrm{Hom}_S(X, Y) \to \mathrm{Hom}_H(fX, fY)$ *is injective.*
(ii) *For* $\alpha \in \Lambda(n, r)$ *and* $Y \in \mathcal{F}(\nabla)$ *restriction*

$$\mathrm{Hom}_S(S^\alpha E, Y) \to \mathrm{Hom}_H(fS^\alpha E, fY)$$

is an isomorphism.

(iii) For $\alpha \in \Lambda(n,r)$ and $X \in \mathcal{F}(\nabla)$ restriction

$$\operatorname{Hom}_S(X, \textstyle\bigwedge^\alpha E) \to \operatorname{Hom}_H(fX, f\textstyle\bigwedge^\alpha E)$$

is an isomorphism.

For (i) we argue (as in [66; Section 4]) using a mixture of the argument given in the classical case [32; (2.3)], and a refinement due to Erdmann, [45; 1.8]. For (i) one reduces (as in the classical case) to $X = \nabla(\lambda)$. However, we have an epimorphism $Se \to E^{\otimes r} \to \bigwedge^{\lambda'} E \to \nabla(\lambda) = X$ (the first map is the map θ of (7), the second map is the natural map, and the third map comes from (15)(i)(a)). We conclude that $\operatorname{Hom}_S(X,Y) \to \operatorname{Hom}_H(fX, fY)$ is injective by part (3) of the argument of the proof of [45; 1.8 Proposition] (or by invoking the Proposition directly).

For (ii) we use the argument of [45; 1.8 Proposition] once again. Certainly $\operatorname{Hom}_S(S^\alpha E, Y) \to \operatorname{Hom}_H(fS^\alpha E, fY)$ is injective by (i). Let $\phi : E^{\otimes r} \to S^\alpha E$ be a good epimorphism (see (15)(ii)(b) and let $K = \operatorname{Ker}(\phi)$. We have a commutative diagram

$$
\begin{array}{ccccccc}
0 & \to & \operatorname{Hom}_S(S^\alpha E, Y) & \to & \operatorname{Hom}_S(E^{\otimes r}, Y) & \to & \operatorname{Hom}_S(K, Y) \\
& & \downarrow & & \downarrow & & \downarrow \\
0 & \to & \operatorname{Hom}_H(fS^\alpha E, fY) & \to & \operatorname{Hom}_H(fE^{\otimes r}, fY) & \to & \operatorname{Hom}_H(fK, fY)
\end{array}
$$

with rows exact and injective vertical maps. Furthermore we have

$$\dim \operatorname{Hom}_S(E^{\otimes r}, Y) = \dim Y^\omega = \dim fY,$$

by (8)(ii), and $\dim \operatorname{Hom}_H(fE^{\otimes r}, fY) = \dim fY$, since $fE^{\otimes r} \cong fSe = H$. Thus the central vertical map is an isomorphism. Now a diagram chase gives that $\operatorname{Hom}_S(S^\alpha E, Y) \to \operatorname{Hom}_H(fS^\alpha, fY)$ is surjective, as required.

We now consider (iii). First take $\alpha = \omega$. We have $\dim \operatorname{Hom}_S(X, E^{\otimes r}) = \dim X^\omega = \dim fX$, by (8), and

$$\dim \operatorname{Hom}_H(fX, fE^{\otimes r}) = \dim \operatorname{Hom}_H(fX, H) = \dim fX$$

since H is Frobenius. Moreover $\operatorname{Hom}_S(X, E^{\otimes r}) \to \operatorname{Hom}_H(fX, fE^{\otimes r})$ is injective, by (i), and hence is an isomorphism, by dimensions. Now let α be arbitrary. We have a short exact sequence $0 \to \bigwedge^\alpha E \to E^{\otimes r} \to Q \to 0$, with $Q \in \mathcal{F}(\nabla)$, by (15)(ii)(b). Thus we have a commutative diagram

$$
\begin{array}{ccccccc}
0 & \to & \operatorname{Hom}_S(X, \bigwedge^\alpha E) & \to & \operatorname{Hom}_S(X, E^{\otimes r}) & \to & \operatorname{Hom}_S(X, Q) \\
& & \downarrow & & \downarrow & & \downarrow \\
0 & \to & \operatorname{Hom}_H(fX, f\bigwedge^\alpha E) & \to & \operatorname{Hom}_H(fX, fE^{\otimes r}) & \to & \operatorname{Hom}_H(fX, fQ)
\end{array}
$$

and the surjectivity of $\operatorname{Hom}_S(X, \bigwedge^\alpha E) \to \operatorname{Hom}_H(fX, f\bigwedge^\alpha E)$ follows from a diagram chase.

We have linear characters ν, ε of the Hecke algebra H given by $\nu(T_w) = q^{l(w)}$ and $\varepsilon(T_w) = \text{sgn}(w)$, for $w \in W$. We call these (respectively) the trivial and sign representations of H. We write simply k (resp. k_s) for the field k regarded as a left H-module via the representation ν (resp. ε).

We set $N = \binom{r}{2}$. The following is an exercise (which we leave to the reader) in the use of the form (,).

(17)   (i) We have $\nu(b_\sigma) = q^{l(\sigma)}$, $\varepsilon(b_\sigma) = \text{sgn}(\sigma)$, for $\sigma \in \text{Sym}(r)$.
(ii) We have $\sum_{\sigma \in \text{Sym}(r)} b_\sigma = \sum_{\pi \in \text{Sym}(r)} q^{N-l(\pi)} b'_\pi = b_\nu$, say, and
$\sum_{\sigma \in \text{Sym}(r)} (-q)^{N-l(\sigma)} b_\sigma = \sum_{\pi \in \text{Sym}(r)} b'_\pi = b_\varepsilon$, say.
(iii) The elements b_ν and b_ε belong to the centre of H and span left H-submodules of H affording representations ν and ε respectively.

We now fix $1 \le a < r$ and calculate the action of T_{s_a} on $(E^{\otimes r})^\omega$. For $\sigma \in \text{Sym}(r)$ we have $v_u b_\sigma = v_{u\sigma}$ (see before (11)). Hence we have

$$v_{u\sigma} b_{s_a} = v_u b_\sigma b_{s_a}$$

$$= \begin{cases} v_u b_{s_a \sigma}, & \text{if } l(s_a \sigma) = l(\sigma) + 1; \\ q v_u b_{s_a \sigma} + (q-1) v_{u\sigma}, & \text{if } l(s_a \sigma) = l(\sigma) - 1 \end{cases}$$

$$= \begin{cases} v_{u s_a \sigma}, & \text{if } l(s_a \sigma) = l(\sigma) + 1; \\ q v_{u s_a \sigma} + (q-1) v_{u\sigma}, & \text{if } l(s_a \sigma) = l(\sigma) - 1. \end{cases}$$

We have $b_{s_a} e_{u\pi} = \sum_{\sigma \in \text{Sym}(r)} \lambda_\sigma e_{u\sigma}$ for scalars λ_σ. Now $\lambda_\sigma = (v_{u\sigma}, b_{s_a} e_{u\pi}) = (v_{u\sigma} b_{s_a}, e_{u\pi})$ from which it is easy to determine λ_σ, $\sigma \in \text{Sym}(r)$ and obtain the following.

(18)   For $1 \le a < r$ and $\sigma \in \text{Sym}(r)$ we have

$$T_{s_a} e_{u\sigma} = \begin{cases} q e_{u s_a \sigma}, & \text{if } l(s_a \sigma) = l(\sigma) + 1; \\ e_{u s_a \sigma} + (q-1) e_{u\sigma}, & \text{if } l(s_a \sigma) = l(\sigma) - 1. \end{cases}$$

By writing $v\sigma$ as $uw_0\sigma$ we deduce the following.

(19)   For $1 \le a < r, \sigma \in \text{Sym}(r)$ we have

$$T_{s_a} e_{v\sigma} = \begin{cases} q e_{v s_{r-a} \sigma}, & \text{if } l(s_{r-a}\sigma) = l(\sigma) - 1; \\ e_{v s_{r-a}\sigma} + (q-1) e_{v\sigma}, & \text{if } l(s_{r-a}\sigma) = l(\sigma) + 1. \end{cases}$$

Let $\alpha = (\alpha_1, \ldots, \alpha_n) \in \Lambda(n,r)$. For $1 \le i < r$ we define a subset $J_i(\alpha)$ of $[1, r-1]$ as follows. We define $J_1(\alpha)$ to be the set of a such that $1 \le a < \alpha_1$ and, for $i > 1$, define $J_i(\alpha)$ to be the set of a such that $\alpha_1 + \cdots + \alpha_{i-1} < a < \alpha_1 + \cdots + \alpha_i$. We define $J(\alpha) = \bigcup_{i=1}^n J_i(\alpha)$. We write $\text{Sym}(\alpha)$ for the

Young subgroup $\mathrm{Sym}[1, \alpha_1] \times \mathrm{Sym}[\alpha_1 + 1, \alpha_1 + \alpha_2] \times \cdots$ of $\mathrm{Sym}(r)$. We write $H(\alpha)$ for the subalgebra of $H(r)$ generated by $\{T_{s_a} \mid a \in J(\alpha)\}$. We have a natural isomorphism $H(\alpha) \cong H(\alpha_1) \otimes H(\alpha_2) \otimes \cdots$. Moreover, H is free as a right $H(\alpha)$-module of rank $r!/(\alpha_1! \alpha_2! \ldots)$.

Notation
We write $x(\alpha)$ for $\sum_{w \in \mathrm{Sym}(\alpha)} T_w$ and $y(\alpha)$ for $\sum_{w \in \mathrm{Sym}(\alpha)} (-q)^{N-l(w)} T_w$.

The following is the q-analogue of [33; (3.5) Lemma].

(20) *(i) For $\alpha \in \Lambda(n, r)$ we have an isomorphism of left $H(r)$-modules $\phi : H(r) \otimes_{H(\alpha)} k \to f S^\alpha E$ taking $(1 \otimes 1)$ to $\bar{e}_u$ and an isomorphism of left $H(r)$-modules $\psi : H(r) \otimes_{H(\alpha)} k \to H(r)x(\alpha)$ taking $1 \otimes 1$ to $x(\alpha)$.*
(ii) Let $\alpha = (\alpha_1, \ldots, \alpha_m)$ be a composition of r and let $\beta = (\alpha_m, \ldots, \alpha_1)$. We have an isomorphism of left $H(r)$-modules $\zeta : H(r) \otimes_{H(\beta)} k_s \to f \bigwedge^\alpha E$ taking $1 \otimes 1$ to $\hat{e}_v$ and an isomorphism of left $H(r)$-modules $\eta : H(r) \otimes_{H(\beta)} k_s \to H(r)y(\beta)$ taking $1 \otimes 1$ to $y(\beta)$.

Proof   (i) We get $T_{s_a} \bar{e}_u = \bar{e}_u$ for all $a \in J(\alpha)$ from (18) so that $k\bar{e}_u \cong k$, as $H(\alpha)$-modules. Thus (by the universal property of induction) the $H(\alpha)$-map $k \to f S^\alpha E$ gives rise to a $H(r)$-module homomorphism $\phi : H(r) \otimes_{H(\alpha)} k \to f S^\alpha E$ taking $1 \otimes 1$ to $\bar{e}_u$. We have $\dim H(r) \otimes_{H(\alpha)} k = \dim f S^\alpha E = [\mathrm{Sym}(r) : \mathrm{Sym}(\alpha)]$; hence we only have to check that ϕ is surjective, i.e. $f S^\alpha E = H(r)\hat{e}_u$. But we have $b'_\pi \bar{e}_u = \bar{e}_{u\pi}$ for $\pi \in \mathrm{Sym}(r)$ (see before (11)) and the result follows. We let $H_i(\alpha)$ be the subalgebra of H generated by the elements T_{s_a}, $a \in J_i(\alpha)$. The subalgebras $H_i(\alpha), H_j(\alpha)$ commute for $i \neq j$ and we have $x(\alpha) = x_1(\alpha)x_2(\alpha) \ldots$. We have a natural isomorphism $H(\alpha_i) \to H_i(\alpha)$ and it follows that $T_{s_a} x_i(\alpha) = q x_i(\alpha)$ for $a \in J_i(\alpha)$. Thus we get $T_{s_a} x(\alpha) = q x(\alpha)$, for $a \in J(\alpha)$. By the universal property of induction, this gives rise to an $H(r)$-module homomorphism $\psi : H(r) \otimes_{H(\alpha)} k \to H(r)x(\alpha)$, taking $1 \otimes 1$ to $x(\alpha)$. Now ψ is surjective and $\dim H(r) \otimes_{H(\alpha)} k = \dim H(r)x(\alpha) = [\mathrm{Sym}(r) : \mathrm{Sym}(\alpha)]$ (this follows from the freeness of $H(r)$ over $H(\alpha)$) and hence ψ is an isomorphism.
(ii) For $a \in J(\beta)$ we have $r - a \in J(\alpha)$ and hence $T_{s_a} \hat{e}_v = \hat{e}_{v s_{r-a}} + (q-1)\hat{e}_v = -q\hat{e}_v + (q-1)\hat{e}_v = -\hat{e}_v$, by (19). Hence $k\hat{e}_v \cong k_s$, as $H(\beta)$-modules. By the universal property of induction we get an $H(r)$-module map $\zeta : H(r) \otimes_{H(\beta)} k_s \to f \bigwedge^\alpha E$ taking $1 \otimes 1$ to $\hat{e}_v$. The image of ζ is $H(r)\hat{e}_v$ and $b'_\pi \hat{e}_v = \hat{e}_{v\pi}$, for $\pi \in \mathrm{Sym}(r)$, so $H(r)\hat{e}_v = f \bigwedge^\alpha E$ and ζ is surjective. Now $\dim H(r) \otimes_{H(\beta)} k_s = [\mathrm{Sym}(r) : \mathrm{Sym}(\beta)] = [\mathrm{Sym}(r) : \mathrm{Sym}(\alpha)] = \dim f \bigwedge^\alpha E$ so that ζ is an isomorphism.
    One gets the isomorphism $\eta : H(r) \otimes_{H(\beta)} k_s \to H(r)y(\beta)$ by arguing as in part (i).

2.2 The 0-Schur algebra

We fix a field k. We shall derive a simple formula for the character of the irreducible $S(n,r)$-modules at $q = 0$. This will emerge as a consequence of the analysis of the the Hecke algebra of a finite Coxeter group at $q = 0$ made by P. N. Norton, [69] (a variation on a theme of Solomon, [73]). The representation theory of the 0-Schur algebra is also discussed by Krob and Thibon, [62].

For $1 \leq i \leq n$ we define ϵ_i to be the element $(0,\ldots,0,1,0,\ldots,0)$ (with 1 in the ith position) of $X(n)$. Thus we have $X(n) = \mathbf{Z}\epsilon_1 \oplus \cdots \oplus \mathbf{Z}\epsilon_n$.

Suppose that $n \geq m$. Let $\xi = \sum_{\alpha \in \Lambda(m,r)} \xi_\alpha$. We shall produce an isomorphism $\psi : S(m,r) \to \xi S(n,r)\xi$. The classical case, $q = 1$, is given in [51, Section 6.5]. The general case is somewhat more complicated, at least from the point of view of notation.

To emphasize dependence on n (resp. m) we write, for the moment, $^n c_{ij}$ (resp. $^m c_{ij}$) for the element of $A(n)$ (resp. $A(m)$) previously denoted c_{ij}, for $i, j \in I(n,r)$ (resp. $i, j \in I(m,r)$) and $r \geq 1$.

Also we write, for the moment, $^n I^-(\alpha)$ (resp. $^m I^-(\alpha)$) for the set previously denoted $I^-(\alpha)$, for $\alpha \in \Lambda(n,r)$ (resp. $\alpha \in \Lambda(m,r)$). We also write $^n \xi_\alpha$ (resp. $^m \xi_\alpha$) and $^n \xi_{i\alpha j}$ (resp. $^m \xi_{i\alpha j}$) for the elements of $S(n,r)$ (resp. $S(m,r)$) previously denoted ξ_α and $\xi_{i\alpha j}$, for $\alpha \in \Lambda(n,r)$ (resp. $\alpha \in \Lambda(m,r)$) and $j \in {}^n I^-(\alpha)$ (resp. $j \in {}^m I^-(\alpha)$). We identify $I(m,r)$ with a subset of $I(n,r)$ and identify $\Lambda(m,r)$ with a subset of $\Lambda(n,r)$ and identify $^m I^-(\alpha)$ with a subset of $^n I^-(\alpha)$, for $\alpha \in \Lambda(m,r)$, in the obvious way.

From the description of $A(m)$ and $A(n)$ by generators and relations we see that there is a k-algebra map $\phi : A(m) \to A(n)$ taking $^m c_{ij}$ to $^n c_{ij}$. Moreover, from the basis of $A(m,r)$ (resp. $A(n,r)$), for $r \geq 0$, given in 2.1(2), we see that ϕ is injective. We write $\bar{A}(m,r)$ for the image of $A(m,r)$ under ϕ. Then we have $\bar{A}(m,r) = \bigoplus_{\alpha,\beta \in \Lambda(m,r)} {}^\alpha A(n,r)^\beta$ and so we have a decomposition $A(n,r) = \bar{A}(m,r) \oplus Z$, where $Z = \bigoplus_{\alpha,\beta} {}^\alpha A(n,r)^\beta$, the sum running over $(\alpha,\beta) \in \Lambda(n,r) \times \Lambda(n,r)$ with either α or β not in $\Lambda(m,r)$. Let $\pi : A(n,r) \to \bar{A}(m,r)$ be the projection. Thus we have

$$\pi(^n c_{ij}) = \begin{cases} {}^n c_{ij}, & \text{if } i, j \in I(m,r); \\ 0, & \text{otherwise.} \end{cases}$$

We define $\bar{\varepsilon} : \bar{A}(m,r) \to k$ to be the restriction of $\varepsilon : A(n,r) \to k$. We define $\bar{\delta} : \bar{A}(m,r) \to \bar{A}(m,r) \otimes \bar{A}(m,r)$ to be $(\pi \otimes \pi) \circ \delta \circ \iota$, where $\iota : \bar{A}(m,r) \to A(n,r)$ is inclusion. Let $\theta : A(m,r) \to \bar{A}(m,r)$ be the restriction of $\phi : A(m) \to A(n)$. Thus θ is a linear isomorphism. We write $^n \varepsilon$ and $^n \delta$ for the augmentation and comultiplication maps for the coalgebra $A(n,r)$ and we write $^m \varepsilon$ and $^m \delta$ for the augmentation and comultiplication maps for the coalgebra $A(m,r)$. For $i, j \in I(m,r)$ we have $\bar{\varepsilon} \circ \theta(^m c_{ij}) = \bar{\varepsilon}(^n c_{ij}) =$

$\varepsilon(^n c_{ij}) = \delta_{ij} = {}^m\varepsilon(^m c_{ij})$ and hence $\bar{\varepsilon} \circ \theta = {}^m\varepsilon$. Moreover, we have

$$\bar{\delta} \circ \theta(^m c_{ij}) = \bar{\delta}(^n c_{ij}) = (\pi \otimes \pi)(\sum_{h \in I(n,r)} {}^n c_{ih} \otimes {}^n c_{hj})$$

$$= (\sum_{h \in I(m,r)} {}^n c_{ih} \otimes {}^n c_{hj})$$

$$= (\theta \otimes \theta) \circ {}^m\delta(^m c_{ij})$$

for $i,j \in I(m,r)$, and so $\bar{\delta} \circ \theta = (\theta \otimes \theta) \circ {}^m\delta$. Thus $(\bar{A}(m,r), \bar{\delta}, \bar{\varepsilon})$ is the coalgebra obtained by transport of structure from the coalgebra $(A(m,r), {}^m\delta, {}^m\varepsilon)$, via the linear isomorphism $\theta : A(m,r) \to \bar{A}(m,r)$. In particular, $(\bar{A}(m,r), \bar{\delta}, \bar{\varepsilon})$ is a k-coalgebra and $\theta : A(m,r) \to \bar{A}(m,r)$ is a coalgebra isomorphism. Let $\bar{S}(m,r)$ the the dual algebra of the coalgebra $\bar{A}(m,r)$.

We define a linear map $\chi : \bar{S}(m,r) \to S(n,r)$ by $\chi(x) = x \circ \pi$, for $x \in \bar{S}(m,r)$. For $i,j \in I(n,r)$ we have

$$\chi(1)(^n c_{ij}) = \bar{\varepsilon}(\pi(^n c_{ij}))$$

$$= \begin{cases} \bar{\varepsilon}(^n c_{ij}), & \text{if } i,j \in I(m,r); \\ 0, & \text{otherwise} \end{cases}$$

$$= \begin{cases} \delta_{ij}, & \text{if } i,j \in I(m,r); \\ 0, & \text{otherwise.} \end{cases}$$

Hence we have $\chi(1) = \sum_{\alpha \in \Lambda(m,r)} {}^n\xi_\alpha = \xi$. We now prove that χ is multiplicative. Let $x,y \in \bar{S}(m,r)$ and $i,j \in I(n,r)$. For $i,j \in I(m,r)$ we have

$$\chi(xy)(^n c_{ij}) = (xy)(\pi(^n c_{ij}))$$

$$= (x \otimes y)(\bar{\delta}(^n c_{ij}))$$

$$= \sum_{h \in I(n,r)} x(\pi(^n c_{ih}))y(\pi(^n c_{hj}))$$

$$= (\chi(x)\chi(y))(^n c_{ij}).$$

Moreover, if i or j is not in $I(n,r)$ then $\chi(xy)(^n c_{ij}) = (xy)(\pi(^n c_{ij})) = 0$ and

$$(\chi(x)\chi(y))(^n c_{ij}) = \sum_{h \in I(n,r)} x(\pi(^n c_{ih}))y(\pi(^n c_{hj})) = 0.$$

Hence $\chi(xy) = \chi(x)\chi(y)$, and χ is multiplicative. Now we have $\chi(x) = \chi(1x1) = \chi(1)\chi(x)\chi(1) \in \xi S(n,r)\xi$ and the restriction $\chi_0 : \bar{S}(m,r) \to \xi S(n,r)\xi$, of $\chi : \bar{S}(m,r) \to S(n,r)$, is an algebra map. Since π is surjective, χ_0 is injective. The dimension of $\bar{S}(m,r)$ is equal to the dimension of $S(m,r)$ and it follows from 2.1(5) that this is also the dimension of $\xi S(n,r)\xi$. Hence χ_0 is an isomorphism. We put $\psi = \chi_0 \circ (\theta^*)^{-1} : S(m,r) \to \xi S(n,r)\xi$, where $\theta^* : \bar{S}(m,r) \to S(m,r)$ is the algebra isomorphism dual to the coalgebra isomorphism $\theta : A(m,r) \to \bar{A}(m,r)$. We claim that

$$\psi(^m\xi_{i\alpha j}) = {}^n\xi_{i\alpha j} \qquad (*)$$

for $\alpha \in \Lambda(m,r)$, $j \in {}^mI^-(\alpha)$. For $\alpha \in \Lambda(m,r)$ and $j \in {}^mI^-(\alpha)$ we write $\bar{\xi}_{i\alpha j}$ for the restriction of $^n\xi_{i\alpha j}$ to $\bar{A}(m,r)$. Then $\{\bar{\xi}_{i\alpha j} \mid \alpha \in \Lambda(m,r), \, j \in {}^mI^-(\alpha)\}$ is the basis of $\bar{S}(m,r)$ dual to the basis $\{^nc_{i\alpha j} \mid \alpha \in \Lambda(m,r), \, j \in {}^mI^-(\alpha)\}$ of $\bar{A}(m,r)$. Thus we have $(\theta^*)(\bar{\xi}_{i\alpha j}) = {}^m\xi_{i\alpha j}$ and, to prove $(*)$, it suffices to show that $\chi_0(\bar{\xi}_{i\alpha j}) = {}^n\xi_{i\alpha j}$. So let $a \in A(n,r)$ and suppose that $a \in \bar{A}(m,r)$ or $a \in Z$. Then we have

$$\chi_0(\bar{\xi}_{i\alpha j})(a) = {}^n\xi_{i\alpha j}(\pi(a))$$
$$= \begin{cases} {}^n\xi_{i\alpha j}(a), & \text{if } a \in \bar{A}(m,r); \\ 0, & \text{otherwise} \end{cases}$$
$$= {}^n\xi_{i\alpha j}(a)$$

(since $^n\xi_{i\alpha j}(a) = 0$ if $a \in Z$) and hence $\chi_0(\bar{\xi}_{i\alpha j}) = {}^n\xi_{i\alpha j}$, completing the proof of $(*)$. To summarize: we have shown the following.

(1)   For $m \leq n$, there is an algebra isomorphism $\psi : S(m,r) \to \xi S(n,r)\xi$ such that $\psi(^m\xi_{i\alpha j}) = {}^n\xi_{i\alpha j}$, for all $\alpha \in \Lambda(m,r)$, $j \in {}^mI^-(\alpha)$.

For the rest of this section we take $q = 0$.

(2)   Suppose $\lambda = (\lambda_1, \ldots, \lambda_n) \in X(n)$ and $\lambda_1, \ldots, \lambda_n \neq 0$. Then there is a 1-dimensional $M(n)$-module of weight λ.

Proof   Let $\mu = (\mu_1, \ldots, \mu_n) \in X(n)$ with $\mu_1, \ldots, \mu_n \geq 0$. We define $c^\mu = c_{11}^{\mu_1} \ldots c_{nn}^{\mu_n}$. We claim that dc^μ is a group-like element of $k[M]$, i.e. that $dc^\mu \neq 0$ and $\delta(dc^\mu) = \delta(dc^\mu) \otimes \delta(dc^\mu)$. Note that $\varepsilon(dc^\mu) = 1$ (where d is the determinant), in particular $dc^\mu \neq 0$. Certainly $d = dc^0$ is group-like (e.g. because of the status of d as coefficient function of $\bigwedge^n E$, cf. [**18**; Section 4.1]). Now suppose that $\mu = (\mu_1, \ldots, \mu_m, 0, \ldots, 0)$ with $\mu_m \neq 0$ and suppose that dc^ν is group-like for $\nu = \mu - \epsilon_m = (\mu_1, \ldots, \mu_m - 1, 0, \ldots, 0)$. Recall that we have $dc_{ij} = q^{j-i}c_{ij}d$, if $1 \leq i < j \leq n$, and $c_{ij}d = q^{i-j}dc_{ij}$, if $1 \leq j \leq i \leq n$, for arbitrary q, see [**18**; 4.1.9 Theorem]. Thus, in our situation, we have $dc_{ij} = 0$, for $1 \leq i < j \leq n$, and $dc_{rr} = c_{rr}d$, for $1 \leq r \leq n$. Thus we have $dc^\nu = c^\nu d$ and

$$\delta(dc^\mu) = \delta(dc^\nu)\delta(c_{mm}) = (dc^\nu \otimes dc^\nu)(\sum_{i=1}^n c_{mi} \otimes c_{im})$$
$$= (c^\nu \otimes c^\nu)(\sum_{i=1}^n dc_{mi} \otimes dc_{im})$$
$$= (c^\nu \otimes c^\nu)(dc_{mm} \otimes dc_{mm}) = dc^\mu \otimes dc^\mu.$$

Thus we get that dc^μ is group-like for all $\mu \in X(n)$, with $\mu_1, \ldots, \mu_n \geq 0$, by induction.

Now let $\mu = \lambda - (1, \ldots, 1)$ and put $f = dc^\mu$. Then $\delta(f) = f \otimes f$ so that kf is a 1-dimensional M-submodule of $k[M]$, of weight λ.

We write $X_0^+(n)$ for the set of $\lambda = (\lambda_1, \ldots, \lambda_n) \in X(n)$ such that, for some $0 \leq m < n$, we have $\lambda_i \neq 0$ for all $1 \leq i \leq m$ and $\lambda_i = 0$ for $i > m$.

(3) (i) For $\lambda = (\lambda_1, \ldots, \lambda_m, 0, \ldots, 0) \in X_0^+(n)$, with $\lambda_m \neq 0$, there exists an irreducible $M(n)$-module $L(\lambda)$, say, such that $L(\lambda)$ has weight λ, which occurs with multiplicity 1, and if $\mu = (\mu_1, \ldots, \mu_n)$ is any other weight then $\mu_i \neq 0$ for some $i > m$.
(ii) The modules $L(\lambda)$, $\lambda \in X_0^+(n)$, are pairwise non-isomorphic.

Proof (i) We take ξ as in (1). By (1) and (2), there is a 1-dimensional simple $\xi S(n,r)\xi$-module with weight λ. By the general theory of the Schur functor, Appendix A1(4)(iv), there exists a simple $S(n,r)$-module L, say, such that ξL is a 1-dimensional $\xi S(n,r)\xi$-module of weight λ. Now we have $\xi L = \sum_{\alpha \in \Lambda(m,r)} \xi_\alpha L = \bigoplus_{\alpha \in \Lambda(m,r)} L^\alpha$, from which we deduce that $\dim L^\lambda = 1$ and $L^\alpha = 0$ for $\lambda \neq \alpha \in \Lambda(m,r)$. Thus $L(\lambda) = L$ has the required properties.
(ii) It follows from the description of weight spaces of $L(\lambda)$ that $\operatorname{ch} L(\lambda) \neq \operatorname{ch} L(\mu)$ and therefore $L(\lambda)$ and $L(\mu)$ are not isomorphic for distinct elements λ, μ of $X_0^+(n)$.

(4) For $\alpha \in \Lambda(n,r)$ the natural map $E^{\otimes r} \to S^\alpha E$ is a split epimorphism.

Proof If $\phi_i : X_i \to Y_i$ is a split epimorphism of M-modules, for $1 \leq i \leq m$, then $\phi_1 \otimes \cdots \otimes \phi_m : X_1 \otimes \cdots \otimes X_m \to Y_1 \otimes \cdots \otimes Y_m$ is a split epimorphism. Hence it suffices to show that the natural map $\eta : E^{\otimes r} \to S^r E$ is split. Let Q denote the k-span of $e_i \in E^{\otimes r}$ with i weakly increasing and let $\tau : E^{\otimes r} \to E^{\otimes r} \otimes k[M(n)]$ denote the structure map. Then we have

$$\tau(e_i) = \sum_{j \in I(n,r)} e_j \otimes c_{ji}$$

for $i \in I(n,r)$. If $i = (i_1, \ldots, i_r)$ is weakly decreasing and $j = (j_1, \ldots, j_r)$ has $j_a > j_{a+1}$ for some $1 \leq a < r$ then, by the defining relations, we have $c_{j_a i_a} c_{j_{a+1} i_{a+1}} = q c_{j_{a+1} i_{a+1}} c_{j_a i_a} = 0$ and hence $c_{ji} = 0$. Thus $\tau(Q) \leq Q \otimes k[M(n)]$ and Q is an M-submodule of $E^{\otimes r}$. Now $\eta : E^{\otimes r} \to S^r E$ restricts to an isomorphism on Q and so η is split.

(5) For $r \leq n$ we have:
(i) $S(n,r)$ is a quasi-Frobenius algebra;

(ii) $E^{\otimes r}$ is a projective generator of $S(n,r)$;
(iii) the algebras $S(n,r)$ and $H(r)$ are Morita equivalent via the equivalence of categories $f : \mathrm{mod}(S(n,r)) \to \mathrm{mod}(H(r))$.

Proof Let $S = S(n,r)$. We have $A(n,r) \cong \bigoplus_{\alpha \in \Lambda(n,r)} S^\alpha E$, by 2.1(1)(ii). Hence any injective indecomposable S-module is isomorphic to a component of $S^\alpha E$ for some $\alpha \in \Lambda(n,r)$ and hence, by (4), isomorphic to a direct summand of $E^{\otimes r}$. Hence, by 2.1(7), every injective S-module is projective and so S is quasi-Frobenius. A projective indecomposable module is therefore injective and hence a component of $E^{\otimes r}$. Hence $E^{\otimes r}$ is a projective generator. Recall the factorization $f = \tilde{\phi} \circ F$ of Remark (ii) of Section 2.1 following (10)'. Certainly $\tilde{\phi}$ is an isomorphism and since $E^{\otimes r}$ is a projective generator we have that $F : \mathrm{mod}(S) \to \mathrm{mod}(\mathrm{End}_S(E^{\otimes r})^{\mathrm{op}})$ is an equivalence of categories, see e.g. [4; Lemma 2.2.3]. Hence f is an equivalence of categories.

Let $r \geq 1$. We put $\Lambda_0^+(n,r) = X_0^+(n) \cap \Lambda(n,r)$. From [69], we have that the irreducible $H(r)$-modules are labelled by the subsets J of $[1, r-1]$. To $\lambda = (\lambda_1, \ldots, \lambda_m, 0, \ldots, 0) \in \Lambda_0^+(n,r)$ (with $\lambda_m \neq 0$) there corresponds the subset $J(\lambda) = [1, \lambda_1 - 1] \bigcup [\lambda_1 + 1, \lambda_1 + \lambda_2 - 1] \bigcup \cdots$. From [69; Section 3], we get a 1-dimensional $H(r)$-module k_λ such that

$$T_{s_a} z = \begin{cases} 0, & \text{if } a \in J(\lambda); \\ -1, & \text{if } a \notin J(\lambda) \end{cases}$$

for $z \in k_\lambda$, and furthermore $\{k_\lambda \mid \lambda \in \Lambda_0^+(n,r)\}$ is a complete set of pairwise non-isomorphic irreducible $H(r)$-modules.

From (5)(iii), the number of isomorphism classes of irreducible $S(n,r)$-modules, for $r \leq n$, is equal to the number of isomorphism classes of irreducible $H(r)$-modules, i.e. 2^{r-1}. But this is also the number of isomorphism classes of irreducible M-modules of degree r, i.e. $S(n,r)$-modules, described in (3). Thus we have the following.

(6) *The modules $\{L(\lambda) \mid \lambda \in \Lambda_0^+(n,r)\}$ (of (3) above) form a complete set of pairwise non-isomorphic irreducible $S(n,r)$-modules, for $r \leq n$.*

Let $P(\lambda)$ denote the projective cover of k_λ and $Q(\lambda)$ denote the injective hull of k_λ, for $\lambda \in \Lambda_0^+(n,r)$. Since $H(r)$ is a Frobenius algebra (there is a non-singular invariant bilinear form on $H(r)$ by 2.1(11)) there is a bijection $\lambda \mapsto \tilde{\lambda}$ such that $P(\lambda) \cong Q(\tilde{\lambda})$, for $\lambda \in \Lambda_0^+(n,r)$. Now by the explicit description of $P(\lambda)$ given by Norton, [69; 4.22 Theorem], and by [69; 4.23 Lemma], we have that $\tilde{\lambda} = (\lambda_m, \ldots, \lambda_1)$, for $\lambda = (\lambda_1, \ldots, \lambda_m) \in \Lambda_0^+(n,r)$ (with $\lambda_m \neq 0$).

We write $I(\lambda)$ for the injective hull of $L(\lambda)$, $\lambda \in \Lambda_0^+(n,r)$. Since f is an equivalence of categories, we have a bijection $\lambda \mapsto \check{\lambda}$ on $\Lambda_0^+(n,r)$ determined by the condition $fI(\lambda) \cong Q(\check{\lambda})$. We claim that in fact $\check{\lambda} = \tilde{\lambda}$, for $\lambda \in \Lambda_0^+(n,r)$.

Now, for $\lambda \in \Lambda_0^+(n,r)$, the module $S^\lambda E$ is injective and hence a direct sum of copies of $I(\mu)$, $\mu \in \Lambda_0^+(n,r)$. Further, since f is an equivalence, the multiplicity of $I(\mu)$ as a summand of $S^\lambda E$ is equal to the multiplicity of $fI(\mu)$ as a summand of $fS^\lambda E$, which is isomorphic to $H(r)x(\lambda)$, by 2.1(20)(i). But we have

$$H(r)x(\lambda) \cong \bigoplus_{\lambda \subseteq \mu} P(\mu) = \bigoplus_{\lambda \subseteq \mu} Q(\tilde{\mu}) \qquad (\dagger)$$

by [69; 4.14 Corollary (2)] (and [69; 4.22 Theorem]), where $\lambda \subseteq \mu$ means that $J(\lambda) \subseteq J(\mu)$. Hence the multiplicity of $I(\mu)$, as a summand of $S^\lambda E$, is at most 1 and we have

$$S^\lambda E \cong \bigoplus_{\mu \in F(\lambda)} I(\mu)$$

for some subset $F(\lambda)$ of $\Lambda_0^+(n,r)$. By 2.1(8), the multiplicity of $I(\mu)$ as a summand of $S^\lambda E$ is $\dim L(\lambda)^\mu$. By (3)(i) we have

$$S^\lambda E \cong I(\lambda) \oplus \left(\bigoplus_{\mu \in F_1(\lambda)} I(\mu) \right)$$

where $F_1(\lambda)$ consists of elements of $\Lambda_0^+(n,r)$ which have more parts than λ. We can assume inductively that $fI(\mu) = Q(\tilde{\mu})$ for all $\mu \in F_1(\lambda)$. Hence we obtain

$$Hx(\lambda) \cong fS^\lambda E = Q(\check{\lambda}) \oplus \left(\bigoplus_{\mu \in F_1(\lambda)} Q(\tilde{\mu}) \right).$$

By $(\dagger)$ (and the Krull–Schmidt theorem) $Q(\tilde{\lambda})$ occurs as a summand of $fS^\lambda E$. But $\tilde{\lambda}$ is not equal to $\tilde{\mu}$ for any $\mu \in F_1(\lambda)$ (since all elements of $F_1(\lambda)$ have more parts than λ). Hence we must have $\check{\lambda} = \tilde{\lambda}$.

Since $L(\lambda)$ is the socle of $I(\lambda)$ and $k_{\tilde{\lambda}}$ is the socle of $Q(\tilde{\lambda})$ we get $fL(\lambda) \cong k_{\tilde{\lambda}}$, for $\lambda \in \Lambda_0^+(n,r)$. Let $\alpha \in \Lambda(n,r)$ and write $\bar{\alpha}$ for the element of $\Lambda_0^+(n,r)$ obtained by deleting the zeros from $\alpha = (\alpha_1, \alpha_2, \ldots, \alpha_n)$. Now we have $S^\alpha E = S^{\bar{\alpha}} E$ and so

$$\dim L(\lambda)^\alpha = \dim \mathrm{Hom}_S(L(\lambda), S^\alpha E) = \dim \mathrm{Hom}_S(L(\lambda), S^{\bar{\alpha}} E)$$
$$= \dim \mathrm{Hom}_H(k_{\tilde{\lambda}}, H(r)x(\bar{\alpha}))$$

which (by $(\dagger)$) is 1 if $\bar{\alpha} \subseteq \lambda$ and 0 otherwise.

We write x_i for the canonical generator $e(\epsilon_i)$ of the character ring $\mathbf{Z}X$ (see 0.12) and write x^α for $x_1^{\alpha_1} \ldots x_n^{\alpha_n}$, for $\alpha \in \Lambda(n,r)$. We have shown that, for $\lambda \in \Lambda_0^+(n,r)$, the character of $L(\lambda)$ is given by the formula

$$\mathrm{ch}\, L(\lambda) = \sum_{\alpha \in \Lambda(n,r), \bar{\alpha} \subseteq \lambda} x^\alpha. \qquad (\ddagger)$$

We now note that $\lambda \subseteq \mu$ implies $\lambda \leq \mu$ (where $\leq$ is the usual partial order) for $\lambda, \mu \in \Lambda_0^+(n,r)$. Let $\lambda = (\lambda_1, \lambda_2, \ldots)$ and $\mu = (\mu_1, \ldots, \mu_b, 0, \ldots, 0)$, with $\mu_b \neq 0$. Then $J(\lambda) = [1, r]\backslash\{\lambda_1, \lambda_1 + \lambda_2, \ldots\}$ and $J(\mu) = [1, r]\backslash\{\mu_1, \mu_1 + \mu_2, \ldots\}$ so that $J(\lambda) \subseteq J(\mu)$ gives $\{\mu_1, \mu_1 + \mu_2, \ldots\} \subseteq \{\lambda_1, \lambda_1 + \lambda_2, \ldots\}$. In particular $\mu_1 = \lambda_1 + \cdots + \lambda_s$, for some $1 \leq s \leq n$, giving $\lambda_1 \leq \mu_1$. Now suppose that $1 \leq h < b$ and we have proved $\lambda_1 + \cdots + \lambda_h \leq \mu_1 + \cdots + \mu_h$. We have $\mu_1 + \cdots + \mu_h + \mu_{h+1} = \lambda_1 + \cdots + \lambda_l$, for some $1 \leq l \leq n$. If $l \leq h$ we get

$$\lambda_1 + \cdots + \lambda_l \leq \lambda_1 + \cdots + \lambda_h \leq \mu_1 + \cdots + \mu_h \leq \mu_1 + \cdots + \mu_{l+1}$$

so we must have equality throughout. But then we have $\mu_{h+1} = 0$, a contradiction. Hence we have $l > h$ and hence $\lambda_1 + \cdots + \lambda_h + \lambda_{h+1} \leq \lambda_1 + \cdots + \lambda_l = \mu_1 + \cdots + \mu_{h+1}$. Thus we get $\lambda_1 + \cdots + \lambda_i \leq \mu_1 + \cdots + \mu_i$ for all $1 \leq i \leq b$ and hence $\lambda \leq \mu$.

Now let $m \geq 1$, $r \geq 0$ and choose $n \gg 0$. For $\lambda \in \Lambda_0^+(n, r)$ the $\xi S(n, r)\xi$-module $\xi L(\lambda)$ is either 0 or a simple $\xi S(n, r)\xi$-module and each irreducible $\xi S(n, r)\xi$-module is isomorphic to some such $\xi L(\lambda)$, by the general theory of the Schur functor, Appendix A1(4)(iv). Now $\xi L(\lambda) = \bigoplus_{\alpha \in \Lambda(m,r)} L(\lambda)^\alpha$ is non-zero if and only if there is some $\alpha \in \Lambda(m, r)$ with $\bar{\alpha} \leq \lambda$, by ($\ddagger$). In that case we have $\bar{\alpha} \leq \lambda$ and so $r = \bar{\alpha}_1 + \cdots + \bar{\alpha}_m \leq \lambda_1 + \cdots + \lambda_m$ (where $\bar{\alpha} = (\bar{\alpha}_1, \bar{\alpha}_2, \ldots)$). This gives $\lambda_1 + \cdots + \lambda_m = r$ so that $\lambda_{m+1} = 0$ and $\lambda \in \Lambda_0^+(m, r)$. Hence the number of isomorphism classes of $\xi S(n, r)\xi$-modules is at most $|\Lambda_0^+(m, r)|$. By (1) and (3)(ii), it is at least this number. Hence $\{\xi L(\lambda) \mid \lambda \in \Lambda_0^+(m, r)\}$ is a complete set of pairwise inequivalent irreducible $\xi S(n, r)\xi$-modules. Applying ξ to a weight space decomposition of $L(\lambda)$, using ($\ddagger$) and (1) and replacing m by n we get the following.

(7) $\{L(\lambda) \mid \lambda \in \Lambda_0^+(n)\}$ is a complete set of pairwise non-isomorphic irreducible $M(n)$-modules and, for $\lambda \in X_0^+(n)$, we have

$$\operatorname{ch} L(\lambda) = \sum_{\alpha \in \Lambda(n,r),\, \bar{\alpha} \subseteq \lambda} x^\alpha.$$

3. Infinitesimal Theory and Steinberg's Tensor Product Theorem

Suppose that q is a primitive lth root of unity. Then the quantum general linear group $G = G(n)$ has a finite subgroup G_1, called the infinitesimal subgroup, and G_1 plays the role that the first infinitesimal subgroup enjoys in the representation theory of reductive groups in positive characteristic (as in [61; I Chapter 9, II Chapter 9]). We establish the main features of the representation theory of G_1 and related subgroups in Section 3.1. Of particular importance is the subgroup $\hat{G}_1$, generated by G_1 and the torus $T = T(n)$. The main attraction of $\hat{G}_1$ is that while its representation differs only trivially from that of G_1 in most respects it has, unlike G_1, a theory of weights, and the weight structure of a G-module is preserved on restriction to $\hat{G}_1$. We call $\hat{G}_1$ and related subgroups "Jantzen subgroups". These correspond to the subgroups introduced into the representation theory of reductive groups in positive characteristic by Jantzen, [59]. Most of the results of 3.1 have been proved for the Manin quantization (and l odd) by Parshall and Wang, [71; Chapter 9].

In Section 3.2 we give the q-analogue of the famous tensor product theorem of Steinberg. The proof we shall give is similar to that given by Parshall and Wang, [71; (9.4.1) Theorem], for the Manin quantization when q is an odd root of unity, and has some elements in common with the proof of the classical Steinberg tensor product theorem given in [5]. Cliff, [9], has given a proof of this theorem for the Manin quantization in the remaining cases. Moreover Dipper and Du, [19; 5.6 Theorem], have given a proof of the tensor product theorem for the quantization we use here. Their proof is quite different from that given here and is based on the representation theory of the Hecke algebra of type A.

In Section 3.3 we record the basic properties of tilting modules for quantum general linear group G and relationships with modules for infinitesimal subgroups of G. Here the arguments of [33] carry over with little change.

In the final section we give the tilting modules for quantum GL_2 and again, for the most part, this is simply a modification of the classical case, [33; Section 2, Example 2].

3.1 Infinitesimal theory

If H is a quantum group and H_1, H_2 are subgroups then we denote by $H_1 \cap H_2$ the subgroup whose defining ideal of $k[H]$ is $I_{H_1} + I_{H_2}$.

Throughout this section q is a primitive lth root of unity, for some positive integer l. We put $N = \binom{l}{2}$, the number of positive roots. We consider the quantum subgroup G_1 of G defined by the ideal of $k[G]$ generated by the elements $c_{ij}^l - \delta_{ij}$, $1 \le i, j \le n$. We define $B_1 = B \cap G_1$, $B_1^+ = B^+ \cap G_1$

and $T_1 = T \cap G_1$. To keep track of weight spaces in the infinitesimal theory we define the Jantzen subgroups $\hat{G}_1$, $\hat{B}_1$ and $\hat{B}_1^+$ of G. The group $\hat{G}_1$ has ideal $I_{\hat{G}_1}$ (of $k[G]$) generated by the elements c_{ij}^l, with $1 \leq i, j \leq n$ and $i \neq j$. We put $\hat{B}_1 = B \cap \hat{G}_1$ and $\hat{B}_1^+ = B^+ \cap \hat{G}_1$. Since the irreducible modules for T, B and B^+ are 1-dimensional we get (cf. the argument of [36; Lemma 2.6]) that any irreducible T_1, B_1 or B_1^+-module is a restriction of a 1-dimensional module for the corresponding "global" quantum group. More precisely, writing X_1 for the set of $\lambda = (\lambda_1, \ldots, \lambda_n) \in X$ such that $\lambda_1 - \lambda_2, \ldots, \lambda_{n-1} - \lambda_n, \lambda_n \in [0, l-1]$, it is easy to derive the following results.

(1) $\{k_\lambda \mid \lambda \in X\}$ is a full set of irreducible H-modules, for $H = T, \hat{B}_1$ and $\hat{B}_1^+$.

(2)   For $\lambda, \mu \in X$ we have $k_\lambda|_{H_1} \cong k_\mu|_{H_1}$ if and only if $\lambda - \mu \in lX$ and indeed $\{k_\lambda|_{H_1} \mid \lambda \in X_1\}$ is a full set of irreducible H_1-modules, for $H = T, B$ and B^+.

Let H be an arbitrary finite quantum group over k. We write $|H|$ for the dimension of $k[H]$ (and call $|H|$ the *order* of H). Let J be a (quantum) subgroup of H. Then $k[H]|_J$ is a direct sum of copies of $k[J]$, by [38] or by (the dual of the main result of) [68]. This, together with [71; (2.9.1) Theorem], gives the following.

(3)   If J is a subgroup of a finite quantum group H then $|J|$ divides $|H|$. For $V \in \mathrm{mod}(J)$ we have $\dim \mathrm{Ind}_J^H V = [H : J] \dim V$. In particular Ind_J^H is exact.

We are writing $[H : J]$ for the integer $|H|/|J|$, which we call the *index* of J in H. For $\lambda \in X$ we write $I_1(\lambda)$ for the injective hull of k_λ as a B_1^+-module and $\hat{I}_1(\lambda)$ for the injective hull of k_λ as a $\hat{B}_1^+$-module. It is easy to check that the images of the elements $c_{11}^{a_{11}} \ldots c_{nn}^{a_{nn}}$, $1 \leq a_{11}, \ldots, a_{nn} < l$, under the restriction map $k[G] \to k[G_1]$, form a k-basis of $k[G_1]$. From this and similar observations we obtain:

(4) $|G_1| = l^{n^2}$, $|B_1| = |B_1^+| = l^{\binom{n+1}{2}}$ and $|T_1| = l^n$.

Let $\rho = (n, n-1, \ldots, 1) \in X$. Let $\Phi^+ = \{\epsilon_i - \epsilon_j \mid 1 \leq i < j \leq n\}$, the set of positive roots. For $\lambda \in X$ let $A(\lambda) = \sum_{\pi \in \mathrm{Sym}(n)} \mathrm{sgn}(\pi) e(\pi\lambda) \in \mathbb{Z}X$. Then $A(\lambda + \rho)$ is divisible by $A(\rho)$ in $\mathbb{Z}X(n)$ (see e.g. [29; (2.2.7)]) and we set $\chi(\lambda) = A(\lambda + \rho)/A(\rho)$, for $\lambda \in X$. Note that $\chi(\lambda)$ is the Schur symmetric function in n variables, for λ a dominant polynomial weight, see e.g. [63; I,(3.1)]. The character formula may also be expressed in the form $\chi(\lambda) = A(\lambda + \rho)e(-\rho)/\prod_{\alpha \in \Phi^+}(1 - e(-\alpha))$, cf. [54; Section 24.2]. Taking $\lambda = 0$, we get $A(\rho)e(-\rho) = \prod_{\alpha \in \Phi^+}(1 - e(-\alpha))$. We consider the case $\lambda = (r-1)\rho$, for $r \geq 1$. For $\psi = \sum_{\mu \in X} m_\mu e(\mu) \in \mathbb{Z}X$, we write $\psi^{(r)}$ for $\sum_{\mu \in X} m_\mu e(r\mu)$. Note the map $\mathbb{Z}X \to \mathbb{Z}X$ sending ψ to $\psi^{(r)}$ is a ring homomorphism. Now we have

$$\chi((r-1)\rho) = \sum_{w \in \mathrm{Sym}(n)} \mathrm{sgn}(w)e(rw\rho)e(-\rho)/\prod_{\alpha \in \Phi^+}(1 - e(-\alpha))$$

$$= e(-\rho)\left(\sum_{w \in \mathrm{Sym}(n)} \mathrm{sgn}(w)e(w\rho)\right)^{(r)}/\prod_{\alpha \in \Phi^+}(1 - e(-\alpha))$$

$$= e(-\rho)(e(\rho)\prod_{\alpha \in \Phi^+}(1 - e(-\alpha)))^{(r)}/\prod_{\alpha \in \Phi^+}(1 - e(-\alpha))$$

$$= e((r-1)\rho)\prod_{\alpha \in \Phi^+}(1 - e(-\alpha)^r)/\prod_{\alpha \in \Phi^+}(1 - e(-\alpha))$$

$$= e((r-1)\rho)\prod_{\alpha \in \Phi^+}(1 - e(-\alpha) - \cdots - e(-(r-1)\alpha)).$$

Now by the argument of the proof of [36; Lemma 2.8(i)] (or compare with the algebraic group case, [61; II, 9.2]) we have that for $\lambda, \mu \in X$, the dimension of $\hat{I}_1(\lambda)^\mu$ is the number of ways of expressing $\lambda - \mu$ as a sum $\sum_{\alpha \in \Phi^+} r_\alpha \alpha$, with $0 \leq r_\alpha < l$ for $\alpha \in \Phi^+$, and that $\hat{I}_1(\lambda) \cong \mathrm{Ind}_{T_1}^{\hat{B}_1^+} k_\lambda$. Thus $\hat{I}_1((l-1)\rho)$ has character given by the Weyl character $\chi((l-1)\rho)$ and $\mathrm{ch}\,\hat{I}_1(\lambda) = e(\lambda - (l-1)\rho)\mathrm{ch}\,\hat{I}_1((l-1)\rho)$.

(5)   For $\lambda \in X$ we have $\hat{I}_1(\lambda) \cong \mathrm{Ind}_T^{\hat{B}_1^+} k_\lambda$ and

$$\mathrm{ch}\,\hat{I}_1(\lambda) = e(\lambda - (l-1)\rho)\chi((l-1)\rho).$$

From (4) we have $\dim \mathrm{Ind}_{T_1}^{B_1^+} k_\lambda = [B_1^+ : T_1] = l^{\binom{n}{2}}$. It follows from Frobenius reciprocity that $\mathrm{Ind}_{T_1}^{B_1^+} k_\lambda$ contains k_λ in its socle and, since induction takes injectives to injectives, must therefore contain a copy of $I_1(\lambda)$. Hence $\dim I_1(\lambda) \leq l^N$. However, we have a B_1^+-module decomposition $k[B_1^+] \cong \bigoplus_{\lambda \in X_1} I_1(\lambda)$ (by [50; (1.5g)(iii)]). It follows from a dimension count that $\dim I_1(\lambda) = l^N$, for $\lambda \in X_1$. Also, the natural map $k[\hat{B}_1^+] \to k[T] \otimes k[B_1^+]$ is injective and it follows that $\hat{I}_1(\lambda) \cong \mathrm{Ind}_T^{\hat{B}_1^+} k_\lambda$ has a simple B_1^+-socle k_λ. Thus $\hat{I}_1(\lambda)|_{B_1^+}$ embeds in $I_1(\lambda)$ and by dimensions we must have:

(6) $\hat{I}_1(\lambda)|_{B_1^+} \cong I_1(\lambda)$, for $\lambda \in X$.

We define $\nabla_1(\lambda) = \mathrm{Ind}_{B_1}^{G_1} k_\lambda$ and $\hat{\nabla}_1(\lambda) = \mathrm{Ind}_{\hat{B}_1}^{\hat{G}_1} k_\lambda$, for $\lambda \in X$. It is easy to prove (cf. [36; Lemma 2.9]) that the natural maps $k[G_1] \to k[B_1] \otimes k[B_1^+]$ and $k[\hat{G}_1] \to k[\hat{B}_1] \otimes k[B_1^+]$ are injective and we get the following.

(7)   For $\lambda \in X$ we have $\mathrm{soc}_{B_1^+} \nabla_1(\lambda) \cong k_\lambda$ and $\mathrm{soc}_{\hat{B}_1^+} \hat{\nabla}(\lambda) \cong k_\lambda$.

Now from (3) and (4) we get $\dim \nabla_1(\lambda) = l^N$ and so, from (7), we get:

(8) $\nabla_1(\lambda)|_{B_1^+} \cong I_1(\lambda)$ for all $\lambda \in X$.

We define a quotient group S of $\hat{G}_1$ by setting $k[S]$ to be the k-subalgebra of $k[\hat{G}_1]$ generated by (the images in $k[\hat{G}_1]$ of) the elements $c_{11}^l, \ldots, c_{nn}^l$ and d^{-l}. Note that S is an n-dimensional split torus and hence every S-module is completely reducible. Note also that $k[S]$ is central in $k[\hat{G}]$ and that the subgroup of $\hat{G}_1$ defined by the ideal generated by $\mathrm{Ker}(\varepsilon_{\hat{G}_1}) \cap k[S]$ is G_1. Indeed $k[\hat{G}_1]$ is faithfully flat (in fact free) over $k[S]$. Similar remarks apply to $\hat{B}_1$ and $\hat{B}_1^+$. Hence we get the following, from [71; (2.10.2) Theorem] and [36; Corollary 1.4].

(9)   The restriction of any injective $\hat{H}_1$-module to H_1 is injective and $\mathrm{Ind}_{H_1}^{\hat{H}_1}$ is exact, for $H = G, B$ and B^+.

The following result may be applied in several cases in the present set-up.

(10)   Let H be a quantum group. Suppose that J and K are subgroups such that the natural map $k[H] \to k[J] \otimes k[K]$ is injective. A subspace of an H-module is an H-submodule if and only if it is both a J-submodule and a K-submodule.

Proof   Let Z be an H-module and let V be a subspace of Z. Certainly if V is an H-submodule then it is a J-submodule and a K-submodule. Now suppose that V is a J-submodule and a K-submodule. Let x_r, $r \in R$, be a k-basis of Z such that $x_r, r \in R_0$, is a k-basis of V for some subset R_0 of R. We have $\tau_Z(x_r) = \sum_{s \in R} x_s \otimes f_{sr}$ for some elements $f_{rs} \in k[G]$. By hypothesis we have $f_{sr} \in I_J \cap I_K$, for $r \in R_0$ and $s \in R \backslash R_0$. However, we have $\delta(f_{sr}) = \sum_{t \in R} f_{st} \otimes f_{tr}$ and it follows that the canonical map $k[H] \to k[J] \otimes k[K]$ takes f_{sr} to 0. Hence $f_{sr} = 0$ for $r \in R_0$, $s \in R \backslash R_0$, and V is a G-submodule.

We leave it to the reader to verify injectivity of the appropriate map (cf. [36; Lemma 2.9]) in each of the cases below.

(11)   In each of the following cases the natural map $k[H] \to k[J] \otimes k[K]$ is injective and hence a subspace of an H-module is an H-submodule if and only if it is both a J-submodule and a K-submodule: (a) $H = G, J = B, K = B^+$; (b) $H = G_1, J = B_1, K = B_1^+$; (c) $H = \hat{G}_1, J = G_1, K = T$; (d) $H = \hat{B}_1, J = B_1, K = T$; (e) $H = \hat{B}_1^+, J = B_1^+, K = T$.

We shall also need:

(12)   Each isotypic component of the B_1-socle (resp. B_1^+-socle) of any B-module (resp. B^+-module) is a B-submodule (resp. B^+-submodule).

Proof   We shall prove the B-version. Let V be a B-module and let V_λ be the k_λ-isotypic component of the B_1-socle of V. We have $V_0 = V^{B_1}$, which is B-stable by [**36**; Proposition 1.5(i)]. For general $\lambda \in X_1$ we note that identifying V with $k_\lambda \otimes (k_{-\lambda} \otimes V)$ identifies V_λ with $k_\lambda \otimes (k_{-\lambda} \otimes V)_0$, which is a B-submodule by the case already considered.

We now consider the simple G_1-modules and simple $\hat{G}_1$-modules. For $\lambda \in X$ the G_1-module $\nabla_1(\lambda)$ has a simple G_1-socle, which we denote by $L_1(\lambda)$, and the $\hat{G}_1$-module $\hat{\nabla}_1(\lambda)$ has a simple $\hat{G}_1$-socle, which we denote $\hat{L}_1(\lambda)$. Let V be a simple G_1-module and let $\lambda \in X$ be such that there is a non-zero B_1-homomorphism $V \to k_\lambda$. By Frobenius reciprocity, we have a non-zero G_1-module homomorphism $V \to \nabla_1(\lambda)$ and hence $V \cong L_1(\lambda)$. Similar remarks apply to $\hat{G}_1$ so we obtain the following.

(13)   (i) For $\lambda \in X$ the G_1-socle (resp. $\hat{G}_1$-socle) of $\nabla_1(\lambda)$ (resp. $\hat{\nabla}_1(\lambda)$) is isomorphic to $L_1(\lambda)$ (resp. $\hat{L}_1(\lambda)$).
(ii) For $\lambda \in X$ the G_1-module $L_1(\lambda)$ has a simple B_1^+-socle k_λ and the $\hat{G}_1$-module $\hat{L}_1(\lambda)$ has a simple B_1^+-socle $k_\lambda|_{B_1}$.
(iii) $\{L_1(\lambda) \mid \lambda \in X_1\}$ is a full set of simple G_1-modules and $\{\hat{L}_1(\lambda) \mid \lambda \in X\}$ is a full set of simple $\hat{G}_1$-modules.

Let K be an extension field of k. For a k-coalgebra C we write C_K for the K-coalgebra $K \otimes_k C$ obtained by base change. For a quantum group H over k we write H_K for the quantum group over K obtained by base change, i.e. H_K is the quantum group whose coordinate algebra is $K \otimes_k k[H]$. Recall that if V is a finite dimensional simple module over a k-algebra A then V is absolutely irreducible if and only if $\mathrm{End}_A(V) = k$ (see [**16**; (29.13) Theorem]). Now let C be a k-coalgebra, let V be an irreducible C-comodule and let D be a finite dimensional subcoalgebra of C containing the coefficient space of V. Then the comodule $V_K = K \otimes_k V$ is irreducible over C_K if and only if it is irreducible over D_K. Furthermore, every (right) D-comodule is naturally a left module for the dual coalgebra $D^* = \mathrm{Hom}_k(D, k)$ and in this way we have an equivalence of categories between (right) D-comodules and (left) D^*-modules. Furthermore we have a natural isomorphism $(D^*)_K \to (D_K)^*$. It follows that V is absolutely irreducible if and only if $\mathrm{End}_C(V) = \mathrm{End}_D(V)$ is equal to k. Thus we have the following.

(14)   A simple module V for a quantum group H over k is absolutely irreducible (i.e. the H_K-module V_K obtained by base change is irreducible for every field extension K of k) if and only if $\mathrm{End}_H(V) = k$.

Now if θ is a G_1-endomorphism of $L_1(\lambda)$ (for $\lambda \in X$) then θ stabilizes the B_1^+-socle of $L_1(\lambda)$ and hence, by (13)(ii), $\theta - cI$ is zero on the B_1^+-socle for some $c \in k$, where I is the identity map on $L_1(\lambda)$. Thus the kernel of $\theta - cI$ is non-zero and therefore equal to $L_1(\lambda)$. Hence $\mathrm{End}_{G_1}(L_1(\lambda)) = k$. The same argument applies to $\hat{G}_1$ and G so we obtain the following.

(15) *Every simple H-module is absolutely irreducible, for $H = G, \hat{G}_1, G_1,$ $B, \hat{B}_1, B_1, B^+, \hat{B}_1^+$ and B_1^+.*

Let H be a quantum group over k and let K be an extension field. If V is an essential submodule of an H-module Z then V_K is an essential submodule of the H_K-module Z_K, e.g. by [37; Lemma 6(i)]. Hence we get the following.

(16) *For any H-module V we have $\mathrm{soc}_H(V)_K = \mathrm{soc}_{H_K}(V_K)$, for $H = G, \hat{G}_1, G_1, B, \hat{B}_1, B_1, B^+, \hat{B}_1^+$ and B_1^+.*

For $\lambda = (\lambda_1, \ldots, \lambda_n) \in X$, the element $c_{11}^{l\lambda_1} \ldots c_{nn}^{l\lambda_n}$ spans a 1-dimensional S-submodule, M_λ, say, of $k[S]$. We regard M_λ as a $\hat{G}_1$-module by ϕ-restriction (see 0.16). The weight of this module is $l\lambda$. Hence we obtain that $\hat{L}_1(l\lambda)$ is 1-dimensional and trivial as a G_1-module. We shall often write simply $k_{l\lambda}$ for $\hat{L}_1(l\lambda)$. We now get the following.

(17) $\hat{L}_1(\lambda + l\mu) \cong \hat{L}_1(\lambda) \otimes k_{l\mu}$ as $\hat{G}_1$-modules, for $\lambda, \mu \in X$.

Note that if k is algebraically closed then T may be identified with the group of its k-rational points and, furthermore, if T acts as k-algebra automorphisms of some finite dimensional k-algebra A then T fixes some decomposition of 1 as a sum of mutually orthogonal primitive idempotents. This is well known but we include a proof for the sake of completeness. Suppose not and that A is a counterexample of minimal dimension. If T fixes a non-trivial idempotent e, say, then by minimality T fixes a primitive decomposition in e and in $(1-e)A(1-e)$, and putting these together gives a primitive decomposition of 1 in A fixed by T. Thus T fixes no non-trivial idempotent. Now there are finitely many central idempotents and these are permuted by T. Since T is connected, T must fix all central idempotents. Hence the only central idempotents are 0 and 1, and A consists of a single block. If the nilpotent radical N of A is non-zero and A/N has non-trivial idempotents, then, by minimality, there is a non-trivial f, say, of A/N fixed by T. But the sequence of fixed points $0 \to N^T \to A^T \to (A/N)^T \to 0$ is exact, since T is linearly reductive, and hence there is an element $x \in A^T$ such that $f = x + N$. But now one gets an idempotent $g \in A$, such that g is a polynomial in x and $x + N = f$, by the usual idempotent lifting procedure, see e.g. [41; Section 44], and this is a contradiction. But then A/N, and hence A, has only trivial idempotents and A is not a counterexample. Thus we must have $N = 0$. Thus A is a product of matrix algebras over k and, since A has only one block, $A = M_r(k)$, the algebra of $r \times r$ matrices, for some positive integer r. Now every algebra automorphism of $M_r(k)$ is inner and hence, for each $t \in T$, there exists some $g_t \in \mathrm{GL}_n(k)$, such that $t \cdot a = g_t a g_t^{-1}$, for all $a \in A$. Moreover g_t is determined up to a scalar multiple so we get a homomorphism of algebraic groups $\theta : T \to \mathrm{PGL}_r(K)$, given by $\theta(t) = c_t Z$, where Z is the centre of $\mathrm{GL}_r(K)$, i.e. the group of non-zero scalar matrices. Now the image of θ is a torus. The maximal tori in $\mathrm{PGL}_r(k)$ are the conjugates of T_0, say, the image in $\mathrm{PGL}_r(k)$ of the group of invertible diagonal matrices in $\mathrm{GL}_r(k)$. Thus there exists some $h \in \mathrm{GL}_r(k)$ such that $h c_t h^{-1}$ is diagonal for all $t \in T$. Now let e_i be the matrix with entry 1 in the (i,i)-position and all other entries 0, for $1 \leq i \leq r$. Then e_i is centralized by $h c_t h^{-1}$ (for $1 \leq i \leq r$, $t \in T$) and T fixes the decomposition $1 = e_1' + \cdots + e_r'$, where $e_i' = h^{-1} e_i h$, for $1 \leq i \leq r$, a contradiction.

(18) Let $H = G, B$ or B^+.
(i) *Suppose that k is algebraically closed and let V, V', V'' be finite dimensional $\hat{H}_1$-modules. Then T acts on $\mathrm{Hom}_{H_1}(V', V'')$ and $\mathrm{Hom}_{\hat{H}_1}(V', V'') = \mathrm{Hom}_{H_1}(V', V'')^T$. The action of T on $\mathrm{End}_{H_1}(V)$ is by k-algebra automorphisms. Furthermore, V is indecomposable as an $\hat{H}_1$-module if and only if it is indecomposable as an H_1-module.*
(ii) *We have $\hat{L}_1(\lambda)|_{G_1} \cong L_1(\lambda)$, for any $\lambda \in X$.*
(iii) *For any $\hat{H}_1$-module V we have $\mathrm{soc}_{H_1}(V) = \mathrm{soc}_{\hat{H}_1}(V)$ and furthermore, each isotypic component of $\mathrm{soc}_{H_1}(V)$ is an $\hat{H}_1$-submodule.*

Proof We have that

$$\mathrm{Hom}_{H_1}(V', V'') = \mathrm{Hom}_k(V', V'')^{H_1}$$

is an $\hat{H}_1$-submodule of $\mathrm{Hom}_k(V', V'')$, by [**36**; Proposition 1.5]. Since $T \leq \hat{H}_1$ we have $\mathrm{Hom}_{\hat{H}_1}(V', V'') \leq \mathrm{Hom}_{H_1}(V', V'')^T$ and the reverse inclusion holds by (11) above. Note that T acts on $\mathrm{End}_k(V)$ naturally as k-algebra automorphisms and hence on the T-stable subalgebra $\mathrm{End}_{H_1}(V)$. If V is a decomposable H_1-module then the length n, say, of an expression for $1 \in \mathrm{End}_{H_1}(V)$ as a sum of mutually orthogonal primitive idempotents is greater than 1. By the remark above, 1 can be written as a sum of n mutually orthogonal idempotents in $\mathrm{End}_{\hat{H}_1}(V)$. In particular $\mathrm{End}_{\hat{H}_1}(V)$ contains a non-trivial idempotent and hence V is a decomposable $\hat{H}_1$-module. This proves that if V is indecomposable then $V|_{H_1}$ is indecomposable. The converse is clear and the proof of (i) is complete.

In proving (ii) and (iii) we may assume, by (16), that k is algebraically closed. We claim that if Z is an $\hat{H}_1$-module, U is an H_1-submodule of Z and $t \in T$ then $t \cdot U$ is also an H_1-submodule. We may assume that Z is finite dimensional. Suppose first that $H = B^+$. Let P be a finite dimensional projective B_1^+-module mapping onto U. Since B_1^+ is finite, P is injective. It follows from (6) that there is an injective $\hat{B}_1$-module $\hat{P}$ such that $\hat{P}|_{B_1} \cong P$. Now $\hat{P}$ is also projective as a $\hat{B}_1$-module. Now T acts naturally on $\mathrm{Hom}_{B_1}(\hat{P}, Z)$. Let $\theta \in \mathrm{Hom}_{B_1^+}(\hat{P}, Z)$ be such that $\mathrm{Im}(\theta) = U$. Then $\mathrm{Im}(t \cdot \theta) = t \cdot U$ and hence $t \cdot U$ is a B_1-submodule.

The case $H = B$ is similar. Now consider the case $H = G$. So let Z be a $\hat{G}_1$-module and let U be a G_1-submodule. Then U is a B_1-submodule and B_1^+-submodule. Hence, by the above, $t \cdot U$ is a B_1-submodule and B_1^+-submodule. Hence $t \cdot U$ is a G_1-submodule, by (11)(a). This proves the claim.

Now let L be a simple H_1-submodule of V. For $t \in T$, $t \cdot L$ is an H_1-submodule of V and must be simple, for if J is an H_1-submodule of $t \cdot L$ then $t^{-1} \cdot J$ is an H_1-submodule of the simple H_1-module L. Thus $\mathrm{soc}_{H_1}(Z)$ is an H_1-submodule and a T-submodule of Z and hence, by (11),(c)–(e), also an $\hat{H}_1$-submodule. By (i), each indecomposable $\hat{H}_1$-summand of $\mathrm{soc}_{H_1}(V)$ is also indecomposable, and hence simple, as an H_1-module. Thus we have $\mathrm{soc}_{H_1}(V) \le \mathrm{soc}_{\hat{H}_1}(V)$. Let L be a simple $\hat{H}_1$-module. Then $\mathrm{soc}_{H_1}(L)$ is an $\hat{H}_1$-submodule and therefore $L = \mathrm{soc}_{H_1}(L)$, i.e. L is semisimple as an H_1-module. Now (i) gives that L is simple as an H_1-module. We get that $\mathrm{soc}_{\hat{H}_1}(V) \le \mathrm{soc}_{H_1}(V)$ and so $\mathrm{soc}_{H_1}(V) = \mathrm{soc}_{\hat{H}_1}(V)$. Furthermore we get that, for $\lambda \in X$, the $\hat{G}_1$-module $\hat{L}_1(\lambda)$ is simple as a G_1-module. Now (13)(ii) implies that $\hat{L}_1(\lambda)|_{G_1}$ is isomorphic to $L_1(\lambda)$.

It only remains to prove the last assertion of (iii). We may assume that V is finite dimensional and, replacing V by $\mathrm{soc}_{H_1}(V)$, that V is semisimple as an H_1-module. Let $V = V_1 \oplus \cdots \oplus V_r$ be the decomposition of V into isotypic components. The idempotent $e_i \in A = \mathrm{End}_{H_1}(V)$, describing projection onto V_i, is central. Moreover, T permutes the central idempotents of A. Since A is finite dimensional it has only finitely many central idempotents and since T is a connected group it must fix each central idempotent. By (i), each e_i belongs to $\mathrm{End}_{\hat{H}_1}(V)$ and hence V_i, the image of e_i, is an $\hat{H}_1$-submodule of V.

We consider now induction from T to $\hat{G}_1$. The coordinate algebra $k[\hat{G}_1]$ has a basis of (right and left) weight vectors $c_{11}^{a_{11}} \ldots c_{nn}^{a_{nn}} c_{12}^{b_{12}} \ldots c_{n-1,n}^{b_{n-1,n}}$, with $a_{11}, \ldots, a_{nn} \in \mathbf{Z}$ and $b_{12}, \ldots, b_{n-1,n} \in [0, l-1]$. (We are writing the restriction of a coordinate function c_{ij} to $\hat{G}_1$ also as c_{ij}.) Let $\lambda \in X$. For each choice of $b_{12}, \ldots, b_{n-1,n} \in [0, l-1]$, there is exactly 1 way of choosing the $a_{11}, \ldots, a_{nn} \in \mathbf{Z}$ in such a way that the basis element $c_{11}^{a_{11}} \ldots c_{nn}^{a_{nn}} c_{12}^{b_{12}} \ldots c_{n-1,n}^{b_{n-1,n}}$ has weight λ. It follows that the λ weight space of $k[\hat{G}_1]$ has dimension l^{2N}. Hence the induced module $\mathrm{Ind}_T^{\hat{G}_1} k_\lambda$ has dimension l^{2N}. By exactness of induction we get the following.

(19)   For $V \in \mathrm{mod}(T)$ we have $\dim \mathrm{Ind}_T^{\hat{G}_1} V = l^{2N} \dim V$.

We shall also need:

(20)   (i) dim $\operatorname{Ind}_{\hat{B}_1}^{\hat{G}_1} V = l^N \dim V$, for all $V \in \operatorname{mod}(\hat{B}_1)$, in particular $\operatorname{Ind}_{\hat{B}_1}^{\hat{G}_1}$ is exact.
(ii) $\hat{\nabla}_1(\lambda)|_{\hat{B}_1^+} \cong \hat{I}_1(\lambda)$, and so $\operatorname{ch} \hat{\nabla}_1(\lambda) = e(\lambda - (l-1)\rho)\chi((l-1)\rho)$, for all $\lambda \in X$.

Proof   From (7) we get that $\hat{\nabla}_1(\lambda)|_{\hat{B}_1^+}$ embeds in $\hat{I}_1(\lambda)$. Furthermore, by (5), we have dim $\hat{I}_1(\lambda) = l^N$, so it will follow from (i) that this is an isomorphism. The second assertion of (ii) follows from the first and (5). Thus it only remains to prove (i).

Let $V \in \operatorname{mod}(\hat{B}_1)$. By left exactness we have dim $\operatorname{Ind}_{\hat{T}}^{\hat{B}_1} V \le l^N \dim V$. Let $Y = \operatorname{Ind}_{\hat{T}}^{\hat{B}_1} V$. Frobenius reciprocity provides us with an embedding $V \to Y$, so we have a short exact sequence of $\hat{B}_1$-modules $0 \to V \to Y \to Q \to 0$. Moreover, we have dim $\operatorname{Ind}_{\hat{B}_1}^{\hat{G}_1} Y = \dim \operatorname{Ind}_{\hat{T}}^{\hat{G}_1} V = l^{2N} \dim V$, by (19) (and transitivity of induction). By left exactness and the first paragraph of the proof we have

$$l^{2N} \dim V \le \dim \operatorname{Ind}_{\hat{B}_1}^{\hat{G}_1} V + \dim \operatorname{Ind}_{\hat{B}_1}^{\hat{G}_1} Q \le l^N \dim V + l^N \dim Q = l^N \dim Y.$$

But dim $Y \le l^N \dim V$. It follows that we have equality throughout and that dim $\operatorname{Ind}_{\hat{B}_1}^{\hat{G}_1} V = l^N \dim V$, as required.

We have the so called dot action of $\operatorname{Sym}(n)$ on X defined by $w \cdot \lambda = w(\lambda + \rho) - \rho$, for $w \in \operatorname{Sym}(r)$, $\lambda \in X$. Let $\lambda \in X$. We have $\operatorname{ch} \hat{\nabla}_1(\lambda) = e(\lambda - (l-1)\rho)\chi((l-1)\rho)$ and therefore $\hat{\nabla}_1(\lambda)$ has unique smallest weight $\lambda + (l-1)w_0 \cdot 0$, and this occurs with multiplicity 1. Hence $\hat{\nabla}_1(\lambda)^*$ has highest weight μ, where $-\mu = \lambda + (l-1)w_0 \cdot 0$. Hence there is a non-zero $\hat{B}_1$-homomorphism $\phi : \hat{\nabla}_1(\lambda)^* \to k_\mu$ and, by Frobenius reciprocity, a $\hat{G}_1$-homomorphism $\tilde{\phi} : \hat{\nabla}_1(\lambda)^* \to \hat{\nabla}_1(\mu)$, such that $\phi = \eta \circ \tilde{\phi}$, where $\eta : \hat{\nabla}_1(\mu) \to k_\mu$ is the natural map. Since B_1^+ is finite, $\nabla_1(\lambda)$ is projective as well as injective. Hence the dual $\hat{\nabla}_1(\lambda)^*$ is an injective indecomposable B_1^+-module. Hence $\hat{\nabla}_1(\lambda)^*$ has a simple B_1^+-socle k_μ. Since $\tilde{\phi}$ is non-zero on $(\hat{\nabla}_1(\lambda)^*)^\mu$ the map must be injective and hence an isomorphism. Thus we have the following important result (the q-analogue of a result sometimes known as Serre duality).

(21)   For $\lambda \in X$ we have $\hat{\nabla}_1(\lambda)^* \cong \hat{\nabla}_1(-\lambda - (l-1)w_0 \cdot 0)$ as $\hat{G}_1$-modules, and hence also $\nabla_1(\lambda)^* \cong \nabla_1(-\lambda - (l-1)w_0 \cdot 0)$ (as G_1-modules).

From (21) we obtain:

(22)   For $\lambda \in X$ the $\hat{G}_1$-module (resp. G_1-module) $\hat{\nabla}_1(\lambda)$ (resp. $\nabla_1(\lambda)$) has simple head, which is isomorphic to $\hat{L}_1(-\lambda - (l-1)w_0 \cdot 0)^*$ (resp. $L_1(-\lambda - (l-1)w_0 \cdot 0)^*$).

3.2 Steinberg's tensor product theorem

Let $a = (a_1, \ldots, a_m)$ be a composition of n. Thus $a_1, \ldots, a_m$ are positive integers whose sum is n. We recall, from [36; Section 2], the construction of various subgroups defined by a. Let $\Psi(a)$ be the subset $[1, a_1]^2 \times [a_1 + 1, a_1 + a_2]^2 \times \cdots \times [a_1 + \cdots + a_{m-1} + 1, n]^2$ of $[1, n]^2$. Let $\Psi^+ = \{(i, j) \mid 1 \le i < j \le n\}$, $\Psi^- = \{(i, j) \mid 1 \le j < i \le n\}$ and let $\Psi^+(a) = \Psi^+ \bigcup \Psi(a)$, $\Psi^-(a) = \Psi^- \bigcup \Psi(a)$. We define $P(a)$ (resp. $P^+(a)$) to be the subgroup of G whose defining ideal is generated by the elements c_{ij} with $(i, j) \in [1, n]^2 \backslash \Psi^-(a)$ (resp. $(i, j) \in [1, n]^2 \backslash \Psi^+(a)$). We define $G(a) = P^+(a) \cap P(a)$, $B(a) = B \cap G(a)$ and $B^+(a) = B(a) \cap G(a)$. We have natural isomorphisms $G(a_1) \times \cdots \times G(a_m) \to G(a)$, $B(a_1) \times \cdots \times B(a_m) \to B(a)$ and $B^+(a_1) \times \cdots \times B^+(a_m) \to B(a)$ (see [36; Section 2]).

We shall need the following.

(1)   For $1 \le r < n$ let $a(r)$ be the composition $(1, \ldots, 1, 2, 1, \ldots, 1)$ (where 2 occurs in the rth position) of n and put $A = \{a(1), \ldots, a(n-1)\}$. For a B-module (resp. B^+-module) Z, a subspace V is a B-submodule (resp. B^+-submodule) if and only if it is a $B(a)$ (resp. $B^+(a)$-submodule) for all $a \in A$.

Proof We shall prove the version involving B. Certainly if V is a B-submodule then it is a $B(a)$-submodule for all $a \in A$. We now prove the converse. By local finiteness we can assume that Z is finite dimensional. Suppose the result is false and choose Z of minimal dimension for which it fails. Let L be a non-zero B-submodule of Z. Now $(L + V)/L$ is a $B(a)$-submodule of Z/L, for all $a \in A$, and hence, by minimality, a B-submodule. If $L + V \ne Z$ then, by minimality, V is a B-submodule of $L + V$ and hence of Z, a contradiction. Thus $L + V = Z$ for every non-zero B-submodule L of Z. Taking, in particular, L to be a 1-dimensional submodule we get that V has codimension 1 in Z. Let N be a B-submodule of Z of codimension 1. Now $V \cap N$ is subspace of N which is $B(a)$-stable, for all $a \in A$, and hence a B-submodule. If $V \cap N \ne 0$ we therefore get $Z = (V \cap N) + N = N$, a contradiction. Hence $V \cap N = 0$. Thus Z is 2-dimensional and $Z = L \oplus V$, for some 1-dimensional submodule L. Tensoring Z by $(Z/L)^*$, we can assume that Z/L is trivial. If L is trivial then Z is a trivial B-module, by [36; Lemma 2.8(ii)]. But then every subspace is a B-submodule and again we have a contradiction. Thus $L \cong k_\theta$, say, for some $0 \ne \theta = (\theta_1, \ldots, \theta_n) \in X$. Now $V = Z^T$ and if Z is semisimple then $V = Z^B$, a B-submodule. Thus Z is a non-split extension of k by k_θ. By [36; Lemma 2.8(ii)], we have $\theta < 0$, in particular we have $\theta_i - \theta_{i+1} < 0$ for some $1 \le i < n$. We take $a = (1, \ldots, 1, 2, 1, \ldots, 1)$, with 2 in the ith position. For any $\nu \in X \backslash X^+(a)$

we have the 5-term exact sequence

$$0 \to H^1(P(a), \mathrm{Ind}_B^{P(a)} k_\nu) \to H^1(B, k_\nu) \to (R^1 \mathrm{Ind}_B^{P(a)} k_\nu)^{P(a)}$$
$$\to H^2(P(a), \mathrm{Ind}_B^{P(a)} k_\nu) \to H^2(B, k_\nu)$$

from [**36**; Proposition 1.1]. Thus we have $H^1(B, k_\nu) \cong (R^1 \mathrm{Ind}_B^{P(a)} k_\nu)^{P(a)}$. It follows from [**36**; Lemmas 3.1(iv) and 2.12] that $\nu = s\alpha$ for some $s < 0$, where $\alpha = \epsilon_i - \epsilon_{i+1}$. Putting $S = \{s < 0 \mid H^1(B, k_{s\alpha}) \neq 0\}$ we get that dim $H^1(B, k_{s\alpha}) = 1$ for all $s \in S$. Another application of the 5-term exact sequence gives that $H^1(B(a), k_{s\alpha}) \cong (R^1 \mathrm{Ind}_{B(a)}^{G(a)} k_{s\alpha})^{G(a)}$, for $s < 0$, and [**36**; Lemma 2.12] gives that $H^1(B(a), k_{s\alpha}) \neq 0$ if and only if $s \in S$ and dim $H^1(B(a), k_{s\alpha}) = 1$ for all $s \in S$. Thus there is precisely 1 non-split B-module extension of k by $k_{s\alpha}$, and precisely 1 non-split $B(a)$-module extension of k by $k_{s\alpha}$ (up to isomorphism) for each $s \in S$. However, we have a natural homomorphism $B \to B(a)$ and each $B(a)$-module extension $0 \to k_{s\alpha} \to E \to k \to 0$ gives rise, via inflation, to a B-module extension and the extension is B-split if and only if it is $B(a)$-split. In particular Z is a non-split $B(a)$-module extension, in contradiction to the $B(a)$-module decomposition $Z = L \oplus V$.

(2)   Let $\lambda \in X^+$. The B-socle of $\nabla(\lambda)$ is isomorphic to $k_{w_0\lambda}$ and the B^+-socle of $\nabla(\lambda)$ is isomorphic to k_λ.

Proof   For $\mu \in X$ we have $\mathrm{Hom}_B(k_\mu, \nabla(\lambda)) \cong (k_{-\mu} \otimes \nabla(\lambda))^B \cong (\mathrm{Ind}_B^G k_{-\mu} \otimes \nabla(\lambda))^G$, by the tensor identity and Frobenius reciprocity. If k_μ appears in the B-socle of $\nabla(\lambda)$ we therefore have $\mathrm{Ind}_B^G k_{-\mu} \neq 0$ and hence, by 0.21(2), we have $-\mu \in X^+$. In that case we get $\mathrm{Hom}_B(k_\mu, \nabla(\lambda)) \cong H^0(G, \nabla(-\mu) \otimes \nabla(\lambda)) = k$, if $\mu = w_0\lambda$, and 0 otherwise, by [**36**; Section 4(2)]. This shows that $k_{w_0\lambda}$ is the B-socle of $\nabla(\lambda)$.

We write the restriction of c_{ij} to B^+ also as c_{ij}. Now $k[B^+]$ has a k-algebra grading $k[B^+] = \bigoplus_{\alpha \in X} k[B^+]_\alpha$, where c_{ij} has degree ϵ_i. Then $k[B^+] = \bigoplus_{\alpha \in X} k[B^+]_\alpha$ is a left B^+-module decomposition and $k[B^+]_\alpha$ is the injective hull of k_α, as a B^+-module, for $\alpha \in X$ (see [**36**; Lemma 2.7] for the B-version). Now $\nabla(\lambda) = \mathrm{Ind}_B^G k_\lambda$ is naturally identified with the submodule of $k[G]$ consisting of the elements f such that $(\pi \otimes \mathrm{id})\delta(f) = f_\lambda \otimes f$, where $\pi : k[G] \to k[B]$ is the natural map and $f_\lambda = \pi(c_{11}^{\lambda_1} c_{22}^{\lambda_2} \dots c_{nn}^{\lambda_n})$. The natural map $(\pi \otimes \pi^+) \circ \delta : k[G] \to k[B] \otimes k[B^+]$ is injective (where $\pi^+ : k[G] \to k[B^+]$ is restriction). It follows that restriction $\nabla(\lambda) \to k[B^+]$ is injective and it is easy to check that the image of this map lies in $k[B^+]_\lambda$. Hence $\nabla(\lambda)$ has simple B^+-socle k_λ.

We now take q to be a primitive lth root of unity and write, as in Section 3.1, X_1 for $\{\lambda = (\lambda_1, \dots, \lambda_n) \in X \mid 0 \le \lambda_1 - \lambda_2, \dots, \lambda_{n-1} - \lambda_n, \lambda_n < l\}$.

We write $\bar{G}$ for the algebraic group scheme $GL_n(k)$. For $1 \leq i, j \leq n$ let x_{ij} denote the (i, j)-coordinate function on $GL_n(k)$. Now c_{ij}^l is a central element of $k[G]$ (see [18; 1.3.2 Corollary]). We have the Frobenius morphism $F : G \rightarrow \bar{G}$ whose comorphism takes x_{ij} to c_{ij}^l, for $1 \leq i, j \leq n$. For $\lambda \in X^+$ let $\bar{L}(\lambda)$ denote the irreducible rational $\bar{G}$-module of highest weight λ. Then the G-module $\bar{L}(\lambda)^F$ is irreducible and has highest weight $l\lambda$ and hence is isomorphic to $L(l\lambda)$.

We can now prove, as in [71; (9.3.4) Proposition], the following.

(3) $\{L(\lambda)|_{G_1} \mid \lambda \in X_1\}$ is a full set of irreducible G_1-modules.

Proof   We shall show that if $\lambda = (\lambda_1, \ldots, \lambda_n) \in X^+$ and $0 \leq \lambda_i - \lambda_{i+1} < l$, for $1 \leq i < n$, then $L(\lambda)|_{G_1}$ is irreducible. It will be important to know that if $\lambda - r\alpha$ is a weight of $L(\lambda)$, for a simple root α and $r \geq 0$, then $r < l$. Since $L(\lambda)$ is a submodule of $\nabla(\lambda)$ it suffices to observe this for $\nabla(\lambda)$. Now the character of $\nabla(\lambda)$ is given by Weyl's character formula and it is easy to deduce the required property from Freudenthal's formula, [54; 22.3 Theorem] (or by reducing to the rank 1 case using [27; p. 230 para. 2]).

We claim that $L(\lambda)$ has simple B_1^+-socle $k_\lambda|_{B_1^+}$. By (2) and the fact that $L(\lambda)$ embeds in $\nabla(\lambda)$, we have that $L(\lambda)$ has simple B^+-socle k_λ. This gives, by 3.2(11), that the B_1^+-socle is sum of copies of $k_\lambda|_{B_1^+}$. Let L be the B^+-socle and let N be the B_1^+-socle of $L(\lambda)$. We assume, for a contradiction, that $L \neq N$. Let μ be a maximal weight of M/L. Then $\mu = \lambda - l\nu$, for some $\nu \in X$. Since λ occurs with multiplicity 1 as a weight of $L(\lambda)$, we have $\nu \neq 0$. By (the B^+ version of) [36; Lemma 2.8(ii)], there is a 1-dimensional B^+ submodule M/L, say, of N/L of weight μ. Let $1 \leq r < n$ and take $a = a(r)$ as in (1). If the extension M, of M/L by L, is non-split as a $B^+(a)$-module then we have, by (the $B^+(a)$ version of) [36; Lemma 2.8(ii)], that $\lambda - \mu = l\nu$ is a multiple of $\alpha = \epsilon_r - \epsilon_{r-1}$. But then ν is a multiple of α and μ has the form $\lambda + ls\alpha$ for some $s \neq 0$, contrary to the paragraph above. Hence the extension of $B^+(a(r))$-modules splits and M^μ is a $B(a(r))$-submodule of $L(\lambda)$, for $1 \leq r < n$. By (1) we get that M^μ is a B^+-submodule of $L(\lambda)$, contrary to the fact that the B^+-socle of $L(\lambda)$ is L. This proves the claim. Similarly we have that $L(\lambda)$ has simple B-socle $k_{w_0\lambda}$.

Hence $L(\lambda)$ has simple G_1-socle R, say, and R contains $L(\lambda)^\lambda$. But now, the dual module $L(\lambda)^* \cong L(\lambda^*)$ has simple B_1-socle $k_{w_0\lambda^*} = k_{-\lambda}$. Hence $L(\lambda)$ has simple head k_λ, as a B_1-module. Hence there is a unique maximal G_1-submodule S, say, and $S \nleq L(\lambda)^\lambda = L$. If $S \neq 0$ then S contains R and hence L, a contradiction. Hence $S = 0$ and $L(\lambda)$ is simple, as a G_1-module.

Let L_1 be an irreducible G_1-module. Then L_1 is a composition factor of some G-module and hence of some irreducible G-module $L(\mu)$, say, with $\mu \in X^+$. We can express μ as $\lambda + l\nu$, where $\lambda \in X_1$ and $\nu \in X^+$. Then $L(\lambda) \otimes \bar{L}(\nu)^F$ has highest weight μ so that $L(\mu)$ is a composition factor of $L(\lambda) \otimes \bar{L}(\nu)^F$ and hence L_1 is a G_1 composition factor of $L(\lambda) \otimes \bar{L}(\nu)^F$. But

$L(\lambda) \otimes \bar{L}(\nu)^F$, as a G_1-module, is a direct sum of the irreducible G_1-module $L(\lambda)|_{G_1}$. Hence L_1 is isomorphic to $L(\lambda)|_{G_1}$. Now by 3.1(13)(iii), the number of isomorphism classes of irreducible G_1-modules is the cardinality of X_1 and therefore we must have that $\{L(\lambda)|_{G_1} \mid \lambda \in X_1\}$ is a complete set of pairwise inequivalent irreducible G_1-modules, as required.

Now we obtain, as in [**71**; bottom of p. 107]:

(4)   For $\lambda \in X_1$ and $V \in \mathrm{mod}(G)$, the natural map $\mathrm{Hom}_{G_1}(L(\lambda), V) \otimes L(\lambda) \to V$ is a morphism of G-modules. In particular the $L(\lambda)$-isotypic component of the G_1-socle of V is a G-submodule.

(5) (**Steinberg's tensor product theorem**)   For $\lambda \in X_1$ and $\mu \in X^+$ we have
$$L(\lambda + l\mu) \cong L(\lambda) \otimes \bar{L}(\mu)^F.$$

Proof   (Compare with [**71**; (9.41)] or [**23**; 2.4(A)].) Let V be a non-zero submodule of $L(\lambda) \otimes \bar{L}(\mu)^F$. Then we have the G-module isomorphism $\mathrm{Hom}_{G_1}(L(\lambda), V) \otimes L(\lambda) \to V$. Moreover we have

$$\mathrm{Hom}_{G_1}(L(\lambda), V) \le \mathrm{Hom}_{G_1}(L(\lambda), L(\lambda) \otimes \bar{L}(\mu)^F) \cong \mathrm{End}_{G_1}(L(\lambda)) \otimes \bar{L}(\mu)^F$$
$$\cong \bar{L}(\mu)^F$$

by Schur's lemma and (3). Since $\bar{L}(\mu)^F$ is irreducible, we get

$$\mathrm{Hom}_{G_1}(L(\lambda), V) \cong \bar{L}(\mu)^F$$

so that $\dim \mathrm{Hom}_{G_1}(L(\lambda), V) = \dim \bar{L}(\mu)^F$ and

$$\dim V = \dim L(\lambda).\dim \bar{L}(\mu)^F$$

and hence $V = L(\lambda) \otimes \bar{L}(\mu)^F$.

We record the following for future use.

(6)   Let $\alpha, \beta \in X_1$ and $\lambda, \mu \in X^+$. If $\mathrm{Ext}^1_G(L(\alpha + l\lambda), L(\beta + l\mu)) \ne 0$ then either $\alpha = \beta$ or $\mathrm{Ext}^1_{G_1}(L(\alpha), L(\beta)) \ne 0$.

Proof   For $\gamma \in X_1$, $\tau \in X^+$ the G-module $\bar{L}(\tau)^F \otimes L(\gamma)$ has highest weight $\gamma + l\tau$ and so has $L(\gamma) \otimes \bar{L}(\tau)^F$ as a composition factor. By dimensions we must have $\bar{L}(\tau)^F \otimes L(\gamma) \cong L(\gamma) \otimes \bar{L}(\tau)^F$. Thus we have $\mathrm{Ext}^1_G(L(\alpha + l\lambda), L(\beta + l\mu)) \cong \mathrm{Ext}^1_G(\bar{L}(\lambda)^F \otimes L(\alpha), \bar{L}(\mu)^F \otimes L(\beta))$, which is isomorphic to $H^1(G, \bar{L}(\mu)^F \otimes L(\beta) \otimes L(\alpha)^* \otimes (\bar{L}(\lambda)^F)^*)$, by [**71**; (2.4.1) Lemma and (2.8.2) Proposition (3)]. If this non-zero then, by the 5-term exact sequence (see [**71**;

(2.11.2) Corollary] and the proof of [36; Proposition 3.10]), we must have
$H^1(\bar{G}, H^0(G_1, \bar{L}(\mu)^F \otimes L(\beta) \otimes L(\alpha)^* \otimes (\bar{L}(\lambda)^F)^*)) \neq 0$ or $H^1(G_1, \bar{L}(\mu)^F \otimes L(\beta) \otimes L(\alpha)^* \otimes (\bar{L}(\lambda)^F)^*)^{\bar{G}} \neq 0$. Since $\bar{L}(\lambda)^F$ and $\bar{L}(\mu)^F$ are trivial, as
G_1-modules, the first condition gives $H^0(G_1, L(\beta) \otimes L(\alpha)^*) \neq 0$ and the
second gives $H^1(G_1, L(\beta) \otimes L(\alpha)^*) \neq 0$. A further application of [71; (2.8.2)
Proposition,(3)] (and (3) above) gives $\alpha = \beta$ or $\mathrm{Ext}^1_{G_1}(L(\alpha), L(\beta)) \neq 0$.

We now consider the effect of the Weyl group on the composition multiplicities $[\hat{\nabla}_1(\lambda) : \hat{L}_1(\mu)]$ (for $\lambda, \mu \in X$).

(7)   For $\lambda, \nu \in X$, $\mu \in X_1$ and $w \in W$ we have $[\hat{\nabla}_1(\lambda) : \hat{L}_1(\mu + l\nu)] =$
$[\hat{\nabla}_1(w \cdot \lambda - lw \cdot 0) : \hat{L}_1(\mu + lw\nu)]$.

Proof We have

$$\mathrm{ch}\,\hat{\nabla}_1(\tau) = e(\tau - (l-1)\rho)\chi((l-1)\rho).$$

for any $\tau \in X$ so that

$$\begin{aligned}
\mathrm{ch}\,\hat{\nabla}_1(w \cdot \lambda - lw \cdot 0) &= e(w \cdot \lambda - lw \cdot 0 - (l-1)\rho)\chi((l-1)\rho) \\
&= w(e(\lambda - (l-1)\rho))\chi((l-1)\rho) \\
&= w(\mathrm{ch}\,\hat{\nabla}_1(\lambda)) \\
&= w(\sum_{\mu \in X_1, \nu \in X} [\hat{\nabla}_1(\lambda) : \hat{L}_1(\mu + l\nu)]e(l\nu)\mathrm{ch}\,\hat{L}_1(\mu)) \\
&= \sum_{\mu \in X_1, \nu \in X} [\hat{\nabla}_1(\lambda) : \hat{L}_1(\mu + l\nu)]e(lw\nu)\mathrm{ch}\,\hat{L}_1(\mu)
\end{aligned}$$

and we also have

$$\mathrm{ch}\,\hat{\nabla}_1(w \cdot \lambda - lw \cdot 0) = \sum_{\mu \in X_1, \nu \in X} [\hat{\nabla}_1(w \cdot \lambda - lw \cdot 0) : \hat{L}_1(\mu + lw\nu)]e(lw\nu)\mathrm{ch}\,\hat{L}_1(\mu).$$

Comparing these two expressions gives the result.

We call $\sigma = (\sigma_1, \ldots, \sigma_n) \in X$ a *Steinberg weight* if $\sigma_1 - \sigma_2, \ldots, \sigma_{n-1} - \sigma_n \equiv -1$ modulo l.

(8)   For a Steinberg weight σ we have $\hat{\nabla}_1(\sigma) \cong \hat{L}_1(\sigma)$ (as $\hat{G}_1$-modules) and
$\nabla_1(\sigma) \cong L_1(\sigma)$ (as G_1-modules).

Clearly the second assertion follows from the first, which we now prove. Since
$\hat{L}_1(\sigma)$ embeds in $\hat{\nabla}_1(\sigma)$ and $\hat{\nabla}_1(\sigma)$ has dimension l^N it suffices to prove
that $\hat{L}_1(\sigma)$ also has this dimension. We can write $\sigma = (l-1)\rho + l\nu + r\omega$
for suitable $\nu \in X$ and $r \in \mathbb{N}_0$ where $\omega = (1, 1, \ldots, 1)$. We have $\hat{L}_1(\sigma) \cong$

$\hat{L}_1((l-1)\rho) \otimes \hat{L}_1(l\nu) \otimes \hat{L}_1(\omega)^{\otimes r}$, and both $\hat{L}_1(l\nu)$ and $L_1(\omega)$ are 1-dimensional so we may assume $\sigma = (l-1)\rho$. By 3.1(21) the head of $\hat{\nabla}_1((l-1)\rho)$ is $\hat{L}_1(-(l-1)\rho - (l-1)w_0 \cdot 0)^* = \hat{L}_1(-(l-1)w_0\rho)^*$. Now $(l-1)\rho \in X_1$ so that $\hat{L}_1((l-1)\rho) = L((l-1)\rho)|_{\hat{G}_1}$ and hence $\hat{L}_1((l-1)\rho)$ has lowest weight $(l-1)w_0\rho$ and $\hat{L}_1((l-1)\rho)^*$ has highest weight $-(l-1)w_0\rho$. Hence we have $\hat{L}_1(-(l-1)w_0\rho)^* \cong L_1((l-1)\rho)$. Thus both the head and socle of $\hat{\nabla}_1((l-1)\rho)$ are isomorphic to $\hat{L}_1((l-1)\rho)$. Since $(l-1)\rho$ occurs with multiplicity 1 as a weight of $\hat{\nabla}_1((l-1)\rho)$ we must have $\hat{\nabla}_1((l-1)\rho) \cong \hat{L}_1((l-1)\rho)$, as required.

We continue with some further "classical" infinitesimal theory. The head of a finite dimensional module Y will be denoted $\operatorname{hd} Y$. We define $\hat{\nabla}_1^+(\lambda) = \operatorname{Ind}_{\hat{B}_1^+}^{\hat{B}_1} k_\lambda$, for $\lambda \in X$. Arguing as in 3.1(13)(i), (20)(ii), (21) and (22) we get the following.

(9)   Let $\lambda \in X$. Then $\hat{\nabla}_1^+(\lambda)|_{\hat{B}_1}$ is the injective hull of k_λ in $\operatorname{mod}(\hat{B}_1)$. The character of $\hat{\nabla}_1^+(\lambda)$ is $e(\lambda - (l-1)\rho)\chi((l-1)\rho)$ and $\hat{\nabla}_1^+(\lambda)^* \cong \hat{\nabla}_1^+(-\lambda + (l-1)w_0 \cdot 0)$. We have $\operatorname{soc}_{\hat{G}_1} \hat{\nabla}_1^+(\lambda) = \hat{L}_1(-\lambda)^*$ and $\operatorname{hd}_{\hat{G}_1} \hat{\nabla}_1^+(\lambda) = \hat{L}_1(\lambda - (l-1)w_0 \cdot 0)$.

For $\lambda \in X$ we write $Q_1(\lambda)$ for the injective envelope of $L_1(\lambda)$ (as a G_1-module) and write $\hat{Q}_1(\lambda)$ for the injective envelope of $\hat{L}_1(\lambda)$ (as a $\hat{G}_1$-module). Note that, for $\lambda \in X$, the simple module $\hat{L}_1(\lambda)$ embeds in the induced module $\operatorname{Ind}_T^{\hat{G}_1} k_\lambda$ and hence the injective envelope $\hat{Q}_1(\lambda)$ of $\hat{L}_1(\lambda)$ occurs as a summand of $\operatorname{Ind}_T^{\hat{G}_1} k_\lambda$. This gives the first part of the following. The second part holds since tensoring with a 1-dimensional module preserves injectivity and indecomposability. The final part may be obtained by arguing as in the proof of 3.1(18).

(10)   Let $\lambda \in X$. Then
(i) $\hat{Q}_1(\lambda)$ is finite dimensional;
(ii) $\hat{Q}_1(\lambda) \otimes k_{l\mu} \cong \hat{Q}_1(\lambda + l\mu)$, for every $\mu \in X$;
(iii) $\hat{Q}_1(\lambda)|_{G_1} \cong Q_1(\lambda)$.

We shall write $V \in \mathcal{F}(\hat{\nabla}_1)$ to indicate that V is a finite dimensional $\hat{G}_1$-module which has a filtration $0 = V_0 \le V_1 \le \cdots \le V_r = V$ such that, for each $1 \le i \le r$, we have V_i/V_{i-1} is either 0 or isomorphic to $\hat{\nabla}_1(\lambda)$ for some $\lambda \in X$. Note that if $V \in \mathcal{F}(\hat{\nabla}_1)$ as above then $\operatorname{ch} V = \sum_{i \in I} e(\lambda_i - (l-1)\rho)\chi((l-1)\rho)$, where I is the set of $i \in [1,r]$ such that $V_i/V_{i-1} \ne 0$ and where $V_i/V_{i-1} \cong \hat{\nabla}_1(\lambda_i)$, for $i \in I$. It follows that, for each $\lambda \in X$, the cardinality of $\{i \in [1,r] \mid V_i/V_{i-1} \cong \hat{\nabla}_1(\lambda)\}$ is independent of the choice of the $\hat{\nabla}_1$-filtration. We denote this cardinality by $(V : \hat{\nabla}_1(\lambda))$.
    For $\lambda \in X$ we define $\hat{\Delta}_1(\lambda) = \hat{\nabla}_1^+(-\lambda)^*$. Then $\operatorname{ch} \hat{\Delta}_1(\lambda) = \operatorname{ch} \hat{\nabla}_1(\lambda)$ and $\hat{\Delta}_1(\lambda)$ has simple $\hat{G}_1$-head $\hat{L}_1(\lambda)$.

We obtain the following as in the "global" case.

(11) (i) For $\lambda, \mu \in X$ we have $\mathrm{Ext}^1_{\hat{G}_1}(\hat{\nabla}_1(\lambda), \hat{\nabla}_1(\mu)) = 0$ for $\lambda \not> \mu$.
(ii) For $i \in \mathbb{N}_0$, $\lambda, \mu \in X$ we have

$$\mathrm{Ext}^i_{\hat{G}_1}(\hat{\Delta}_1(\lambda), \hat{\nabla}_1(\mu)) = \begin{cases} k, & \text{if } i = 0, \lambda = \mu; \\ 0, & \text{otherwise.} \end{cases}$$

(iii) For $V \in \mathrm{mod}(\hat{G}_1)$ we have $V \in \mathcal{F}(\hat{\nabla}_1)$ if and only if $\mathrm{Ext}^1_{\hat{G}_1}(\hat{\Delta}_1(\lambda), V)) = 0$ for all $\lambda \in X$.
(iv) For $V \in \mathcal{F}(\hat{\nabla}_1)$ and $\mu \in X$ we have $(V : \hat{\nabla}_1(\mu)) = \dim \mathrm{Hom}_{\hat{G}_1}(\hat{\Delta}_1(\mu), V)$.
(v) For all $\lambda \in X$ we have $\hat{Q}_1(\lambda) \in \mathcal{F}(\hat{\nabla}_1)$ and $(\hat{Q}_1(\lambda) : \hat{\nabla}_1(\mu)) = [\hat{\nabla}_1(\mu) : \hat{L}_1(\lambda)]$.

Proof We have

$$\mathrm{Ext}^1_{\hat{G}_1}(\hat{\nabla}_1(\lambda), \hat{\nabla}_1(\mu)) \cong \mathrm{Ext}^1_{\hat{B}_1}(\hat{\nabla}_1(\lambda), k_\mu)$$

by Shapiro's lemma (valid in this context since $\mathrm{Ind}^{\hat{G}_1}_{\hat{B}_1}$ is exact). If this is non-zero then, by the long exact sequence, we have $\mathrm{Ext}^1_{\hat{B}_1}(k_\tau, k_\lambda) \neq 0$ for some weight τ of $\hat{\nabla}_1(\lambda)$. Hence we have $\tau > \mu$, by [36; Lemma 2.8(ii)]. But λ is the unique highest weight of $\hat{\nabla}_1(\lambda)$ so we get $\lambda > \mu$, proving (i).
We have $\mathrm{Ext}^i_{\hat{G}_1}(\hat{\Delta}_1(\lambda), \hat{\nabla}_1(\mu)) \cong \mathrm{Ext}^i_{\hat{B}_1}(\hat{\Delta}_1(\lambda), k_\mu)$. This is 0 if $i > 0$ since $\hat{\nabla}^+_1(\lambda)$ is injective and hence projective as a $\hat{B}_1$-module. For $i = 0$ we get $\mathrm{Hom}_{\hat{B}_1}(\hat{\Delta}_1(\lambda), k_\mu)$. Now λ is the highest weight of $\hat{\Delta}_1(\lambda)$ and occurs with multiplicity 1 so we get $\dim \mathrm{Hom}_{\hat{B}_1}(\hat{\Delta}_1(\lambda), k_\lambda) = 1$. Furthermore $\hat{\Delta}_1(\lambda)|_{\hat{B}_1}$ is an injective, hence projective, indecomposable $\hat{B}_1$-module (by (9)) and so has a simple head. Thus the head of $\hat{\Delta}_1(\lambda)$ is k_λ and we get $\mathrm{Hom}_{\hat{B}_1}(\hat{\Delta}_1(\lambda), k_\mu) = 0$ for $\lambda \neq \mu$.
One now obtains (iii) from the argument of [26; Corollary 1.3]. Part (iv) follows directly from (ii).
Since $\hat{Q}_1(\lambda)$ is injective it satisfies the condition of (iii) and so has a $\hat{\nabla}_1$-filtration. Since $\hat{\nabla}_1(\lambda)$ has socle $\hat{L}_1(\lambda)$ the first term in a $\hat{\nabla}_1$-filtration of $\hat{Q}_1(\lambda)$ must be $\hat{\nabla}_1(\lambda)$. Hence we have a short exact sequence $0 \to \hat{\nabla}_1(\lambda) \to \hat{Q}_1(\lambda) \to Y \to 0$, where $Y \in \mathcal{F}(\hat{\nabla}_1(\lambda))$. We now get by dimension shifting that $\mathrm{Ext}^i_{\hat{G}_1}(\hat{\nabla}^+_1(\nu), \hat{\nabla}(\lambda)) = 0$ for all $i > 0$. The filtration multiplicity $(\hat{Q}_1(\lambda) : \hat{\nabla}_1(\mu))$ is equal to $\dim \mathrm{Hom}_{\hat{G}_1}(\hat{\Delta}_1(\mu), \hat{Q}_1(\lambda))$, which is the composition multiplicity $[\hat{\Delta}_1(\mu) : \hat{L}_1(\lambda)] = [\hat{\nabla}_1(\mu) : \hat{L}_1(\lambda)]$ (since $\hat{\Delta}_1(\mu)$ and $\hat{\nabla}_1(\mu)$ have the same character), proving the last part of (v).

Let σ be a Steinberg weight. Then we have $(\hat{Q}_1(\sigma) : \hat{\nabla}_1(\mu)) = [\hat{\nabla}_1(\mu) : \hat{L}_1(\sigma)]$, which is 1 if $\sigma = \mu$ and 0 otherwise (since $\hat{\nabla}_1(\sigma) = \hat{L}_1(\sigma)$ is simple). Thus we have $\hat{Q}_1(\sigma) = \hat{\nabla}_1(\sigma)$. This gives part (i) of the following.

(12)   Let σ be a Steinberg weight.
(i) We have $\hat{Q}_1(\sigma) \cong \hat{\nabla}_1(\sigma) \cong \hat{L}_1(\sigma)$, as $\hat{G}_1$-modules.
(ii) If $\sigma \in X^+$ then $\nabla(\sigma)|_{\hat{G}_1} \cong \hat{L}_1(\sigma)$.

For part (ii) we note that, by Weyl's dimensional formula (or the argument after 3.1(4)) we have dim $\nabla(\sigma) = l^N$. Since $\nabla(\sigma)$ has highest weight σ it has $\hat{L}_1(\sigma)$ as a $\hat{G}_1$-composition factor and since dim $\hat{L}_1(\sigma) = $ dim $\hat{\nabla}_1(\sigma) = l^N$ we must have $\nabla(\sigma)|_{\hat{G}_1} \cong \hat{L}_1(\sigma)$, as required.

Now fix a Steinberg weight $\sigma \in X_1$ and let $\pi_\sigma = \{\sigma + l\lambda \mid \lambda \in X^+\}$. For any subset π of X^+ and $V \in \text{mod}(G)$ we say that V belongs to π if all composition factors of V belong to $\{L(\lambda) \mid \lambda \in \pi\}$. We write $\text{mod}(\pi)$ for the full subcategory of $\text{mod}(G)$ whose objects are the G-modules belonging to π. The arguments of for example [**35**; Section 4, Theorem], combined with (6) above, give the following.

(13) π_σ is a union of blocks of G and the functor $g : \text{mod}(\bar{G}) \to \text{mod}(\pi_\sigma)$, taking $V \to L(\sigma) \otimes V^F$, is an equivalence of categories and $g(\bar{\nabla}(\lambda)) = L(\sigma) \otimes \bar{\nabla}(\lambda)^F$ is isomorphic to $\nabla(\sigma + l\lambda)$, for $\lambda \in X^+$.

We now consider the effect of the operation of the Weyl group on filtration multiplicities of the injective $\hat{G}_1$-modules.

(14)   Let $\lambda \in X_1$, $\mu, \nu \in X$ and $w \in W$.
(i) $(\hat{Q}_1(\lambda + l\nu) : \hat{\nabla}_1(\mu)) = (\hat{Q}_1(\lambda + lw\nu) : \hat{\nabla}_1(w \cdot \mu - lw \cdot 0))$.
(ii) The module $\hat{Q}_1(\lambda)$ has unique smallest weight $\lambda - (l-1)w_0 \cdot 0$, and this weight occurs with multiplicity 1.
(iii) ch $\hat{Q}_1(\lambda) \in (\mathbb{Z}X)^W$.
(iv) $\hat{Q}_1(\lambda)$ has unique highest weight $w_0\lambda - (l-1)w_0 \cdot 0$ and this weight occurs with multiplicity 1.
(v) $(\hat{Q}_1(\lambda) : \hat{\nabla}_1(w \cdot \lambda - lw \cdot 0)) = 1$ for all $w \in W$ and if $(\hat{Q}_1(\lambda) : \hat{\nabla}_1(\tau)) \neq 0$ for some $\tau \in X$ then we have $\lambda \leq \tau \leq w_0\lambda - (l-1)w_0 \cdot 0$.
(vi) $\hat{Q}_1(\lambda)$ has a simple head $\hat{L}_1(\lambda)$.

Proof   (i) This follows from (7) and (11)(iii).
(ii) For $\nu \in X$ we have

$$\text{dim } \hat{Q}_1(\lambda)^\nu = \sum_{\mu \in X} (\hat{Q}_1(\lambda) : \hat{\nabla}_1(\mu)) \text{ dim } \hat{\nabla}_1(\mu)^\nu$$

$$= \sum_{\mu \in X} [\hat{\nabla}_1(\mu) : \hat{L}_1(\lambda)] \text{ dim } \hat{\nabla}_1(\mu)^\nu$$

$$= \sum_{\mu \geq \lambda} [\hat{\nabla}_1(\mu) : \hat{L}_1(\lambda)] \text{ dim } \hat{\nabla}_1(\mu)^\nu.$$

Now (v) follows from the fact that $\hat{\nabla}_1(\mu)$ has unique lowest weight $\mu - (l - 1)w_0 \cdot \rho$, and this weight occurs with multiplicity 1, for $\mu \in X$.

(iii) We have

$$\text{ch}\,\hat{Q}_1(\lambda) = \sum_{\mu \in X} (\hat{Q}_1(\lambda) : \hat{\nabla}_1(\mu))\text{ch}\,\hat{\nabla}_1(\mu)$$

$$= \sum_{\mu \in X} (\hat{Q}_1(\lambda) : \hat{\nabla}_1(\mu))e(\mu - (l-1)\rho)\chi((l-1)\rho)$$

$$= \sum_{\mu \in X} (\hat{Q}_1(\lambda) : \hat{\nabla}_1(\mu + (l-1)\rho))e(\mu)\chi((l-1)\rho).$$

Hence $\text{ch}\,\hat{Q}_1(\lambda) \in (\mathbb{Z}X)^W$ provided that $(\hat{Q}_1(\lambda) : \hat{\nabla}_1(\mu + (l-1)\rho) = (\hat{Q}_1(\lambda) : \hat{\nabla}_1(w\mu + (l-1)\rho)$, for $w \in W$, and this is true by (i).

Now part (iv) follows from (iii) and (11)(v) and part (v) follows from (iv) and (11)(v).

(vi) The module $\hat{Q}_1(\lambda)$ is an injective indecomposable G_1-module and hence a projective indecomposable G_1-module. Thus $\hat{Q}_1(\lambda)$ has a simple head as a G_1-module and therefore as a $\hat{G}_1$-module. It follows from (v) and (11)(i) that there is an epimorphism $\hat{Q}_1(\lambda) \to \hat{\nabla}_1(w_0 - (l-1)w_0 \cdot 0)$. Hence the head of $\hat{Q}_1(\lambda)$ is the head of $\hat{\nabla}_1(w_0\lambda - (l-1)w_0 \cdot 0)$, which by 3.1(22) is the dual of $\hat{L}_1(-w_0\lambda)$, i.e. $\hat{L}_1(\lambda)$.

3.3 Tilting modules

In this section and the next we briefly describe the application of infinitesimal methods to the calculation of tilting modules and work out explicitly the tilting modules for quantum GL_2. This calculation gives the decomposition numbers for 2-rowed partitions for the Hecke algebra, as we shall see in Section 4.4. We write $\mathcal{F}(\Delta)$ for the class of finite dimensional G-modules which have a filtration with sections isomorphic to modules of the form $\Delta(\lambda)$, $\lambda \in X^+$. We recall that $S(n,r)$ is a quasihereditary algebra with standard modules $\Delta(\lambda)$ and costandard modules $\nabla(\lambda)$, for $\lambda \in \Lambda^+(n,r)$ (and the dominance order on partitions), see [36; Section 4]. Thus we have, by results of Ringel (see the Appendix, Section A4), for each $\lambda \in X^+$, an indecomposable finite dimensional $S(n,r)$-module $T(\lambda) \in \mathcal{F}(\nabla) \cap \mathcal{F}(\Delta)$ such that $(T(\lambda) : \nabla(\lambda)) = 1$ and, for $\mu \in \Lambda^+(n,r)$, $(T(\lambda) : \nabla(\mu)) = 0$ unless $\mu \leq \lambda$. Thus we have $\dim T(\lambda)^\lambda = 1$ and, for $\mu \in \Lambda^+(n,r)$, $\dim T(\lambda)^\mu = 0$ unless $\mu \leq \lambda$. (See also [36; Section 4(6)] and compare with [33, Section 1].) Furthermore, every module in $\mathcal{F}(\nabla) \cap \mathcal{F}(\Delta)$ is isomorphic to a direct sum of the modules $T(\lambda)$, $\lambda \in X^+$. Modules in $\mathcal{F}(\nabla) \cap \mathcal{F}(\Delta)$ are here called the tilting modules.

We note that, by [36; Section 4(3)(ii)], if $T, T' \in \mathcal{F}(\Delta) \cap \mathcal{F}(\nabla)$ then $T \otimes T' \in \mathcal{F}(\Delta) \cap \mathcal{F}(\nabla)$. Moreover we have $\bigwedge^r E = \nabla(1^r) = \Delta(1^r)$, for $1 \leq r \leq n$, so that $\bigwedge^r E \in \mathcal{F}(\Delta) \cap \mathcal{F}(\nabla)$. Hence we get $\bigwedge^{a_1} E \otimes \cdots \otimes \bigwedge^{a_m} E \in$

$\mathcal{F}(\Delta) \cap \mathcal{F}(\nabla)$. By examining the highest weight of a tensor product of exterior powers of the natural module we get the following (as in the classical case [**33**; Section 3]).

(1) *The indecomposable tilting modules for $S(n,r)$ are precisely the indecomposable summands of $\bigwedge^{\alpha'} E$, for $\alpha \in \Lambda^+(n,r)$. Furthermore, for $\alpha \in \Lambda^+(n,r)$, the module $T(\alpha)$ occurs exactly once as a summand of $\bigwedge^{\alpha'} E$ and if $T(\lambda)$ is a summand of $\bigwedge^{\alpha'} E$ then $\lambda \leq \alpha$ (for $\lambda \in \Lambda^+(n,r)$).*

(Here α' denotes the transpose of the partition α.)

Remarks (i) From the classification of tilting modules discussed in the first paragraph above it follows that tilting modules T, T' are isomorphic if and only if they have the same character. If $\alpha, \beta \in P(n)$ and β is obtained by permuting the parts of α then we have ch $\bigwedge^\alpha E =$ ch $\bigwedge^\beta E$ and therefore $\bigwedge^\alpha E$ and $\bigwedge^\beta E$ are isomorphic. Similar remarks apply to right M-modules so that the classification of partial tilting modules by highest weight gives a short, calculation-free proof of Lemma 1.3.3.

We also have the corresponding results for symmetric powers. Let $\pi \in \Lambda^+(n,r)$ and, for $\lambda \in \pi$, let $I_\pi(\lambda)$ denote the injective hull of $L(\lambda)$ as an $S(n,r)$-module. It follows from the reciprocity principle, [**36**, Section 4(4)], that the characters of the $I_\pi(\lambda)$, $\lambda \in \pi$, are linearly independent and that injective finite dimensional $S(n,r)$-modules I and I' are isomorphic if and only if ch $I =$ ch I'. It follows that the symmetric powers $S^\alpha E$ and $S^\beta E$ are isomorphic if α is obtained by permuting the parts of β (for $\alpha, \beta \in \Lambda(n,r)$). Similar remarks apply to the symmetric powers of the natural right module V.

(ii) Note that (1) describes the polynomial tilting modules. We say an arbitrary finite dimensional G-module T is a tilting module if $T \in \mathcal{F}(\Delta) \cap \mathcal{F}(\nabla)$. Suppose T is an indecomposable tilting module. Then $T \otimes L(s\omega)$ is polynomial, for $s \gg 0$. Hence we have $T \otimes L(s\omega) \cong T(\lambda)$ for $\lambda \in \Lambda^+(n,r)$ (and some $r \geq 0$). Hence we have $T \cong T(\lambda) \otimes L(-s\omega)$ and therefore (1) describes all indecomposable tilting modules for G.

We assume, in the next 2 statements, that q is a root of unity. One has the following, by the arguments of [**33**; Section 2].

(2) *For $\lambda \in X^+$, the tilting module $T(\lambda)$ is projective as a $\hat{G}_1$-module (equivalently as a G_1-module) if and only if $\lambda_i - \lambda_{i+1} \geq l - 1$, for $i = 1, \ldots, n-1$.*

For $\lambda = (\lambda_1, \ldots, \lambda_n) \in X$ we define $f(\lambda) = \lambda_1 - \lambda_n$. We obtain the following by the arguments of the proof of [**33**; (2.5) Theorem] (an observation of Cornelius Pillen).

(3)   Let $\sigma, \mu \in X_1$ and suppose that σ is a Steinberg weight.
(i) Let $V = L(\sigma) \otimes L(\mu)$ or $T(\sigma + \mu)$. Then $\mathrm{Hom}_{G_1}(L(\sigma + w_0\mu), V) = \mathrm{Hom}_G(L(\sigma + w_0\mu), V) = k$ and moreover whenever $\mathrm{Hom}_{G_1}(L(\lambda), V) \neq 0$, for $\sigma + w_0\mu \neq \lambda \in X_1$, we have $f(\lambda) > f(\sigma + w_0\mu)$.
(ii) $T(\sigma + \mu)$ occurs exactly once as a component of $L(\sigma) \otimes L(\mu)$ and contains a copy of $L(\sigma + w_0\mu)$ in its G-socle.
(iii) $\hat{Q}_1(\sigma + w_0\mu)$ occurs exactly once as a $\hat{G}_1$-component of $T(\sigma + \mu)$.

The next 2 statements are made in order to justify a q-analogue of [33; (2.1) Proposition]. We start with some generalities. Let H be a quantum group over k and $Z \in \mathrm{mod}(H)$. We call Z absolutely indecomposable if the module Z_K, obtained by base change from Z, is an indecomposable G_K-module (where G_K is the K-quantum group obtained by base change), for every field extension K of k. For $V, Z \in \mathrm{mod}(H)$ the natural map $K \otimes_k \mathrm{Hom}_G(V, Z) \to \mathrm{Hom}_{G_K}(V_K, Z_K)$ is an isomorphism (where K is an extension of k) and we get the following result.

(4)   A finite dimensional G-module Z is absolutely indecomposable if and only if the nilpotent radical of $\mathrm{End}_G(Z)$ has codimension 1.

We assume again that q is a root of unity. The following is an analogue of [25; Section 2, Lemma].

(5)   Let $V, Z \in \mathrm{mod}(G)$ and suppose that $V|_{G_1}$ is absolutely indecomposable, that G_1 acts trivially on Z and that Z is absolutely indecomposable as a $\bar{G}$-module. Then $V \otimes Z$ is an absolutely indecomposable G-module.

Proof   Let $Q, R \in \mathrm{mod}(H)$, for a quantum group H over k; then we regard $\mathrm{Hom}_k(Q, R)$ as an H-module via the canonical isomorphism $R \otimes Q^* \to \mathrm{Hom}_k(Q, R)$ (as in [71; (2.4.2) Theorem]). Suppose also $S \in \mathrm{mod}(H)$. The natural map $Q^* \otimes Q \to k$ is a G-module homomorphism. Thus we get a G-module homomorphism $S \otimes Q^* \otimes Q \otimes R^* \to S \otimes R^*$ and so the linear map $\mathrm{Hom}_k(Q, R) \otimes \mathrm{Hom}_k(R, S) \to \mathrm{Hom}_k(Q, S)$, taking $\alpha \otimes \beta$ to $\beta \circ \alpha$ (for $\alpha \in \mathrm{Hom}_k(Q, R)$, $\beta \in \mathrm{Hom}_k(R, S)$) is a G-module map. In particular, multiplication $\mathrm{End}_k(Q)^{\mathrm{op}} \otimes \mathrm{End}_k(Q)^{\mathrm{op}} \to \mathrm{End}_k(Q)^{\mathrm{op}}$ is a G-module map.
    By (4) we can assume that k is algebraically closed. From the hypotheses, the natural map $\mathrm{End}_{G_1}(V) \otimes \mathrm{End}_k(Z) \to \mathrm{End}_{G_1}(V \otimes Z)$ is a G-module isomorphism. Also, from the hypotheses, we have a k-algebra map $\phi : \mathrm{End}_{G_1}(V)^{\mathrm{op}} \to k$ with nilpotent kernel. Such a map must be given by the formula $\phi(cI + x) = c$, for $c \in k$ and x nilpotent, where $I : V \to V$ is the identity map. Since $x \in \mathrm{End}_{G_1}(V)$ is nilpotent if and only if it is nilpotent as an element of $\mathrm{End}_{G_1}(V)^{\mathrm{op}}$, the linear map ϕ is also an algebra map when $\mathrm{End}_{G_1}(V)$ is regarded as a k-algebra in the natural manner. Hence we get a k-algebra map $\Phi = \phi \otimes \mathrm{id} : \mathrm{End}_{G_1}(V \otimes Z) = \mathrm{End}_{G_1}(V) \otimes \mathrm{End}_k(Z) \to \mathrm{End}_k(Z)$ with nilpotent kernel. By restriction we have a k-algebra map $\Psi : \mathrm{End}_G(V \otimes$

$Z) = \mathrm{End}_{G_1}(V \otimes Z)^G \to \mathrm{End}_k(Z)^G = \mathrm{End}_G(Z)$. But Z is indecomposable so that $\mathrm{End}_G(Z)$ is local and hence there is a k-algebra $\Theta : \mathrm{End}_G(Z) \to k$ with nilpotent kernel. The composite $\Theta \circ \Psi : \mathrm{End}_G(V \otimes Z) \to k$ has nilpotent kernel. Hence $\mathrm{End}_G(V \otimes Z)$ is a local k-algebra and $V \otimes Z$ is indecomposable.

Let $\lambda \in X_1$ and suppose there exists a module $Q(\lambda) \in \mathrm{mod}(G)$ such that $Q(\lambda)|_{G_1} \cong \hat{Q}_1(\lambda)$ and assume furthermore that $Q(\lambda)$ is a tilting module. The top weight of $\hat{Q}_1(\lambda)$ (for $\lambda \in X_1$) is $w_0\lambda - (l-1)w_0 \cdot 0$ so we have:

(6) $Q(\lambda) \cong T(w_0\lambda - (l-1)w_0 \cdot 0)$, for $\lambda \in X_1$.

For $\lambda \in X^+$ we write $\bar{T}(\lambda)$ for the indecomposable tilting module and $\bar{\nabla}(\lambda)$ for the induced $\bar{G}$-module of highest weight λ. Using (5), we obtain as in [33; (2.1) Proposition]:

(7) $Q(\lambda) \otimes \bar{T}(\mu)^F \cong T(w_0\lambda - (l-1)w_0 \cdot 0 + l\mu)$, for $\lambda \in X_1$ and $\mu \in X^+$.

In particular, for a Steinberg weight σ we have $T(\sigma) = L(\sigma)$ so we have:

(8) $L(\sigma) \otimes \bar{T}(\mu)^F \cong T(\sigma + l\mu)$, for all $\mu \in X^+$.

Now a $\bar{G}$-module filtration $0 = T_0 < T_1 < \cdots < T_r = \bar{T}(\mu)$ with $T_i/T_{i-1} \cong \bar{\nabla}(\lambda_i)$, for $1 \le i \le r$, gives rise to a filtration $0 = L(\sigma) \otimes T_0^F < L(\sigma) \otimes T_1^F < \cdots < L(\sigma) \otimes T_r^F = L(\sigma) \otimes \bar{T}(\mu)^F$. Hence, using 2.1(13) we get a q-analogue of a result of Erdmann, [47; (2.2) Corollary].

(9)   *For a Steinberg weight* $\sigma \in X_1$ *and* $\lambda, \mu \in X^+$ *we have*

$$(T(\sigma + l\lambda) : \nabla(\mu)) = \begin{cases} (\bar{T}(\lambda) : \bar{\nabla}(\tau)), & \text{if } \mu = \sigma + l\tau, \text{ for some } \tau \in X^+; \\ 0, & \text{otherwise.} \end{cases}$$

In particular, defining $t : X^+ \to X^+$ by $t(\lambda) = (l-1)\delta + l\lambda$, where $\delta = (n-1, \ldots, 1, 0)$, we have $(T(t(\lambda)) : \nabla(t(\mu)) = (\bar{T}(\lambda) : \bar{\nabla}(\mu))$.

Here, for a finite dimensional $\bar{G}$-module V which has a filtration with sections of the form $\bar{\nabla}(\alpha)$, $\alpha \in X^+$, (i.e. a good $\bar{G}$-module filtration) and $\tau \in X^+$, we are using $(V : \bar{\nabla}(\tau))$ to denote the number of occurrences of $\bar{\nabla}(\tau)$ as a section in such a filtration.

3.4 Tilting modules for quantum GL$_2$

We are now in a position to calculate the characters of the tilting modules for quantum GL$_2$. The tilting modules were described in the classical case for SL$_2$ in [33]. If q is not a root of unity then all G-modules are completely reducible and it follows that $T(\lambda) = \nabla(\lambda)$, for all $\lambda \in X^+$. We assume therefore that q is a primitive lth root of unity, and $l > 1$. Let π denote the subset of X^+ consisting of elements (a, b) with $a, b \geq 0$ and $a - b \leq 2(l-1)$. Let $\mathrm{mod}(\pi)$ denote the category of finite dimensional G-modules V such that every composition factor of V comes from $\{L(\lambda) \mid \lambda \in \pi\}$. Note that π is a saturated subset of X^+, i.e. π has the property that whenever $\mu \in \pi$ and λ is a dominant weight such that $\lambda \leq \mu$ then $\lambda \in \pi$. Hence $\mathrm{mod}(\pi)$ is equivalent to the category of finite dimensional modules for the generalized Schur algebra $S(\pi)$ (see [36; Section 4]). We have $L(\lambda) = \nabla(\lambda)$ if $\lambda_1 - \lambda_2 < l$, e.g. by [75]. Hence we have $L(\lambda) = \nabla(\lambda) = \Delta(\lambda)$, in that case, and so $T(\lambda) = \nabla(\lambda)$. If $\lambda_1 - \lambda_2 = l + r$ with $0 \leq r \leq l - 2$ then $\nabla(\lambda)$ has composition factors $L(\lambda) = L(\lambda - l\epsilon_1) \otimes \bar{L}(1,0)^F$ and $L(\lambda - (r+1)(\epsilon_1 - \epsilon_2))$. We denote by $I_\pi(\lambda)$ the injective hull of $L(\lambda)$ in $\mathrm{mod}(\pi)$, for $\lambda \in \pi$. By reciprocity, [36; Section 4,(6)], we have $(I_\pi(\lambda) : \nabla(\mu)) = [\nabla(\mu) : L(\lambda)]$, for $\lambda, \mu \in \pi$. Specifically, we have $I_\pi(\lambda) = \nabla(\lambda)$ if $\lambda_1 - \lambda_2 \geq l - 1$. For $\lambda_1 - \lambda_2 = l + r$ we put $\mu = \lambda - (r+1)(\epsilon_1 - \epsilon_2)$. Then we have $(I_\pi(\mu) : \nabla(\tau)) = 1$ if $\tau = \lambda$ or μ and is 0 otherwise (for $\tau \in \pi$). Thus the block of λ in $\mathrm{mod}(\pi)$ is $\{\lambda\}$ if $\lambda_1 - \lambda_2 = (l-1)$, and if $\lambda_1 - \lambda_2 = l + r$ with $0 \leq r \leq l - 2$ then the block containing λ is $\{\lambda, \lambda - (r+1)(\epsilon_1 - \epsilon_2)\}$. Each block of $\mathrm{mod}(\pi)$ (or $S(\pi)$) is of one of the kinds just specified.

An explicit description of the blocks of the q-Schur algebra $S(n, r)$, for arbitrary n, r, q, has been given by Cox, [15; Chapter 4].

Now suppose $\lambda = (\lambda_1, \lambda_2) \in \pi$ and that $\lambda_1 - \lambda_2 = l + r$, for some $0 \leq r \leq l - 2$. We have the Steinberg weight $\sigma = (l - 1 + \lambda_2, \lambda_2)$ and $\lambda = \sigma + \mu$, where $\mu = (\lambda_1 - \lambda_2 - l + 1, 0)$. We have the Clebsch–Gordan expansion $\chi(\sigma)\chi(\mu) = \sum_{i=0}^{r+1} \chi(\lambda - i(\epsilon_1 - \epsilon_2))$. We write $T'(\lambda)$ for the block component of $\nabla(\sigma) \otimes \nabla(\mu)$ which has highest weight λ. Then $T'(\lambda)$ is a tilting module and has character $\chi(\lambda) + \chi(\lambda - (r+1)(\epsilon_1 - \epsilon_2))$. The dimension of $T'(\lambda)$ is $\lambda_1 - \lambda_2 + 1 + \lambda_1 - \lambda_2 - 2(r+1) + 1 = 2l$. Now $T(\lambda)$ is a direct summand of $T'(\lambda)$ and hence we get $\dim T(\lambda) \leq 2l$. Furthermore, by 3.3(3)(iii), $\dim Q_1(l - 1 + \lambda_2, \lambda_1 - l + 1) \leq 2l$ and therefore we have $\dim Q_1(\nu) \leq 2l$ for any non-Steinberg weight $\nu \in X_1$. Now, as left G_1-modules, we have $k[G_1] \cong \bigoplus_{\nu \in X_1} Q_1(\nu)^{(d_\nu)}$, where d_ν is the dimension of $L(\nu)$ (for $\nu \in X_1$). Hence we have $l^4 = \dim k[G_1] = \sum_{\nu \in X_1} d_\nu \dim Q_1(\nu)$. For $\nu = (\nu_1, \nu_2) \in X_1$ we have $d_\nu = \dim L(\nu) \leq \dim \nabla(\nu) = \nu_1 - \nu_2 + 1$. If $\nu_1 - \nu_2 = l - 1$ then ν is a Steinberg weight and so $\dim Q(\nu) = \dim \nabla(\nu) = l$. Now we have $X_1 = \{(a + b, b) \mid 0 \leq a, b < l\}$ giving

$$l^4 \leq \sum_{\nu \in X_1, \nu_1 - \nu_2 \neq l-1} (\nu_1 - \nu_2 + 1)2l + \sum_{\nu \in X_1, \nu_1 - \nu_2 = l-1} l^2$$

$$= \sum_{a=0}^{l-2} \sum_{b=0}^{l-1} 2(a+1)l + \sum_{a=0}^{l-1} l^2 = l(l-1)l^2 + l^3 = l^4.$$

It follows that we must have $\dim Q_1(\nu) = 2l$ for all non-Steinberg weights in X_1. Further we deduce that $T'(\lambda) = T(\lambda)$ and (from 3.3(3)(iii)) deduce that $T(\lambda)|_{\hat{G}_1} \cong \hat{Q}_1(l - 1 + \lambda_2, \lambda_1 - l + 1)$. For $\nu = (\nu_1, \nu_2) \in X^+$ with $\nu_1 - \nu_2 < l$ we write $Q(\nu)$ for $T(\tilde{\nu})$, where $\tilde{\nu} = (l - 1 + \nu_2, \nu_1 - l + 1)$. By the above discussion we have $Q(\nu)|_{\hat{G}_1} \cong \hat{Q}_1(\nu)$ if $\tilde{\nu} \in \pi$. We write w for the non-identity element of W. By tensoring with a suitable power of the (1-dimensional) determinant representation $L(1, 1)$, it follows that for any $\lambda = (\lambda_1, \lambda_2) \in X$ with $0 \le \lambda_1 - \lambda_2 < l$, writing $Q(\lambda)$ for $T(w\lambda - (l-1)w \cdot 0)$, we have:

(1) $Q(\lambda)|_{\hat{G}_1} \cong \hat{Q}_1(\lambda)$ and ch $Q(\lambda) = s(w\lambda - (l-1)w\rho)\chi((l-1)\rho)$.

Here, for $\nu \in X$, we are writing $s(\nu)$ for the orbit sum, i.e. $s(\nu) = e(\nu)$ if $\nu = w\nu$ and $s(\nu) = e(\nu) + e(w\nu)$ if $w\nu \ne \nu$.

For $\psi = \sum_\mu a_\mu e(\mu) \in \mathbb{Z}X$, we put $\psi^F = \sum_\mu a_\mu e(l\mu) \in \mathbb{Z}X$. Thus for $V \in \mathrm{mod}(G)$ we have ch $V^F = (\mathrm{ch}\,V)^F$. We denote by $\bar{F}$ the (ordinary) Frobenius morphism on $\bar{G}$-modules and similarly define $\bar{F}$ on $\mathbb{Z}X$.

We denote by p the characteristic of k. We shall give the character of the tilting module $T(\lambda)$ in terms of the (l, p) expansion of a non-negative integer associated with $\lambda \in X^+$. Let a be a non-negative integer. Then a has a unique expression in the form $a = a_{-1} + la_0$, with $0 \le a_{-1} < l$. If $p = 0$ we call this expression the (l,p)-expansion of a. If $p > 0$ then a has a unique expression of the form $a_{-1} + l\sum_{i=0}^\infty p^i a_i$, with $0 \le a_{-1} < l$ and $0 \le a_0, a_1, \ldots < p$, and we call this expression the (l,p)-expansion of a. Assume first that $p > 0$. For $\mu = (\mu_1, \mu_2) \in X^+$ with $\mu_1 - \mu_2 < p$ we write $\bar{Q}(\mu)$ for the tilting module for $\bar{G}$ of highest weight $w\mu - (p-1)w \cdot 0$. By [33; Section 2, Example 1], we have that $\bar{Q}(\mu) \otimes \bar{T}(\nu)^{\bar{F}}$ is an indecomposable tilting module, for any $\nu \in X^+$. (Actually in [33], this remark is made for the group SL_2 but the same argument applies to GL_2.) Hence, inductively for $\lambda(-1), \lambda(0), \ldots, \lambda(m-1), \mu \in X^+$ with $\lambda(-1)_1 - \lambda(-1)_2 < l$, $\lambda(0)_1 - \lambda(0)_2, \ldots, \lambda(m-1)_1 - \lambda(m-1)_2, \mu_1 - \mu_2 < p$, we get that $Q(\lambda(-1)) \otimes (\bar{Q}(\lambda(0)) \otimes \cdots \otimes \bar{Q}(\lambda(m-1))^{\bar{F}^{m-1}} \otimes \bar{\nabla}(\mu)^{\bar{F}^m})^F$ is an indecomposable tilting module. This has highest weight

$$w\lambda(-1) - (l-1)w \cdot 0 + l\sum_{i=0}^{m-1}(p^i(w\lambda(i) - (p-1)w \cdot 0) + p^m\mu)$$

so we get

$$T(w\lambda(-1) - (l-1)w \cdot 0 + l\sum_{i=0}^{m-1}(p^i(w\lambda(i) - (p-1)w \cdot 0) + p^m\mu))$$

$$\cong Q(\lambda(-1)) \otimes (\bar{Q}(\lambda(0)) \otimes \cdots \otimes \bar{Q}(\lambda(m-1)^{\bar{F}^{m-1}}) \otimes \bar{\nabla}(\mu)^{\bar{F}^m})^F.$$

Replacing $\lambda(-1)$ by $(l-1)\rho + w\lambda(-1)$ and $\lambda(i)$ by $(p-1)\rho + w\lambda(i)$, for

$0 \leq i \leq m - 1$, we get

$$T(\lambda(-1) + (l-1)\rho + l \sum_{i=0}^{m-1} (p^i(\lambda(i) + (p-1)\rho) + p^m\mu))$$

$$= Q((l-1)\rho + w\lambda(-1)) \otimes (\bar{Q}((p-1)\rho + w\lambda(0))$$

$$\otimes \cdots \otimes \bar{Q}((p-1)\rho + w\lambda(m-1))^{\bar{F}^{m-1}} \otimes \bar{\nabla}(\mu)^{\bar{F}^m})^F.$$

Now for $\nu \in X^+$ with $\nu_1 - \nu_2 < p$, the character of $Q((l-1)\rho + w\nu)$ is $\chi((l-1)\rho)s(\nu)$ and for $\nu \in \bar{X}_1$ the character of $\bar{Q}((p-1)\rho + w\nu)$ is $\chi((p-1)\rho)s(\nu)$ (by the argument of (1) or the GL$_2$ version of [**33**; Section 2, Example 1]). Thus we get

$$\operatorname{ch} T(\lambda(-1) + (l-1)\rho + l \sum_{i=0}^{m-1} (p^i(\lambda(i) - (p-1)\rho) + p^m\mu))$$

$$= \chi((l-1)\rho)s(\lambda(-1))(\chi(\mu)^{\bar{F}^m} \prod_{i=0}^{m-1} \chi((p-1)\rho)^{\bar{F}^i}s(\lambda(i))^{\bar{F}^i})^F$$

$$= \chi((l-1)\rho)s(\lambda(-1))(\chi((p^m - 1)\rho)\chi(\mu)^{\bar{F}^m} \prod_{i=0}^{m-1} s(\lambda(i))^{\bar{F}^i})^F$$

$$= \chi((lp^m - 1)\rho + lp^m\mu)s(\lambda(-1))(\prod_{i=0}^{m-1} s(\lambda(i))^{\bar{F}^i})^F$$

using the fact that $\chi((l-1)\rho)\chi(\mu)^F = \chi((l-1)\rho + l\mu)$ and $\chi((p-1)\rho)\chi(\mu)^F = \chi((p-1)\rho + p\mu)$, for $\mu \in X^+$, by e.g. [**58**; equation (7′)] (or direct calculation).
Setting $\lambda = \lambda(-1) + l \sum_{i=0}^{m-1} p^i\lambda(i)$ we get

$$\operatorname{ch} T(\lambda + (lp^m - 1)\rho + lp^m\mu) = \chi((lp^m - 1)\rho + lp^m\mu)s(\lambda(-1))(\prod_{i=0}^{m-1} s(\lambda(i))^{\bar{F}^i})^F.$$

To recapitulate, we have the following.

(2)   Suppose that $\lambda(-1), \lambda(0), \dots, \lambda(m-1), \mu \in X^+$ with $\lambda(-1)_1 - \lambda(-1)_2 < l$, with $\lambda(i)_1 - \lambda(i)_2 < p$ for $0 \leq i \leq m$, and with $\mu_1 - \mu_2 < p$. Then we have

$$\operatorname{ch} T(\lambda + (lp^m - 1)\rho + lp^m\mu) = \chi((lp^m - 1)\rho + lp^m\mu)s(\lambda(-1))(\prod_{i=0}^{m-1} s(\lambda(i))^{\bar{F}^i})^F.$$

Let a be a non-negative integer with (l, p) expansion

$$a = a_{-1} + \sum_{i=0}^{m-1} lp^i a_i.$$

For a subset J of $[-1, m-1]$ we define a_J to be the sum of the terms indexed by J, i.e. we define a_J to be $a_{-1} + \sum_{-1 \neq i \in J} lp^i a_i$ if $-1 \in J$, and to be $\sum_{i \in J} lp^i a_i$ if $-1 \notin J$. Now let $\lambda = (\lambda_1, \lambda_2)$, suppose that $\lambda_1 - \lambda_2 \geq l - 1$ and let m be the non-negative integer such that $lp^m - 1 \leq \lambda_1 - \lambda_2 < lp^{m+1} - 1$. We write $\lambda_1 - \lambda_2$ in the form $(lp^m - 1) + a + lp^m b$ for (uniquely determined) integers a, b with $0 \leq a \leq lp^m - 1$, $0 \leq b < p$. Let $D = L(1,1)$ be the determinant module. Then we have

$$
\begin{aligned}
T(\lambda) = T(\lambda_1, \lambda_2) &\cong D^{\otimes \lambda_2} \otimes T((\lambda_1 - \lambda_2)\epsilon_1) \\
&\cong D^{\otimes \lambda_2} \otimes T((lp^m - 1)\epsilon_1 + a\epsilon_1 + lp^m b\epsilon_1) \\
&\cong D^{\otimes(\lambda_2 - (lp^m - 1))} \otimes T((lp^m - 1)\rho + a\epsilon_1 + lp^m b\epsilon_1)
\end{aligned}
$$

and hence, from (2), we have

$$
\begin{aligned}
\operatorname{ch} T(\lambda) &= \sum_{J \subseteq [-1, m-1]} \chi((\lambda_2 - (lp^m - 1))\omega)\chi((lp^m - 1)\rho + a\epsilon_1 - a_J(\epsilon_1 - \epsilon_2) + lp^m b\epsilon_1) \\
&= \sum_{J \subseteq [-1, m-1]} \chi(\lambda_2 \omega)\chi((lp^m - 1)\epsilon_1 + a\epsilon_1 - a_J(\epsilon_1 - \epsilon_2) + lp^m b\epsilon_1) \\
&= \sum_{J \subseteq [-1, m-1]} \chi(\lambda - a_J(\epsilon_1 - \epsilon_2)).
\end{aligned}
$$

Thus we have proved the following.

(3)   Let $\lambda = (\lambda_1, \lambda_2) \in X^+$. If $\lambda_1 - \lambda_2 \leq l - 1$ then $T(\lambda) = \nabla(\lambda)$. If $\lambda_1 - \lambda_2 \geq l - 1$ and $\lambda_1 - \lambda_2 = lp^m - 1 + a + lp^m b$, with $a \leq lp^m - 1$ and $0 \leq b < p$, then for $\mu \in X^+$ we have

$$
(T(\lambda) : \nabla(\mu)) = \begin{cases} 1, & \text{if } \mu = \lambda - a_J(\epsilon_1 - \epsilon_2) \text{ for some } J \subseteq [-1, m-1]; \\ 0, & \text{otherwise.} \end{cases}
$$

Now suppose that $p = 0$. A much shortened version of the above discussion gives the following.

(4)   Let $\lambda = (\lambda_1, \lambda_2) \in X^+$. If $\lambda_1 - \lambda_2 \leq l - 1$ then $T(\lambda) = \nabla(\lambda)$. If $\lambda_1 - \lambda_2 \geq l - 1$ and $\lambda_1 - \lambda_2 = l - 1 + a + lb$, with $0 \leq a \leq l - 1$, then for $\mu \in X^+$ we have

$$
(T(\lambda) : \nabla(\mu)) = \begin{cases} 1, & \text{if } \mu = \lambda \text{ or } \lambda - a(\epsilon_1 - \epsilon_2); \\ 0, & \text{otherwise.} \end{cases}
$$

4. Further Topics

This chapter is concerned with diverse topics in the representation theory of q-Schur algebra and connections with the Hecke algebra. We shall not give a general introduction to the chapter but rather treat each section separately.

4.1 The Ringel dual of the Schur algebra

In [33] we determined the Morita type of the "Ringel dual" of the ordinary Schur algebra $S(n,r)$. The result was obtained with the aid of the Schur functor from $S(n,r)$-modules to $\mathrm{Sym}(r)$-modules. Another treatment, valid in somewhat more generality, was given in [35; Section 5(2)], by making the exterior algebra on the space of $n \times n$-matrices into an $(S(n,r), S(n,r))$-bimodule. This is the approach we take here. So the main point is to produce an appropriate q-exterior algebra which is an $(S(n,r), S(n,r))$-bimodule. This we do by brute calculation.

Note that $B = V \otimes E$ is naturally an (M, M)-bimodule with left M-module structure map structure map $\lambda : B \to B \otimes A$ and right M-module structure map $\rho : B \to A \otimes B$ given on $b_{is} = v_i \otimes e_s$ by

$$\lambda(b_{is}) = \sum_{t=1}^{n} b_{it} \otimes c_{ts}$$

and

$$\rho(b_{is}) = \sum_{j=1}^{n} c_{ij} \otimes b_{js}$$

for $1 \leq i, s \leq n$. We have an induced (M, M)-bimodule structure on the tensor algebra $T(B) = \bigoplus_{r=0}^{\infty} B^{\otimes r}$. Let $J = J(n)$ be the ideal of $T(B)$ generated by the elements

TI b_{is}^2 for all $1 \leq i, s \leq n$;

TII $b_{is}b_{jt} + b_{jt}b_{is}$ for all $1 \leq i < j \leq n, 1 \leq s \leq t \leq n$;

TIII $b_{is}b_{it} + qb_{it}b_{is}$ for all $1 \leq i \leq n, 1 \leq s < t \leq n$;

TIV $b_{js}b_{it} + qb_{it}b_{js} + (q-1)b_{jt}b_{is}$ for all $1 \leq i < j \leq n, 1 \leq s < t \leq n$.

Note that J is generated by all elements of the form TI–TIII together with those of the form

TIV′ $b_{is}b_{jt} + qb_{jt}b_{is} + b_{js}b_{it} + qb_{it}b_{js}, \quad 1 \leq i \leq j \leq n, 1 \leq s < t \leq n.$

Lemma 4.1.1 J is a subbimodule of $T(B)$.

Proof   We need to check that J is both a left and right submodule. This may done in eight stages by checking that the generators of type TI–TIV map into $J \otimes A$ under λ and into $A \otimes J$ under ρ. We label the calculation showing that generators of type TI–TIV map into $J \otimes A$ under λ by RI–RIV, and the calculation showing that generators of type TI–TIV map into $A \otimes J$ under ρ by LI-LIV. In RI–RIV the symbol $\equiv$ denotes congruence modulo $J \otimes A$ and in LI–LIV denotes congruence modulo $A \otimes J$.

LI   For $1 \leq i, s \leq n$ we have

$$\lambda(b_{is}^2) = \sum_{k,l} b_{ik}b_{il} \otimes c_{ks}c_{ls}$$

$$\equiv \sum_{k<l} b_{ik}b_{il} \otimes c_{ks}c_{ls} + \sum_{k<l} b_{il}b_{ik} \otimes c_{ls}c_{ks}$$

$$= \sum_{k<l} (b_{ik}b_{il} + qb_{il}b_{ik}) \otimes c_{ks}c_{ls}$$

$$\equiv 0$$

using AII from Section 1.2 and the fact that elements of types TI,TIII belong to J.

RI   For $1 \leq i, s \leq n$ we have

$$\rho(b_{is}^2) = \sum_{k,l} c_{ik}c_{il} \otimes b_{ks}b_{ls}$$

$$\equiv \sum_{k<l} c_{ik}c_{il} \otimes b_{kl}b_{ls} + \sum_{k<l} c_{il}c_{ik} \otimes b_{ls}b_{ks}$$

$$= \sum_{k<l} c_{ik}c_{il} \otimes (b_{kl}b_{ls} + b_{ls}b_{ks})$$

$$\equiv 0$$

using AI and the fact that elements of types TI,TII belong to J.

LII   Let $1 \leq i < j \leq n$ and $1 \leq s \leq t \leq n$. We have

$$\lambda(b_{is}b_{jt} + b_{jt}b_{is}) = \sum_{k,l} b_{ik}b_{jl} \otimes c_{ks}c_{lt} + \sum_{k,l} b_{jk}b_{il} \otimes c_{kt}c_{ls}$$

$$= \sum_{k<l} b_{ik}b_{jl} \otimes c_{ks}c_{lt} + \sum_{k} b_{ik}b_{jk} \otimes c_{ks}c_{kt}$$

$$+ \sum_{k<l} b_{il}b_{jk} \otimes c_{ls}c_{kt} + \sum_{k<l} b_{jk}b_{il} \otimes c_{kt}c_{ls}$$

$$+ \sum_{k} b_{jk}b_{ik} \otimes c_{kt}c_{ks} + \sum_{k<l} b_{jl}b_{ik} \otimes c_{lt}c_{ks}.$$

Moreover we have $b_{ik}b_{jk} \otimes c_{ks}c_{kt} + b_{jk}b_{ik} \otimes c_{kt}c_{ks} \in A \otimes J$, by AI and TII. Thus we have

$$\lambda(b_{is}b_{jt} + b_{jt}b_{is}) \equiv \sum_{k<l} b_{ik}b_{jl} \otimes c_{ks}c_{lt} + \sum_{k<l} b_{il}b_{jk} \otimes c_{ls}c_{kt}$$
$$+ \sum_{k<l} b_{jk}b_{il} \otimes c_{kt}c_{ls} + \sum_{k<l} b_{jl}b_{ik} \otimes c_{lt}c_{ks}.$$

If $s = t$ this is

$$\sum_{k<l}(b_{ik}b_{jl} + qb_{il}b_{jk} + b_{jk}b_{il} + qb_{jl}b_{il}) \otimes c_{ks}c_{ls}$$

by AII, and this belongs to $J \otimes A$, since elements of type TIV' belong to J. Now suppose $s < t$. We get

$$\lambda(b_{is}b_{jt} + b_{jt}b_{is}) \equiv \sum_{k<l} b_{ik}b_{jl} \otimes c_{ks}c_{lt} + q\sum_{k<l} b_{il}b_{jk} \otimes c_{kt}c_{ls}$$
$$+ \sum_{k<l} b_{jk}b_{il} \otimes c_{kt}c_{ls} + \sum_{k<l} b_{jl}b_{ik} \otimes c_{ks}c_{lt}$$
$$+ (q-1)\sum_{k<l} b_{jl}b_{ik} \otimes c_{kt}c_{ls}$$

by AII and AIII. In this the "coefficient" of $c_{ks}c_{lt}$ is $b_{ik}b_{jl} + b_{jl}b_{ik}$, which is an element of type TII and so belongs to J, and the "coefficient" of $c_{kt}c_{ls}$ is $qb_{il}b_{jk} + b_{jk}b_{il} + (q-1)b_{jl}b_{ik}$, which is an element of type TIV and so belongs to J. Hence we have $\lambda(b_{is}b_{jt} + b_{jt}b_{is}) \in J \otimes A$, as required.

RII   Let $1 \leq i < j \leq n$ and $1 \leq s \leq t \leq n$. We have

$$\rho(b_{is}b_{jt} + b_{jt}b_{is}) = \sum_{k,l} c_{ik}c_{jl} \otimes b_{ks}b_{lt} + \sum_{k,l} c_{jk}c_{il} \otimes b_{kt}b_{ls}$$
$$= \sum_{k<l} c_{ik}c_{jl} \otimes b_{ks}b_{lt} + \sum_{k} c_{ik}c_{jk} \otimes b_{ks}b_{kt}$$
$$+ \sum_{k<l} c_{il}c_{jk} \otimes b_{ls}b_{kt} + \sum_{k<l} c_{jk}c_{il} \otimes b_{kt}b_{ls}$$
$$+ \sum_{k} c_{jk}c_{ik} \otimes b_{kt}b_{ks} + \sum_{k<l} c_{jl}c_{ik} \otimes b_{lt}b_{ks}$$
$$= \sum_{k<l} c_{ik}c_{jl} \otimes b_{ks}b_{lt} + \sum_{k} c_{ik}c_{jk} \otimes b_{ks}b_{kt}$$
$$+ \sum_{k<l} c_{il}c_{jk} \otimes b_{ls}b_{kt} + q\sum_{k<l} c_{il}c_{jk} \otimes b_{kt}b_{ls}$$
$$+ q\sum_{k} c_{ik}c_{jk} \otimes b_{kt}b_{ks}$$
$$+ \sum_{k<l} c_{ik}c_{jl} \otimes b_{lt}b_{ks} + (q-1)\sum_{k<l} c_{il}c_{jk} \otimes b_{lt}b_{ks}$$

using AII and AIII. However, we have $c_{ik}c_{jl} \otimes b_{ks}b_{lt} + c_{ik}c_{jl} \otimes b_{lt}b_{ks} \in A \otimes J$ by TII and $c_{ik}c_{jk} \otimes b_{ks}b_{kt} + qc_{ik}c_{jk} \otimes b_{kt}b_{ks} \in A \otimes J$ by TIII. Hence we have

$$\rho(b_{is}b_{jt} + b_{jt}b_{is}) \equiv \sum_{k<l} c_{il}c_{jk} \otimes (b_{ls}b_{kt} + qb_{kt}b_{ls} + (q-1)b_{lt}b_{ks})$$

which belongs to $A \otimes J$ by TIV.

By similar, but somewhat more complicated calculations, which we leave to the reader, one obtains:

LIII For $1 \leq i \leq n$ and $1 \leq s < t \leq n$ we have $\lambda(b_{is}b_{it} + qb_{it}b_{is}) \in J \otimes A$.

RIII For $1 \leq i \leq n$ and $1 \leq s < t \leq n$ we have $\rho(b_{is}b_{it} + qb_{it}b_{is}) \in A \otimes J$.

LIV For $1 \leq i < j \leq n$ and $1 \leq s < t \leq n$ we have $\lambda(b_{is}b_{jt} + qb_{it}b_{js} + (q-1)b_{jt}b_{is}) \in J \otimes A$.

RIV For $1 \leq i < j \leq n$ and $1 \leq s < t \leq n$ we have $\rho(b_{is}b_{jt} + qb_{it}b_{js} + (q-1)b_{jt}b_{is}) \in J \otimes A$.

We have shown that elements of the form TI–TIV span an (M, M) subbimodule of $(V \otimes E)^{\otimes 2}$. However, $\rho : T(B) \to A \otimes T(B)$ and $\lambda : T(B) \to T(B) \otimes A$ are algebra maps sothat $\rho(J) \leq A \otimes J$ and $\lambda(J) \leq J \otimes A$, as required.

We write $\bigwedge(V \otimes E)$ for $T(V \otimes E)/J$. We write $x \wedge y$ for the product xy in $\bigwedge(V \otimes E)$ (for $x, y \in \bigwedge(V \otimes E)$). Notice that J is a homogeneous ideal for the natural grading of $T(V \otimes E)$. Thus $\bigwedge(V \otimes E)$ inherits a grading. We write $\bigwedge^r(V \otimes E)$ for the rth component of $\bigwedge(V \otimes E)$, $r \geq 0$. We order the basis $v_i \otimes e_s$ for $V \otimes E$ by $v_i \otimes e_s < v_j \otimes e_t$ if $i < j$ or $i = j$ and $s < t$. Let $\{v_i \otimes e_s \mid 1 \leq i \leq n, 1 \leq s \leq n\} = \{m_1, m_2, \ldots, m_{n^2}\}$, with $m_1 < m_2 < \cdots < m_{n^2}$. We leave it to the reader to check the following.

Lemma 4.1.2 *Let r be a positive integer. The set $\{m_{h_1} \wedge \cdots \wedge m_{h_r} \mid 1 \leq h_1 < h_2 < \cdots < h_r \leq n^2\}$ is a K-basis of $\bigwedge^r(V \otimes E)$.*

Let $\alpha = (\alpha_1, \alpha_2, \ldots) \in P(n, r)$ (see Section 1.2). We define a map $\phi_\alpha : E^{\otimes r} \to \bigwedge^r(E \otimes V)$ to be the k-map taking e_i to $(v_1 \otimes e_{i_1}) \wedge \cdots \wedge (v_1 \otimes e_{i_{\alpha_1}}) \wedge (v_2 \otimes e_{i_{(\alpha_1+1)}}) \wedge \cdots \wedge (v_2 \otimes e_{i_{(\alpha_1+\alpha_2)}}) \wedge \cdots$. It is easy to check that ϕ_α is a morphism of left M-modules. Furthermore, it follows from TI that $\phi_\alpha(e_i) = 0$ if we have $i_a = i_{a+1}$ for some $1 \leq a < \alpha_1$ or $\alpha_1 + \cdots + \alpha_m < a < \alpha_1 + \cdots + \alpha_m + \alpha_{m+1}$ with $m \geq 1$. It follows from TIII that $\phi_\alpha(e_i + qe_{is_a}) = 0$ if $i_a < i_{a+1}$, for some $1 \leq a < \alpha_1$ or $\alpha_1 + \cdots + \alpha_m < a < \alpha_1 + \cdots + \alpha_m + \alpha_{m+1}$ with $m \geq 1$. Hence ϕ_α induces an M-module homomorphism $\tilde{\phi}_\alpha : \bigwedge^\alpha E \to \bigwedge^r(V \otimes E)$, satisfying $\tilde{\phi}_\alpha(\hat{e}_i) = \phi_\alpha(e_i)$, for all $i \in I(n, r)$. It follows from Lemma 4.1.1 that $\tilde{\phi}_\alpha$ is injective and that $\bigwedge^r(V \otimes E) = \bigoplus_{\alpha \in P(n,r)} \mathrm{Im}(\tilde{\phi}_\alpha)$. In particular we get part (i) of the following. The second part is obtained by the mirror argument.

Lemma 4.1.3 (i) $\bigwedge^r(E \otimes V) \cong \bigoplus_{\alpha \in P(n,r)} \bigwedge^\alpha E$, as left M-modules.
(ii) $\bigwedge^r(E \otimes V) \cong \bigoplus_{\alpha \in P(n,r)} \bigwedge^\alpha V$, as right M-modules.

Suppose now that $r \leq n$. By 3.3(1), the left module $T = \bigwedge^r(V \otimes E)$ is a full tilting module for $S(n,r)$. We write $S(n,r)'$ for $\operatorname{End}_{S(n,r)}(T)$, the Ringel dual of $S(n,r)$. It follows, e.g. from 2.1(13), that the dimension of $S(n,r)'$ is independent of q and the characteristic of k (cf. the proof of [35; Section 5(2)]). Hence, by [33; (3.7) Proposition], we have dim $S(n,r)' =$ dim $S(n,r)$. Now T is a polynomial right M-module of degree r, and hence a right $S(n,r)$-module. We thus get a homomorphism $\eta : S(n,r)^{\operatorname{op}} \to S(n,r)'$, given by $\eta(s)(x) = xs$, for $x \in T$, $s \in S(n,r)$. It follows from the fact that the coefficient space of $V^{\otimes r}$ is $A(n,r)$ that $V^{\otimes r}$ is a faithful right $S(n,r)$-module. Moreover $V^{\otimes r}$ is a right M-module summand of $\bigwedge^r(V \otimes E)$, by Lemma 4.1.3, so that $\bigwedge^r(V \otimes E)$ is faithful and $\eta : S(n,r)^{\operatorname{op}} \to S(n,r)'$ is injective. Now by dimensions we have:

(1) $\eta : S(n,r)^{\operatorname{op}} \to S(n,r)'$ is an isomorphism.

We have the involutory antiautomorphism J of the Hecke algebra $H(r)$ defined on the generators by $J(T_s) = T_s$, for a simple basic reflection s. For $i = (i_1, \ldots, i_r) \in I(n,r)$ we define $d(i)$ to be the number of the set of pairs (a,b) of elements of $[1,r]$ such that $a < b$ and $i_a < i_b$. We define a bilinear form $(\,,\,)$ on $E^{\otimes r}$ by $(e_i, e_j) = \delta_{ij} q^{d(i)}$, for $i,j \in I(n,r)$. Note that $(\,,\,)$ is symmetric and non-degenerate (since $q \neq 0$) so we have an involutory antiautomorphism $J' : \operatorname{End}_k(E^{\otimes r}) \to \operatorname{End}_k(E^{\otimes r})$, given by $(\xi e_i, e_j) = (e_i, J'(\xi)e_j)$, for $\xi \in \operatorname{End}_k(E^{\otimes})$ and $i,j \in I(n,r)$. Now the natural representation $S(n,r) \to \operatorname{End}_k(E^{\otimes r})$ is faithful and the image is exactly the centralizer of the action $\rho : H(r) \to \operatorname{End}_k(E^{\otimes r})$, given in [18; Section 3.1]. Moreover, it is easy to check that $(\rho(T_s)e_i, e_j) = (e_i, \rho(T_s)e_j)$, for simple reflections s and all $i,j \in I(n,r)$. It follows that if $\xi \in \operatorname{End}_H(E^{\otimes r})$ then $J'(\xi) \in \operatorname{End}_H(E^{\otimes r})$. Thus we obtain, by restriction, an involutory antiautomorphism of $S(n,r)$, also denoted J'.

We claim that, for $r \leq n$, we have $J'(\xi) = J(\xi)$, for all $\xi \in H(r) \leq S(n,r)$. It suffices to prove this for $\xi = T_{s_a}$, $1 \leq a < r$. We put $T = T_{s_a}$. We must show that $(Te_i, e_j) = (e_i, Te_j)$ for all $i,j \in I(n,r)$. We have

$$(Te_i, e_j) = \sum_{h \in I(n,r)} T(c_{hi})(e_h, e_j) = q^{d(j)} T(c_{ji})$$

and similarly

$$(e_i, Te_j) = q^{d(i)} T(c_{ij}).$$

Furthermore, for $x, y \in I(n,r)$, we have $T(c_{xy}) = \xi_{u, us_a}(c_{xy}) = 0$ unless both x and y have content ω. Thus it suffices to show that $q^{d(u\sigma)} T(c_{u\sigma, u\pi}) = q^{d(u\pi)} T(c_{u\pi, u\sigma})$, i.e.

$$q^{l(\pi)} T(c_{u\sigma, u\pi}) = q^{l(\sigma)} T(c_{u\pi, u\sigma}) \qquad (*)$$

for all $\pi, \sigma \in \mathrm{Sym}(r)$.

We have

$$v_{u\sigma}T = v_u T_\sigma T_{s_a}$$
$$= \begin{cases} v_u T_{\sigma s_a}, & \text{if } l(\sigma s_a) > l(\sigma); \\ q v_u T_{\sigma s_a} + (q-1) v_u T_\sigma, & \text{if } l(\sigma s_a) < l(\sigma). \end{cases}$$

Equating coefficients of $v_{u\pi}$ we get

$$T(c_{u\sigma, u\pi}) = \begin{cases} \xi_{u, u\sigma s_a}(c_{u, u\pi}), & \text{if } l(\sigma s_a) > l(\sigma); \\ q\xi_{u, u\sigma s_a}(c_{u, u\pi}) + (q-1)\xi_{u, u\sigma}(c_{u, u\pi}), & \text{if } l(\sigma s_a) < l(\sigma), \end{cases}$$

i.e.

$$T(c_{u\sigma, u\pi}) = \begin{cases} \delta_{\sigma s_a, \pi}, & \text{if } l(\sigma s_a) > l(\sigma); \\ q\delta_{\sigma s_a, \pi} + (q-1)\delta_{\sigma, \pi}, & \text{if } l(\sigma s_a) < l(\sigma). \end{cases}$$

Hence we also have

$$T(c_{u\pi, u\sigma}) = \begin{cases} \delta_{\pi s_a, \sigma}, & \text{if } l(\pi s_a) > l(\pi); \\ q\delta_{\pi s_a, \sigma} + (q-1)\delta_{\pi, \sigma}, & \text{if } l(\pi s_a) < l(\pi). \end{cases}$$

Thus both sides of $(*)$ are 0 unless $\sigma s_a = \pi$ or $\sigma = \pi$, and both sides are equal in the case $\sigma = \pi$. If $\sigma s_a = \pi$ and $l(\sigma) < l(\pi)$ then both sides of $(*)$ are $q^{l(\pi)}$. If $\sigma s_a = \pi$ and $l(\sigma) < l(\pi)$ then both sides of $(*)$ are $q^{l(\sigma)}$. This completes the proof of the claim. By abuse of notation we now write J for J'.

For $\xi \in S(n, r)$ and $i, j \in I(n, r)$ from $(\xi e_i, e_j) = (e_i, J(\xi) e_j)$ we obtain $c_{ji}(\xi) q^{d(j)} = c_{ij}(J(\xi)) q^{d(i)}$. In particular, for $h \in I(n, r)$ we have $c_{ij}(\xi_{hh}) = 0$ for $i \neq j$ and $c_{ii}(\xi_{hh}) = \delta_{hi}$. It follows that $J(\xi_\alpha) = \xi_\alpha$, for all $\alpha \in \Lambda(n, r)$.

Remarks (i)   Given a finite dimensional left $H(r)$-module U we write U° for the dual space $\mathrm{Hom}_k(U, k)$ regarded as an $H(r)$-module via the action $(h\theta)(u) = \theta(J(h)u)$, for $h \in H(r)$, $\theta \in \mathrm{Hom}_k(U, k)$ and $u \in U$.
(ii)   Given a finite dimensional left $S(n, r)$-module U we write U° for the dual space $\mathrm{Hom}_k(U, k)$ regarded as an $S(n, r)$-module via the action $(h\theta)(u) = \theta(J(h)u)$, for $h \in S(n, r)$, $\theta \in \mathrm{Hom}_k(U, k)$ and $u \in U$. We call U° the *contravariant dual* (see [51; Section 2.7] for the classical case). Since $J(\xi_\alpha) = \xi_\alpha$, for $\alpha \in \Lambda(n, r)$, the character of U° is equal to the character of U. Hence we have $L(\lambda)^\circ \cong L(\lambda)$, for all $\lambda \in \Lambda^+(n, r)$. Moreover, since $J : H(r) \to H(r)$ is the restriction of $J : S(n, r) \to S(n, r)$, we have $(fU)^\circ \cong fU^\circ$, for $U \in \mathrm{mod}(S(n, r))$. Since J is an involution on $S(n, r)$ the natural linear map $U \to (U^\circ)^\circ$ is an $S(n, r)$-module isomorphism. Note also that the natural isomorphism $\psi : E^{\otimes r} \to (E^{\otimes r})^\circ$, given by $\psi(e_i)(e_j) = (e_i, e_j)$, for $i, j \in I(n, r)$, is an $S(n, r)$-module isomorphism.

Combining the antiautomorphism J with η we get, from (1), the following.

Proposition 4.1.4   We have a k-algebra isomorphism $\theta : S(n,r) \rightarrow S(n,r)'$ given by $\theta(\xi)(x) = xJ(\xi)$, for $\xi \in S(n,r)$, $x \in E^{\otimes r}$.

We have the functor $F : \text{mod}\, S(n,r) \rightarrow \text{mod}\, S(n,r)'$, satisfying $F(V) = \text{Hom}_{S(n,r)}(T,V)$, for $V \in \text{mod}\, S(n,r)$. We thus have the functor $\tilde{F} : \text{mod}\, S(n,r) \rightarrow \text{mod}\, S(n,r)$ satisfying $\tilde{F}(V) = (FV)^\theta$ (for $V \in \text{mod}\, S(n,r)$), the $S(n,r)$-module obtained from FV by composing the action $S(n,r)' \rightarrow \text{End}_k(FV)$ with $\theta : S(n,r) \rightarrow S(n,r)'$. For $\alpha \in \Lambda(n,r)$ we put $\xi'_\alpha = \theta(\xi_\alpha)$. An easy check reveals that ξ'_α is the identity map on $\text{Im}(\phi_\alpha) \cong \bigwedge^\alpha E$ and is 0 on $\text{Im}(\phi_\beta) \cong \bigwedge^\beta E$, for $\beta \neq \alpha$. Now the arguments of the proof of [**33**; (3.8) Corollary and (3.9) Corollary] give the following.

Proposition 4.1.5   Assume $r \leq n$.
(i) We have $\tilde{F}(\nabla(\lambda)) \cong \Delta(\lambda')$, for $\lambda \in \Lambda^+(n,r)$.
(ii) For $\lambda, \mu \in \Lambda^+(n,r)$ we have $(T(\lambda) : \nabla(\mu)) = [\Delta(\mu') : L(\lambda')]$.
(iii) $\text{Ext}_G^i(\nabla(\lambda), \nabla(\mu)) \cong \text{Ext}_G^i(\Delta(\lambda'), \Delta(\mu'))$, for all $\lambda, \mu \in \Lambda^+(n,r)$ and $i \geq 0$.

Finally we record the following results for use later.

Proposition 4.1.6   For $\lambda \in \Lambda^+(n,r)$ we have $\Delta(\lambda)^\circ \cong \nabla(\lambda)$ and $\nabla(\lambda)^\circ \cong \Delta(\lambda)$.

Proof   Since $\Delta(\lambda)$ has head $L(\lambda)$ the contravariant dual $\Delta(\lambda)^\circ$ has simple socle $L(\lambda)$ and the same character as $\nabla(\lambda)$, and hence $\Delta(\lambda)^\circ \cong \nabla(\lambda)$ (cf. the argument of [**36**; Section 4(1)]).

Proposition 4.1.7   For $\alpha \in \Lambda(n,r)$ we have $(\bigwedge^\alpha E)^\circ \cong \bigwedge^\alpha E$.

Proof   From Proposition 4.1.6 we get that $\alpha \in \Lambda(n,r)$ and $(\bigwedge^\alpha E)^\circ \cong \bigwedge^\alpha E$ are tilting modules with the same character and are hence isomorphic.
     One can also construct a more elementary proof by showing that the image of the natural embedding $\bigwedge^\alpha E \rightarrow E^{\otimes r}$ is orthogonal to the kernel of the natural map $E^{\otimes r} \rightarrow \bigwedge^\alpha E$ and hence the form on $E^{\otimes r}$ induces a non-singular bilinear form $(\,,\,)_\alpha : \bigwedge^\alpha E \times \bigwedge^\alpha E \rightarrow k$ such that $(\xi x, y)_\alpha = (x, J(\xi)y)_\alpha$, for all $x, y \in \bigwedge^\alpha E$ and $\xi \in S(n,r)$.

4.2 Truncation to Levi subgroups

We now consider the effect of truncation to Levi subgroups on modules and decomposition numbers. Let $a = (a_1, \dots, a_m)$ be a composition of n. We associate to a a subset Σ_a of the set $\Pi = \{\alpha_1, \dots, \alpha_{n-1}\}$ of simple roots. We define Σ_a to be the set $\{\alpha_1, \dots, \alpha_{a_1-1}, \alpha_{a_1+1}, \dots, \alpha_{a_1+a_2-1}, \alpha_{a_1+a_2+1}, \dots\}$. The assignment $a \mapsto \Sigma_a$ defines a bijection between the set of compositions of n and the set of subsets of Π. We now fix a subset Σ of Π with corresponding composition $a = (a_1, \ldots, a_m)$. We write P_Σ, G_Σ and B_Σ for the subgroups of G denoted $P(a), G(a)$ and $B(a)$ in [**36**; Section 2]. We shall study truncation $\mathrm{mod}(G) \to \mathrm{mod}(G_\Sigma)$. We write X_Σ^+ for the set denoted $X^+(a)$, in [**36**; Section 2], we write $L_\Sigma(\lambda)$ for the G_Σ-module denoted $L_a(\lambda)$ in [**36**; Section 2], and write $\nabla_\Sigma(\lambda)$ for the induced module $\mathrm{Ind}_{B_\Sigma}^{G_\Sigma} k_\lambda$, for $\lambda \in X_\Sigma^+$. We shall now describe the character of $\nabla_\Sigma(\lambda)$. For $\alpha = \epsilon_a - \epsilon_{a+1} \in \Pi$ we write s_α for the transposition $(a, a+1)$ and W_Σ for the subgroup of $W = \mathrm{Sym}(n)$ generated by s_α, $\alpha \in \Sigma$. For $\lambda \in X$ we write $A_\Sigma(\lambda)$ for $\sum_{w \in W_\Sigma} \mathrm{sgn}(w) e^{w\lambda}$. Then $A_\Sigma(\lambda)$ is divisible by $A_\Sigma(\rho)$ in $\mathbf{Z}X$ and we write $\chi_\Sigma(\lambda)$ for the quotient $A_\Sigma(\lambda + \rho)/A_\Sigma(\rho)$. For $\Sigma = \Pi$ and $\lambda \in X^+$ we have $\chi_\Sigma(\lambda) = \chi(\lambda)$ and by [**36**; Theorem 3.6] we have $\mathrm{ch}\,\nabla(\lambda) = \chi(\lambda)$ (Weyl's character formula). For $\lambda(1) \in X^+(a_1)$, $\lambda(2) \in X^+(a_2)$, $\ldots$, and $\lambda = (\lambda(1), \lambda(2), \ldots)$ (the concatenation) we have $\lambda \in X_\Sigma^+$. Moreover, via the natural isomorphism $G(a) \to G(a_1) \times G(a_2) \times \cdots$ the module $\mathrm{Ind}_{B_\Sigma}^{G_\Sigma} k_\lambda$ identifies with $\mathrm{Ind}_{B(a_1)}^{G(a_1)} k_{\lambda(1)} \otimes \mathrm{Ind}_{B(a_2)}^{G(a_2)} k_{\lambda(2)} \otimes \cdots$. From this (or arguing directly from [**36**; Theorem 3.4] (Kempf's vanishing theorem) as in the classical case [**29**; Section 2.2]) one obtains:

(1)   For any $\lambda \in X_\Sigma$ we have $\mathrm{ch}\,\nabla_\Sigma(\lambda) = \chi_\Sigma(\lambda)$.

For $\lambda, \mu \in X$ we write $\lambda \geq_\Sigma \mu$ to mean that $\lambda - \mu$ has an expression $\sum_{\alpha \in \Sigma} r_\alpha \alpha$ with $r_\alpha \geq 0$, for all $\alpha \in \Sigma$. We write $\lambda >_\Sigma \mu$ to indicate that $\lambda \geq_\Sigma \mu$ and $\lambda \neq \mu$. It follows from (1) that $\lambda >_\Sigma \mu$ for every weight μ of $\nabla_\Sigma(\lambda)$ different from λ, and since $L_\Sigma(\lambda)$ embeds in $\nabla_\Sigma(\lambda)$ we also have $\lambda >_\Sigma \mu$ for every weight μ of $L_\Sigma(\lambda)$ different from λ.

(2)   Let $\lambda, \mu \in X_\Sigma$.
(i) If $\mathrm{Ext}_{P_\Sigma}^1(L_\Sigma(\lambda), \nabla_\Sigma(\mu)) \neq 0$ then $\lambda > \mu$.
(ii) If $\mathrm{Ext}_{G_\Sigma}^1(L_\Sigma(\lambda), L_\Sigma(\mu)) \neq 0$ then $\lambda - \mu \in \mathbf{Z}\Sigma$.

Proof   (i) We have $R^i \mathrm{Ind}_B^{P_\Sigma} k_\mu = 0$ for all $i > 0$, by [**36**; Theorem 3.4]. It follows from [**36**; Proposition 1.1] that $\mathrm{Ext}_B^1(L_\Sigma(\lambda), k_\mu) \neq 0$. Thus we get $\nu > \mu$ for some weight ν of $L_\Sigma(\lambda)$, by [**36**; Lemma 2.8(ii)], and since $\lambda \geq \nu$ we have $\lambda > \mu$.
(ii) We have a short exact sequence of G_Σ-modules

$$0 \to L_\Sigma(\mu) \to \nabla_\Sigma(\mu) \to Q \to 0.$$

Applying $\mathrm{Hom}_{G_\Sigma}(L_\Sigma(\lambda), -)$ we obtain that either $\mathrm{Hom}_{G_\Sigma}(L_\Sigma(\lambda), Q) \neq 0$ or $\mathrm{Ext}_{G_\Sigma}^1(L_\Sigma(\lambda), \nabla_\Sigma(\mu)) \neq 0$. In the first case we have that λ is a weight of $\nabla_\Sigma(\mu)$ and hence $\lambda - \mu \in \mathbf{Z}\Sigma$. So we suppose $\mathrm{Ext}_{G_\Sigma}^1(L_\Sigma(\lambda), \nabla_\Sigma(\mu)) \neq 0$. We have $R^i \mathrm{Ind}_{B_\Sigma}^{G_\Sigma} k_\mu = 0$ for all $i > 0$, by [**36**; Theorem 3.4 and Lemma 2.12]. It follows from [**36**; Proposition 1.1] that $\mathrm{Ext}_{B_\Sigma}^1(L_\Sigma(\lambda), k_\mu) \neq 0$. Thus we get $\nu \geq_\Sigma \mu$, in particular $\nu - \mu \in \mathbb{Z}\Sigma$, for some weight ν of $L_\Sigma(\lambda)$, by [36; Lemma 2.8(ii)], and since $\lambda - \nu \in \mathbb{Z}\Sigma$ we have $\lambda - \mu \in \mathbb{Z}\Sigma$.

We now define submodules of G_Σ-modules and filtrations of P_Σ-modules as in [29; Section 3.3]. We define an $\mathbf{R}$-linear map $f : \mathbf{R} \otimes_{\mathbb{Z}} X \to \mathbf{R}$ by setting $f(\alpha) = 0$ for $\alpha \in \Sigma$ and $f(\alpha) = 1$ for $\alpha \in \Pi \backslash \Sigma$. Let $r \in \mathbf{R}$. For a T-module V we define $F_\Sigma^{(r)}V$ to be the sum of the weight spaces V^μ with $f(\mu) = r$. We define $F^r V$ to be the sum of all weight spaces V^μ with $f(\mu) \leq f(\lambda)$.

(3)   (i) If V is a G_Σ-module then $F^{(r)}V$ is a G_Σ-module summand.
(ii) If V is a P_Σ-module then $F^r V$ is a P_Σ-submodule.

Proof In both (i) and (ii) we may assume that V is finite dimensional. For $\lambda \in X_\Sigma$ the G_Σ-module $L_\Sigma(\lambda)$ embeds in $\nabla_\Sigma(\lambda)$. It follows from (1) that f is constant on the weights of $\nabla_\Sigma(\lambda)$ and hence on the weights of $L_\Sigma(\lambda)$. In proving (i) we may assume that V is indecomposable. But then (2)(ii) gives that f is constant on the weights which occur as composition factors of V and hence, by the above remarks, constant on all weights of V. Thus $F^{(r)}V$ is either 0 or V and so certainly a module summand.

We now turn to (ii). If V is simple we have $V \cong L_\Sigma(\lambda)$ for some $\lambda \in X_\Sigma$. Then we have $F^r V = V$ if $f(\lambda) \leq r$ and $F^r V = 0$ if $f(\lambda) > 0$. Now assume that V is not simple and the result holds for all modules of smaller dimension. Clearly we may (and do) assume that V is indecomposable. Let S be a simple submodule of V. We have $F^r(V/S) = U/S$ for some submodule U of V and we get $F^r V = F^r U$. Thus we are done unless $F^r(V/S) = V/S$. Thus we may assume that $f(\lambda) \leq r$ for every composition factor $L_\Sigma(\lambda)$ of V/S. By indecomposability we thus have $\mathrm{Ext}^1_{P_\Sigma}(L_\Sigma(\lambda), L_\Sigma(\mu)) \neq 0$, for some $\lambda \in X_\Sigma$ with $f(\lambda) \leq r$, where $S \cong L_\Sigma(\mu)$. We have a short exact sequence $0 \to L_\Sigma(\mu) \to \nabla_\Sigma(\mu) \to Q \to 0$. Applying $\mathrm{Hom}_{P_\Sigma}(L_\Sigma(\lambda), -)$ we deduce that either $L_\Sigma(\lambda)$ is a composition factor of $\nabla_\Sigma(\mu)$ or that $\mathrm{Ext}^1_{P_\Sigma}(L_\Sigma(\lambda), \nabla_\Sigma(\mu)) \neq 0$. The first possibility gives $\lambda - \mu \in \mathbb{Z}\Sigma$ and hence $f(\lambda) = f(\mu)$. The second possibility gives $\lambda > \mu$ and hence $f(\lambda) \geq f(\mu)$. Hence we get $S \leq F^r V$ and so $f(\nu) \leq r$ for every weight ν of V, and therefore $F^r V = V$.

Fix an element $\lambda \in X_\Sigma^+$. It follows from (2)(ii) that if V is a finite dimensional indecomposable G_Σ-module which has λ as a weight then for each weight μ of V we have $\lambda - \mu \in \mathbb{Z}\Sigma$. It follows that if V is any rational G_Σ-module then $\bigoplus_{\mu \in X, \lambda - \mu \in \mathbb{Z}\Sigma} V^\mu$ is a G_Σ-submodule. We thus obtain an exact functor $\mathrm{Tr}_\Sigma^\lambda : \mathrm{Mod}(G) \to \mathrm{Mod}(G_\Sigma)$ such that $\mathrm{Tr}_\Sigma^\lambda V = \bigoplus_{\mu \in X, \lambda - \mu \in \mathbb{Z}\Sigma} V^\mu$, for $V \in \mathrm{Mod}(G)$. We call $\mathrm{Tr}_\Sigma^\lambda$ the (λ, Σ)-*truncation functor*.

(4)   For $\mu \in X^+$ we have:
(i) $\mathrm{Tr}_\Sigma^\lambda \nabla(\mu) = \begin{cases} \nabla_\Sigma(\mu), & \text{if } \lambda - \mu \in \mathbb{Z}\Sigma; \\ 0, & \text{otherwise} \end{cases}$

and

(ii) $\mathrm{Tr}_\Sigma^\lambda L(\mu) = \begin{cases} L_\Sigma(\mu), & \text{if } \lambda - \mu \in \mathbf{Z}\Sigma; \\ 0, & \text{otherwise.} \end{cases}$

Proof (i) By the argument of [27; p. 230] we have that the character of $F_\Sigma^\lambda \nabla(\mu)$ is ch $\chi_\Sigma(\mu)$ if $\lambda - \mu \in \mathbf{Z}\Sigma$ and is 0 otherwise. Thus we get $F_\Sigma^\lambda \nabla(\mu) = 0$ if $\lambda - \mu \notin \mathbf{Z}\Sigma$. Suppose that $\lambda - \mu \in \mathbf{Z}\Sigma$. Then by [36; Section 4, (3)(ii)], $\nabla(\mu)$ has a G_Σ-module section isomorphic to $\nabla_\Sigma(\mu)$. It follows that $\nabla_\Sigma(\mu)$ is a section of $F_\Sigma^\lambda \nabla(\mu)$ and since ch $F_\Sigma^\lambda \nabla(\mu) = \chi_\Sigma(\mu)$ we get $F_\Sigma^\lambda \nabla(\mu) \cong \nabla_\Sigma(\mu)$.
(ii) We have an embedding $L(\mu) \to \nabla(\mu)$. Hence $\mathrm{Tr}_\Sigma^\lambda L(\mu)$ embeds in $\mathrm{Tr}_\Sigma^\lambda \nabla(\mu)$ and so $\mathrm{Tr}_\Sigma^\lambda L(\mu) = 0$ if $\lambda - \mu \notin \mathbf{Z}\Sigma$. Now suppose $\lambda - \mu \in \mathbf{Z}\Sigma$. Then $\mathrm{Tr}_\Sigma^\lambda = \mathrm{Tr}_\Sigma^\mu$ so may assume $\lambda = \mu$. Since $\nabla_\Sigma(\lambda) = \mathrm{Tr}_\Sigma^\lambda$ has a simple G_Σ-socle $L_\Sigma(\lambda)$ so has $F_\Sigma^\lambda L(\mu)$. Let $r = f(\lambda)$. Since λ is the unique highest weight of $L(\lambda)$ we have $F^r L(\lambda) = \mathrm{Tr}_\Sigma^\lambda L(\lambda)$ and $L(\lambda) = \bigoplus_{s \leq r} F^{(s)} L(\lambda)$. Thus we have $L(\lambda) = U \oplus \mathrm{Tr}_\Sigma^\lambda L(\lambda)$, where $U = \sum_{s < r} F^s L(\lambda)$. Since $\mathrm{Tr}_\Sigma^\lambda L(\lambda)$ has a simple G_Σ-socle $L_\Sigma(\lambda)$ we have that $L(\lambda)/U$ has a simple P_Σ-socle $L_\Sigma(\lambda)$. Suppose that $L_\Sigma(\nu)$ occurs in the P_Σ-module head of $L(\lambda)/U$. Then we have $\mathrm{Hom}_{P_\Sigma}(L(\lambda), L_\Sigma(\nu)) \neq 0$ and hence $\mathrm{Hom}_{P_\Sigma}(L(\lambda), \nabla_\Sigma(\nu)) \neq 0$. Thus, by Frobenius reciprocity and transitivity of induction, we have

$$\mathrm{Hom}_G(L(\lambda), \mathrm{Ind}_B^G k_\nu) = \mathrm{Hom}_G(L(\lambda), \mathrm{Ind}_{P_\Sigma}^G \mathrm{Ind}_B^{P_\Sigma} k_\nu)$$
$$= \mathrm{Hom}_{P_\Sigma}(L(\lambda), \mathrm{Ind}_B^{P_\Sigma} k_\nu) \neq 0.$$

Thus $\mathrm{Ind}_B^G k_\nu \neq 0$, giving $\nu \in X^+$, and $\mathrm{Hom}_G(L(\lambda), \nabla(\nu)) \neq 0$, giving $\lambda = \nu$. Thus $L(\lambda)/U$ has a simple socle isomorphic to $L_\Sigma(\lambda)$ and moreover $L_\Sigma(\lambda)$ also occurs in the head of $L(\lambda)/U$. Since λ occurs with multiplicity 1 as a weight of $L(\lambda)$ we must have $L(\lambda)/U \cong L_\Sigma(\lambda)$. Hence we have $\mathrm{Tr}_\Sigma^\lambda L(\lambda) = L_\Sigma(\lambda)$.

We write $[\nabla_\Sigma(\lambda) : \nabla_\Sigma(\mu)]$ for the multiplicity of $L_\Sigma(\mu)$ as a G_Σ-module composition factor of $\nabla_\Sigma(\lambda)$. Applying $\mathrm{Tr}_\Sigma^\lambda$ to a composition series of $\nabla(\lambda)$ we deduce the following analogue of [27; Section 2, Theorem].

(5) *We have* $[\nabla(\lambda) : L(\mu)] = [\nabla_\Sigma(\lambda) : L_\Sigma(\mu)]$, *for* $\lambda, \mu \in X^+$ *with* $\lambda - \mu \in \mathbf{Z}\Sigma$.

Now suppose $n = h + 1$ and put $\Sigma = \{\alpha_1, \ldots, \alpha_{h-1}\}$. Let λ, μ be partitions of r into at most h parts. Then we have $\lambda - \mu \in \mathbf{Z}\Sigma$ and hence we have $[\nabla(\lambda) : L(\mu)] = [\nabla_\Sigma(\lambda) : L_\Sigma(\mu)]$. Now we have the natural isomorphism $G(h) \times G(1) \to G_\Sigma = G(h, 1)$. Writing now $\nabla_h(\nu)$ for the induced $G(h)$-module and $L_h(\nu)$ for the simple $G(h)$-module of highest weight $\nu \in X^+(h)$ we have that $\nabla_\Sigma(\lambda)$ identifies with $\nabla_h(\lambda) \otimes k$ and $L_\Sigma(\mu)$ identifies with $L_h(\mu) \otimes k$ (where k denotes the trivial $G(1)$-module). It follows that the multiplicity of $L_\Sigma(\mu)$ as a composition factor of $\nabla_\Sigma(\lambda)$ is equal to the multiplicity of $L_h(\mu)$ as a composition factor of $\nabla_h(\lambda)$. Thus we have the following analogue of [51; (6.6e) Theorem (i)].

(6) Let λ and μ be fixed partitions of r. The composition multiplicity $[\nabla(\lambda) : L(\mu)]$ is the same for all quantum groups $G(n)$ such that λ and μ have at most n parts.

We denote this stable composition multiplicity $[\lambda : \mu]$.

Let $\lambda \in X_\Sigma$ have non-negative entries. We can write λ uniquely as the concatenation $\lambda = (\lambda(1), \ldots, \lambda(m))$, with $\lambda(i) \in X^+(a_i)$, for $1 \leq i \leq m$. We similarly write $\mu = (\mu(1), \ldots, \mu(m)) \in X_\Sigma$, where $\mu(1), \ldots, \mu(m)$ also have non-negative entries. Identifying $\nabla(\lambda)$ with $\nabla_{a_1}(\lambda(1)) \otimes \cdots \otimes \nabla_{a_m}(\lambda(m))$ and $L(\mu)$ with $L_{a_1}(\mu(1)) \otimes \cdots \otimes L_{a_m}(\mu(m))$ we get:

(7) $[\nabla_\Sigma(\lambda) : \nabla_\Sigma(\mu)] = \prod_{i=1}^m [\lambda(i) : \mu(i)]$.

Thus from (5) we have:

(8) If $\lambda - \mu \in \mathbb{Z}\Sigma$ then $[\lambda : \mu] = \prod_{i=1}^m [\lambda(i) : \mu(i)]$.

Suppose that $\lambda = (\lambda_1, \ldots)$ and $\mu = (\mu_1, \ldots)$ are partitions of r. We say that the partitions (λ, μ) admit a horizontal h-cut if we have $\lambda_1 + \cdots + \lambda_h = \mu_1 + \cdots + \mu_h$. Put $\lambda^t(h) = (\lambda_1, \ldots, \lambda_h)$, $\lambda^b(h) = (\lambda_{h+1}, \lambda_{h+2}, \ldots)$ and $\mu^t(h) = (\mu_1, \ldots, \mu_h)$, $\mu^b(h) = (\mu_{h+1}, \mu_{h+2}, \ldots)$ (the top and bottom parts of λ and μ). Taking Σ corresponding to the composition $(h, n - h)$ of n we get:

(9) If λ and μ are a pair of partitions admitting a horizontal h-cut then we have $[\lambda : \mu] = [\lambda^t(h) : \mu^t(h)].[\lambda^b(h) : \mu^b(h)]$.

This is a q-analogue of [28; Theorem 1] (generalizing James's result on row removal for decomposition numbers,[55]). The result was also proved for q a prime power in [57], and the argument given there is valid for arbitrary (non-zero) q.

We now discuss the effect of truncation on tilting modules. We first note that $\mathrm{mod}(G_\Sigma)$ has a tilting theory. For $\lambda \in X^+(n)$ we write $T_n(\lambda)$ for the tilting module for $G(n)$ of highest weight λ if we wish to emphasize the role of n. We write $V \in \mathcal{F}(\nabla_\Sigma)$ to denote that V is a finite dimensional G_Σ-module which has a filtration with sections isomorphic to $\nabla_\Sigma(\lambda)$, for various $\lambda \in X_\Sigma$. We write w_Σ for the longest element of the Weyl group W_Σ. For $\lambda \in X_\Sigma$ we write $\Delta_\Sigma(\lambda)$ for the dual module $\nabla_\Sigma(-w_\Sigma\lambda)^*$. It follows from (1) that $\mathrm{ch}\,\Delta_\Sigma(\lambda) = \mathrm{ch}\,\nabla_\Sigma(\lambda) = \chi_\Sigma(\lambda)$. We write $V \in \mathcal{F}(\Delta_\Sigma)$ to indicate that V is a finite dimensional G_Σ-module which has a filtration with sections $\Delta_\Sigma(\lambda)$, for various $\lambda \in X_\Sigma$. We write $V \in \mathcal{F}(\nabla_\Sigma) \cap \mathcal{F}(\Delta_\Sigma)$ to indicate that $V \in \mathcal{F}(\nabla_\Sigma)$ and $V \in \mathcal{F}(\Delta_\Sigma)$. It follows from the natural isomorphism $G(a_1) \times \cdots \times G(a_m) \to G(a) = G_\Sigma$ and the first paragraph of Section 3.3, that for each $\lambda \in X_\Sigma$ there is a unique (up to isomorphism) indecomposable module $T_\Sigma(\lambda)$ such that $T_\Sigma(\lambda) \in \mathcal{F}(\nabla_\Sigma) \cap \mathcal{F}(\Delta_\Sigma)$ and that $T_\Sigma(\lambda)$ has unique highest weight

λ and this weight occurs with multiplicity 1. We call $T_\Sigma(\lambda)$ the tilting module for G_Σ of highest weight λ. We write $(T_\Sigma(\lambda) : \nabla_\Sigma(\mu))$ for the multiplicity of $\nabla_\Sigma(\mu)$ as a section in a filtration of $T_\Sigma(\lambda)$ with all sections of the form $\nabla_\Sigma(\nu)$ (for various $\nu \in X_\Sigma$). Note that $(T_\Sigma(\lambda) : \nabla_\Sigma(\mu))$ is independent of the choice of such a filtration as it is the coefficient of $\chi_\Sigma(\mu)$ in an expression of $\operatorname{ch} T_\Sigma(\lambda)$ as a linear combination of $\chi_\Sigma(\nu)$'s ($\nu \in X_\Sigma$). It follows from the natural isomorphism $G(a_1) \times \cdots \times G(a_m) \to G(a) = G_\Sigma$ that we may identify $T_\Sigma(\lambda)$ with $T_{a_1}(\lambda(1)) \otimes \cdots \otimes T_{a_m}(\lambda(m))$, where $\lambda = (\lambda(1), \ldots, \lambda(m))$, with $\lambda(1) \in X^+(a_1), \ldots, \lambda(m) \in X^+(a_m)$. Writing $\mu \in X_\Sigma$ also in the form $\mu = (\mu(1), \ldots, \mu(m))$, with $\mu(1) \in X^+(a_1), \ldots, \mu(m) \in X^+(m)$ we obtain:

(10) $\quad (T_\Sigma(\lambda) : \nabla_\Sigma(\mu)) = \prod_{i=1}^m (T_{a_i}(\lambda(i)) : \nabla_{a_i}(\mu(i)))$.

Furthermore, the arguments of [33; (1.5) Proposition and (1.6) Corollary] go through essentially unchanged and we obtain we obtain the following results.

(11) $\quad$ Let $\lambda \in X^+$.
(i) If $U \in \mathcal{F}(\Delta)$, $V \in \mathcal{F}(\nabla)$ and every weight of U and V is $\leq \lambda$ then the map $\operatorname{Hom}_G(U, V) \to \operatorname{Hom}_{G_\Sigma}(\operatorname{Tr}_\Sigma^\lambda U, \operatorname{Tr}_\Sigma^\lambda V)$ is surjective.
(ii) For $\lambda \in X^+$ we have $\operatorname{Tr}_\Sigma^\lambda T(\lambda) \cong T_\Sigma(\lambda)$.

(12) $\quad$ For $\lambda \in X^+$ and $\mu \in X_\Sigma$ we have

$$(T_\Sigma(\lambda) : \nabla_\Sigma(\mu)) = \begin{cases} (T(\lambda) : \nabla(\mu)), & \text{if } \mu \in X^+ \text{ and } \lambda - \mu \in \mathbb{Z}\Sigma; \\ 0, & \text{otherwise.} \end{cases}$$

In particular, if we have that $n = h + 1$ and λ, μ are partitions of h, taking $\Sigma = \{\alpha_1, \ldots, \alpha_{h-1}\}$ and arguing as in the paragraph preceding (6) we get that $(T_n(\lambda) : \nabla_n(\mu)) = (T_h(\lambda) : \nabla_h(\mu))$. Thus we have:

(13) $\quad$ Let λ and μ be fixed partitions of r. The filtration multiplicity $(T(\lambda) : \nabla(\mu))$ is the same for all quantum groups $G(n)$ such that λ and μ have at most n parts.

This gives the following version of Proposition 4.1.5(i) without restrictions on n and r:

(14) $\quad$ For $\lambda, \mu \in \Lambda^+(n, r)$ we have $(T(\lambda) : \nabla(\mu)) = [\mu' : \lambda']$.

Suppose that λ, μ are partitions of r and let $\lambda' = \eta$, $\mu' = \zeta$. We say that the pair (λ, μ) admits a vertical h-cut if the pair (η, ζ) admits a horizontal h-cut. We put $\lambda^l(h) = \eta^t(h)'$, $\lambda^r(h) = \eta^b(h)'$ and $\mu^l(h) = \zeta^t(h)'$, $\mu^r(h) = \zeta^b(h)'$ (the left and right parts of λ and μ); for a pictorial representation see [28;

Section 1]). Now we have $[\lambda : \mu] = (T(\mu') : \nabla(\lambda'))$, by (14), and applying (12) and (10) with $\Sigma = \{\alpha_1, \ldots, \alpha_{h-1}, \alpha_{h+1}, \ldots, \alpha_{n-1}\}$ we get $(T(\mu') : \nabla(\lambda')) = (T_h(\mu^l(h)') : \nabla_h(\lambda^l(h)').(T_{n-h}(\mu^r(h)') : T_{n-h}(\lambda^r(h)')$. Reformulating this equation via (14) we obtain the following.

(15) *If λ and μ are a pair of partitions admitting a vertical h-cut then we have $[\lambda : \mu] = [\lambda^l(h) : \mu^l(h)].[\lambda^r(h) : \mu^r(h))]$.*

This is a q analogue of [28; Theorem 2] (generalizing James's result on column removal for decomposition numbers,[55]). It was proved, in case the case in which q is a prime power, in [57]. It may also be deduced from (9) by dualizing, as in [28].

We now give (in (17) below) a homological analogue of (5). This will be not be used in this book, so we shall be brief with the details. For the case $q = 1$ (for arbitrary reductive groups) see [46; (4.3) Corollary]. For the moment let G be an arbitrary quantum group over k and let $\bar{G}$ be a quotient group (i.e. $k[\bar{G}]$ is a sub Hopf algebra of $k[G]$). For $V \in \text{Mod}(\bar{G})$ we obtain inflation maps $H^i(\bar{G}, V) \to H^i(G, V)$ in the following manner. Choose an injective $\bar{G}$-module resolution $0 \to k \to E_0 \to E_1 \to \cdots$ of the trivial $\bar{G}$-module and an injective resolution $0 \to k \to F_0 \to F_1 \to \cdots$ of the trivial G-module. Then (by injectivity) there are G-module homomorphisms $E_i \to F_i$, $i \geq 0$, such that

$$
\begin{array}{ccccccccc}
0 & \to & k & \to & E_0 & \to & E_1 & \to & \cdots \\
 & & \downarrow & & \downarrow & & \downarrow & & \downarrow \\
0 & \to & k & \to & F_0 & \to & F_1 & \to & \cdots
\end{array}
$$

commutes. Let $V \in \text{Mod}(\bar{G})$. Calculating $H^*(\bar{G}, V)$ as the homology of the complex $0 \to V \otimes E_0 \to V \otimes E_1 \to \cdots$ and $H^*(G, V)$ as the homology of the complex $0 \to V \otimes F_0 \to V \otimes F_1 \to \cdots$ we obtain induced maps $H^i(\bar{G}, V) \to H^i(G, V)$, for $i \geq 0$.

We now identify G_Σ with a quotient of P_Σ via the natural map $P_\Sigma \twoheadrightarrow G_\Sigma$.

(16) *Suppose that all weights of the G_Σ-module V lie in $\mathbb{Z}\Sigma$. Then we have $H^i(G_\Sigma, V) \cong H^i(P_\Sigma, V)$, for all $i \geq 0$.*

The morphisms $H^i(G_\Sigma, V) \to H^i(P_\Sigma, V)$ constructed above commute with direct limits so we may suppose V finite dimensional. Certainly we have $H^0(G_\Sigma, V) = H^0(P_\Sigma, V)$ so the result is true in degree 0. We suppose now that $i > 0$ and the result holds in degree $i - 1$ for all finite dimensional G_Σ-modules (satisfying the hypotheses). Consider first the special case $V = \nabla_\Sigma(\lambda)$, where $\lambda \in X_\Sigma$ and $\lambda \in \mathbb{Z}\Sigma$. We get $H^i(G_\Sigma, \nabla_\Sigma(\lambda)) \cong H^i(B_\Sigma, k_\lambda)$ and $H^i(P_\Sigma, \nabla_\Sigma(\lambda)) \cong H^i(B, k_\lambda)$, by [36; Theorem 3.4] (Kempf's vanishing theorem) and Lemma 2.12. If either of these cohomology spaces is non-zero then we get $\lambda < 0$, by [36; Lemma 2.8]. Since $\lambda \in \mathbb{Z}\Sigma$ we get $\lambda = -\sum_{\alpha \in \Sigma} r_\alpha \alpha$, where $r_\alpha \geq 0$ for $\alpha \in \Sigma$ and $r_\beta \neq 0$ for some $\beta \in \Sigma$. However, this is incompatible with the condition $\lambda \in X_\Sigma$. Thus we have $H^i(G_\Sigma, \nabla_\Sigma(\lambda)) = H^i(P_\Sigma, \nabla_\Sigma(\lambda)) = 0$. Hence we also have $H^i(G_\Sigma, V) = H^i(P_\Sigma, V) = 0$, for every G_Σ-module which is filtered by $\nabla_\Sigma(\lambda)$, $\lambda \in X_\Sigma \cap \mathbb{Z}\Sigma$. Now let V be a finite dimensional G_Σ-module such that all weights of Σ belong to $\mathbb{Z}\Sigma$. We claim that V embeds in a finite dimensional G_Σ-module which has a filtration with sections of the form $\nabla_\Sigma(\lambda)$, $\lambda \in X_\Sigma \cap \mathbb{Z}\Sigma$. If U_1, U_2 are submodules of V and V/U_1 embeds in Z_1 and V/U_2 embeds in Z_2 then V embeds in $Z_1 \oplus Z_2$. We may therefore suppose that V has a simple socle $L_\Sigma(\lambda)$ (with $\lambda \in X_\Sigma \cap \mathbb{Z}\Sigma$). Let $\mu(1), \ldots, \mu(r) \in X_\Sigma$ be the set of maximal weights of V and let π be the set of weights $\nu \in X_\Sigma$ such that $\nu \leq \mu(i)$ for some $1 \leq i \leq r$. We say that a G_Σ-module Y belongs to π if all composition factors of Y come from $\{L_\Sigma(\nu) \mid \nu \in \pi\}$. For an arbitrary G_Σ-module Y we write $O_\pi(Y)$ for the largest submodule belonging to π. We write $I_\Sigma(\lambda)$ for the G_Σ-module injective hull of $L_\Sigma(\lambda)$. Let $Z = O_\pi(I(\lambda))$. Then Z is a finite dimensional submodule of $I_\Sigma(\lambda)$ and has a filtration with sections belonging to $\{\nabla_\Sigma(\nu) \mid \nu \in X_\Sigma\}$. Moreover for each section $\nabla_\Sigma(\nu)$ occurring we have $\lambda \leq_\Sigma \mu(i)$, for some i, which gives $\nu \in X_\Sigma \cap \mathbb{Z}\Sigma$. Now we have an embedding $L_\Sigma(\lambda) \to V$ and this extends (by injectivity) to an embedding $V \to I_\Sigma(\lambda)$ and hence to an embedding $V = O_\pi(V) \to O_\pi(I_\Sigma(\lambda)) = Z$. This proves the claim. Thus, for V a finite dimensional G_Σ-module with weights in $\mathbb{Z}\Sigma$, let Z be a finite dimensional G_Σ-module containing V such that Z has a filtration with sections $\nabla_\Sigma(\nu)$, $\nu \in X_\Sigma \cap \mathbb{Z}\Sigma$. Let $Q = Z/V$. Now we have $H^i(G_\Sigma, Y) = H^i(P_\Sigma, Y)$ for all $i > 0$ by the case already considered. Thus for $i = 1$ we have a commutative diagram

$$
\begin{array}{ccccccccc}
0 & \to & H^0(G_\Sigma, V) & \to & H^0(G_\Sigma, Y) & \to & H^0(G_\Sigma, Q) & \to & H^1(G_\Sigma, V) & \to & 0 \\
& & \downarrow & & \downarrow & & \downarrow & & \downarrow & & \\
0 & \to & H^0(P_\Sigma, V) & \to & H^0(P_\Sigma, Y) & \to & H^0(P_\Sigma, Q) & \to & H^1(P_\Sigma, V) & \to & 0
\end{array}
$$

with rows exact and the first three vertical maps isomorphisms. Hence the map $H^1(G_\Sigma, V) \to H^1(P_\Sigma, V)$ is also an isomorphism. For $i > 1$ we have, by dimension shifting, $H^i(G_\Sigma, V) \cong H^{i-1}(G_\Sigma, Q)$ and $H^i(P_\Sigma, V) \cong H^i(P_\Sigma, Q)$, from which the result follows by induction.

Now the arguments of [46; Section 4] (including the arguments used in the results cited from [29]) go through without change and we get:

(17)   For $\lambda, \mu \in X^+$ with $\lambda - \mu \in \mathbb{Z}\Sigma$ and all $i \geq 0$ we have

$$
\mathrm{Ext}^i_{G_\Sigma}(\nabla_\Sigma(\lambda), \nabla_\Sigma(\mu)) \cong \mathrm{Ext}^i_{P_\Sigma}(\nabla_\Sigma(\lambda), \nabla_\Sigma(\mu)) \cong \mathrm{Ext}^i_G(\nabla(\lambda), \nabla(\mu)).
$$

We conclude this section by giving a description of $S(n, r)'$, for general n, r, as a generalized Schur algebra (cf. [33; (3.11) Proposition]). Suppose that $\pi \subseteq \sigma$ are finite saturated subsets of X^+. Then we have generalized Schur coalgebras $A(\pi) \leq A(\sigma)$ (see [36; Section 4]) and the restriction map Res : $S(\sigma) \to S(\pi)$ between the generalized Schur algebras. We have the (T,T)-weight space decomposition $A(\sigma) = \bigoplus_{\lambda,\mu \in X} {}^\lambda A(n,r)^\mu$. For $\alpha \in W\sigma$ we define $\xi_\alpha \in A(\sigma)^* = S(\sigma)$ to be the projection onto ${}^\alpha A(\sigma)^\alpha$ followed by the evaluation map. (Note that this coincides with our use of the symbol ξ_α in Chapter 2, in the case $\sigma = \Lambda^+(n,r)$, i.e. $S(\sigma) = S(n,r)$.) Then $1 = \sum_{\alpha \in W\sigma} \xi_\alpha$ is decomposition of 1 as a sum of mutually orthogonal idempotents. By the argument of [33; (3.3) Lemma] we get the q-analogue of that result:

(18)   Res : $S(\sigma) \to S(\pi)$ induces an isomorphism $S/SeS \to S(\pi)$, where $S = S(\sigma)$ and $e = \sum_{\alpha \in W\sigma \backslash W\pi} \xi_\alpha$.

Now let n,r be arbitrary and choose $N \geq r$ and n. Let E denote the the natural left $M(N)$-module and let V denote the natural right $M(N)$-module. Let $T = \bigwedge^r(E \otimes V)$ and $S(N,r)' = \operatorname{End}_{G(N)}(T)$. Let $\Sigma = \{\alpha_1,\ldots,\alpha_{m-1}\}$, $\lambda = N\epsilon_1$ and $\operatorname{Tr}_\Sigma^\lambda$. The natural map $S(N,r)' \to \operatorname{End}_{G_\Sigma}(\operatorname{Tr}_\Sigma^\lambda T)$ is surjective, by (11)(i). Moreover we have (as in [33; p. 57]) $\operatorname{End}_{G_\Sigma}(\operatorname{Tr}_\Sigma^\lambda T) = \operatorname{End}_{G(n)}(\operatorname{Tr}_\Sigma^\lambda T)$ and T is a full tilting module for $G = G(n)$. Thus, we obtain a surjective algebra map $\phi : S(N,r) \to S(N,r)' \to S(n,r)' = \operatorname{End}_G(\operatorname{Tr}_\Sigma^\lambda T)$, where the first map is the map θ of Proposition 4.1.4. Arguing as in the proof of [33; (3.7) Proposition], we obtain that the kernel of ϕ is exactly $S(N,r)eS(N,r)$, where $e = \sum_{\alpha \in \Lambda(N,r) \backslash \operatorname{Sym}(N)\Lambda^+(n,r)'} \xi_\alpha$ (and $\Lambda^+(n,r)' = \{\alpha' \mid \alpha \in \Lambda^+(n,r)\}$). Hence, by (17), we obtain the following general form of Proposition 4.1.4.

(19)   Let n,r be arbitrary and $N \geq n$ and r. Then the Ringel dual $S(n,r)'$ is naturally isomorphic to the generalized Schur algebra $S(\pi)$, determined by the quantum group $G(N)$ and the saturated subset $\pi = \Lambda^+(n,r)'$ of $\Lambda^+(N,r)$.

Remark   Using the cell structure of the Hecke algebra, Du, Parshall and Scott have recently also obtained a similar result, and also a proof of (14) above in the case $n = r$, see [44; (7.9) Theorem and (8.3) Proposition].

4.3 Components of $E^{\otimes r}$

We now wish to consider some connections between representations of the Schur algebra and the Hecke algebra. It is convenient to first consider which modules occur as components of $E^{\otimes r}$. Throughout this section we assume $r \leq n$. Let $\omega = (1,1,\ldots,1,0,\ldots,0) \in \Lambda(n,r)$.. Recall the definition of $\Lambda^+(n,r)_{\text{row}}$ and $\Lambda^+(n,r)_{\text{col}}$ from 0.18.

(1)   Let $\lambda \in \Lambda^+(n,r)$. If $L(\lambda)^\omega \neq 0$ then $\lambda \in \Lambda^+(n,r)_{\text{col}}$.

Proof If q is not a root of unity or k has characteristic 0 and $q = 1$ then there is nothing to prove.

We now assume that q is a primitive lth root of unity and suppose first that $l > 1$. We write $\lambda = \mu + l\nu$, with $\mu \in X_1$ and $\nu \in X^+$. By Steinberg's tensor product theorem we have $L(\lambda) \cong L(\mu) \otimes \bar{L}(\nu)^F$, where $F : G \to \bar{G}$ is the (quantum) Frobenius map. Then $\mu \in \Lambda^+(n, s)$ and $\nu \in \Lambda^+(n, t)$ for some s, t with $r = s + lt$. The weights of $L(\lambda)$ thus have the form $\alpha + l\beta$, where α is a weight of $L(\mu)$ and β is a weight of $\bar{L}(\nu)$. If $\omega = \alpha + l\beta$ we deduce that $\beta = 0$ and $\alpha = \omega$. Thus $t = 0$ and $\nu = 0$. Hence $\lambda \in \Lambda^+(n, r) \cap X_1 = \Lambda^+(n, r)_{\mathrm{col}}$. Similarly, if $l = 1$ and k has characteristic $p > 0$, writing $\lambda = \mu + p\nu$ and using the (ordinary) Steinberg tensor product theorem we get $\lambda \in \Lambda^+(n, r)_{\mathrm{col}}$.

For $\lambda \in \Lambda^+(n, r)$ we denote by $I(\lambda)$ the injective envelope of $L(\lambda)$ and by $P(\lambda)$ the projective cover of $L(\lambda)$, as $S(n, r)$-modules. Recall that for $V \in \mathrm{mod}\, S(n, r)$, we have the contravariant dual V°, defined in Section 4.1.

(2)   (i) *For* $X, Y \in \mathrm{mod}\, S(n, r)$ *we have*

$$\mathrm{Hom}_{S(n,r)}(X, Y) \cong \mathrm{Hom}_{S(n,r)}(Y^\circ, X^\circ).$$

(ii) *We have* $P(\lambda)^\circ \cong I(\lambda)$ *and* $I(\lambda)^\circ \cong P(\lambda)$, *for each* $\lambda \in \Lambda^+(n, r)$.
(iii) *We have* $(E^{\otimes r})^\circ \cong E^{\otimes r}$.

Proof (i) This is true because $J : S(n, r) \to S(n, r)$ is an antiautomorphism.
(ii) It follows from (i) and the fact that $J : S(n, r) \to S(n, r)$ is an involution that $P \in \mathrm{mod}\, S(n, r)$ is projective if and only if P° is injective. Now $L(\lambda)^\circ$ is a simple $S(n, r)$-module and $\mathrm{ch}\, L(\lambda)^\circ = \mathrm{ch}\, L(\lambda)$ (see Remark (ii) of Section 4.1). Hence we have $L(\lambda)^\circ \cong L(\lambda)$. Now $P(\lambda)$ is a projective module with socle $L(\lambda)$ and hence $P(\lambda)^\circ$ is an injective module with socle $L(\lambda)^\circ \cong L(\lambda)$ and so we have $P(\lambda)^\circ \cong I(\lambda)$.
(iii) See Remark (ii) of Section 4.1.

To make further progress we need the following, which is a well known property of symmetric functions.

(3)   *We have* $(\bigwedge^\alpha E : \nabla(\lambda')) = (S^\alpha E : \nabla(\lambda))$ *for all* $\alpha \in \Lambda(n, r)$ *and* $\lambda \in \Lambda^+(n, r)$.

Proof Since $\mathrm{ch}\, S^\alpha E = \mathrm{ch}\, S^\beta E$ and $\mathrm{ch} \bigwedge^\alpha E = \mathrm{ch} \bigwedge^\beta E$ if $\beta \in \Lambda(n, r)$ is obtained by rearranging the parts of α, we may assume that $\alpha \in \Lambda^+(n, r)$. Let $e_1, e_2, \dots$ denote the elementary symmetric functions and $h_1, h_2, \dots$ the complete symmetric functions in infinitely many variables $x_1, x_2, \dots$, as in [63; I]. For any partition $\lambda = (\lambda_1, \lambda_2, \dots)$ we have symmetric functions $e_\lambda = e_{\lambda_1} e_{\lambda_2} \cdots$, $h_\lambda = h_{\lambda_1} h_{\lambda_2} \cdots$ and the Schur function s_λ (see [63; I, Sections

2 and 3]). We identify the character of an $S(n,r)$-module as a symmetric function in the n variables $x_1 = e(\epsilon_1), \ldots, x_n = e(\epsilon_n)$. Restricting the Schur function s_λ to n variables we obtain the character $\chi(\lambda)$, for $\lambda \in \Lambda^+(n,r)$ (see [63; I, (3.1)]). We write e_α and h_α as linear combinations of Schur functions:

$$e_\alpha = \sum_{\lambda \in \Lambda^+(n,r)} a_\lambda s_\lambda \quad \text{and} \quad h_\alpha = \sum_{\lambda \in \Lambda^+(n,r)} b_\lambda s_\lambda.$$

Now there is the ring automorphism $\bar{\omega}$ (denoted ω in [63]) on the ring of symmetric functions, defined by $\bar{\omega}(e_r) = h_r$, for $r \geq 1$. Furthermore we have $\bar{\omega}(s_\lambda) = s_{\lambda'}$, for all partitions λ, by [63; (5.6)]. Applying $\bar{\omega}$ to the above equations we get $b_\lambda = a_{\lambda'}$, for all $\lambda \in \Lambda^+(n,r)$. Restricting e_α to n variables gives $\text{ch}\bigwedge^\alpha E$ and restricting h_α gives $\text{ch}\, S^\alpha E$ so we get $\text{ch}\bigwedge^\alpha E = \sum_{\lambda \in \Lambda^+(n,r)} a_\lambda \chi(\lambda)$ and $\text{ch}\, S^\alpha E = \sum_{\lambda \in \Lambda^+(n,r)} a_{\lambda'} \chi(\lambda)$. Thus we get $(\bigwedge^\alpha E : \nabla(\lambda')) = (S^\alpha E : \nabla(\lambda))$, for all $\lambda \in \Lambda^+(n,r)$, as required.

For modules Y, Z of finite length with Z indecomposable we write $(Y \mid Z)$ for the number of components of Y isomorphic to Z in a decomposition of Z as a direct sum of indecomposable modules.

(4)   Let $\lambda \in \Lambda^+(n,r)$.
(i) We have $(S^\alpha E \mid I(\lambda)) = (\bigwedge^\alpha E \mid T(\lambda'))$ for all $\alpha \in \Lambda(n,r)$. In particular $I(\lambda)$ occurs as a component of $E^{\otimes r}$ if and only if $T(\lambda')$ occurs as a component of $E^{\otimes r}$.
(ii) The following are equivalent:
(a) $I(\lambda)$ is projective;
(b) $I(\lambda)$ occurs as a component of $E^{\otimes r}$;
(c) $I(\lambda)$ is a tilting module;
(d) $L(\lambda)^\omega \neq 0$.

Proof (i) As in the proof of (3) we may assume that α is dominant. For $\alpha, \mu \in \Lambda^+(n,r)$ we have

$$\sum_{\nu \in \Lambda^+(n,r)} (I(\nu) : \nabla(\mu))(S^\alpha E \mid I(\nu)) = \sum_{\nu \in \Lambda^+(n,r)} (T(\nu') : \nabla(\mu'))(\bigwedge^\alpha E \mid T(\nu')),$$

i.e.

$$\sum_{\nu \in \Lambda^+(n,r)} [\mu : \nu](S^\alpha E \mid I(\nu)) = \sum_{\nu \in \Lambda^+(n,r)} [\mu : \nu](\bigwedge^\alpha E \mid T(\nu')),$$

by Proposition 4.1.5 and [36; Section 4, (6)]. The matrix $([\mu : \nu])_{\mu,\nu \in \Lambda^+(n,r)}$ is invertible (in fact unitriangular) so we get $(I(\mu) \mid S^\alpha E) = (\bigwedge^\alpha E \mid T(\mu'))$, for all $\mu \in \Lambda^+(n,r)$.
(ii) Suppose that $I(\lambda)$ is projective. We have $\text{Hom}_G(L(\lambda), S^\lambda E) \cong L(\lambda)^\lambda \neq 0$ and hence $L(\lambda)$ embeds in $S^\lambda E$. Thus $I(\lambda)$ is a component of $S^\lambda E$ so

there is an epimorphism $S^\lambda E \to I(\lambda)$. Composing with the natural map $E^{\otimes r} \to S^\lambda E$ we get an epimorphism $E^{\otimes r} \to I(\lambda)$. By projectivity this map splits and so $I(\lambda)$ occurs as a component of $E^{\otimes r}$. If $I(\lambda)$ occurs as a component of $E^{\otimes r}$ then it is tilting module, by 3.3(1). If $I(\lambda)$ is tilting module then $I(\lambda)$ embeds in $\bigwedge^\alpha E$, for some $\alpha \in \Lambda^+(n,r)$, by 3.3(1). Tensoring together natural embeddings $\bigwedge^a E \to E^{\otimes a}$, [36; Lemma 3.3(i)], (where a runs over the components of α) we obtain an embedding $\bigwedge^\alpha E \to E^{\otimes r}$ and hence an embedding $I(\lambda) \to E^{\otimes r}$. By injectivity this splits, so $I(\lambda)$ occurs as a component of $E^{\otimes r}$. If $I(\lambda)$ occurs as a component of $E^{\otimes r}$ then it is projective, by 2.1(7). Thus (a), (b) and (c) are equivalent. Now $I(\lambda)$ occurs as a component of $E^{\otimes r}$ if and only if $\mathrm{Hom}_G(L(\lambda), E^{\otimes r}) \neq 0$. By 2.1(8) this holds if and only if $L(\lambda)^\omega \neq 0$ and so (b) and (d) are equivalent.

We now describe explicitly the set of $\lambda \in \Lambda^+(n,r)$ which satisfy (4)(ii). We shall need a couple of preliminary remarks.

(5)   (i) Let H be a quantum group over k. Let g be a group-like element of $k[H]$ and let L_g be the corresponding 1-dimensional H-module. (Thus the structure map takes $x \in L_g$ to $x \otimes g \in L_g \otimes k[H]$.) Suppose further that $k[H]$ has a k-basis f_i, $i \in I$, such that $g f_i g^{-1}$ is a scalar multiple of f_i, for each $i \in I$. If $U, V \in \mathrm{Mod}\,H$, if $\phi : U \otimes V \to V \otimes U$ is an H-module monomorphism and if L is a submodule of U isomorphic to L_g then $\{v \in V \mid \phi(L \otimes v) \le v \otimes L\}$ is an H-submodule of V.
(ii) For $H = B$ or B^+ every group-like element of $k[H]$ has the property described in (i).

Proof   (i) Let $V_0 = \{v \in V \mid \phi(L \otimes v) \le v \otimes L\}$. We have $g f_i g^{-1} = \lambda_i f_i$, for scalars λ_i, $i \in I$. We have $L = k u_0$, for some $u_0 \in U$ satisfying $\tau_U(u_0) = u_0 \otimes g$. Now suppose that $v \in V_0$, $\phi(u_0 \otimes v) = \alpha(v \otimes u_0)$ (with $\alpha \in k$) and that $\tau_V(v) = \sum_i v_i \otimes f_i$. We have

$$\tau_{V \otimes U}\phi(u_0 \otimes v) = \alpha \sum_i v_i \otimes u_0 \otimes f_i g$$

and

$$(\phi \otimes \mathrm{id})\tau_{U \otimes V}(u_0 \otimes v) = \alpha \sum_i \phi(u_0 \otimes v_i) \otimes g f_i = \alpha \sum_i \lambda_i \phi(u_0 \otimes v_i) \otimes f_i g.$$

Since ϕ is a module homomorphism these two expressions are equal and since $f_i g$, $i \in I$, is a basis of $k[H]$ we have $\phi(u_0 \otimes v_i) = \lambda_i(v_i \otimes u_0)$, for all $i \in I$. Thus all $v_i \in V_0$ and V_0 is an H-submodule of V.
(ii) We consider $H = B$ and leave B^+ to the reader. By abuse of notation we write simply c_{ij} for the restriction of $c_{ij} \in k[G]$ to B. The modules k_λ, $\lambda \in X$, form a complete set of 1-dimensional B-modules and it follows

that the elements $c^\lambda = c_{11}^{\lambda_1} \dots c_{nn}^{\lambda_n}$, with $\lambda = (\lambda_1, \dots, \lambda_n) \in X$, are exactly the group-like elements of $k[B]$ (see the proof of [36; Lemma 2.6]). Let $1 \leq i \leq n$. It follows from the defining relations, given in Section 1.2, that $c_{ii}c_{rs} \in k[B]$ is a scalar multiple of $c_{rs}c_{ii} \in k[B]$. Thus we have that $c_{ii}fc_{ii}^{-1}$ is a scalar multiple of f, and therefore that $c^\lambda f c^{-\lambda}$ is a scalar multiple of f, for any monomial f in the elements c_{rs} ($1 \leq s \leq r \leq n$) and any $\lambda \in X$. Since $k[B]$ has a basis consisting of monomials in the c_{rs}, [36; Corollary 2.4], we get that each c^λ (with $\lambda \in X$) has the property described in (i).

Now let $a, b \geq 1$ with $a + b \leq n$. Recall, we have the isomorphism $\phi : \bigwedge^a E \otimes \bigwedge^b E \to \bigwedge^b E \otimes \bigwedge^a E$ of Lemma 1.3.3. To save on notation, for the rest of this section we shall indicate multiplication in the exterior algebra on E simply by juxtaposition.

(6) $\phi(e_1 \dots e_a \otimes v)$ is a scalar multiple of $v \otimes e_1 \dots e_a$, for all $v \in \bigwedge^b E$.

Proof This holds for $v = \hat{e}_j$ with $j = (n - b + 1, \dots, n)$ by Lemma 1.3.3(i). Moreover, $u_0 = e_1 \dots e_a$ spans a 1-dimensional B^+-submodule of $\bigwedge^a E$. Thus, by (5), the set V_0 of elements v of $\bigwedge^b E$ such that $\phi(u_0 \otimes v)$ is a multiple of $v \otimes u_0$ is a B^+-submodule of $\bigwedge^b E$. However, it is easy to check that $\hat{e}_j$ generates $\bigwedge^b E$, as a B^+-module, so that $V_0 = \bigwedge^b E$, as required.

We shall need the fact that the natural map $E^{\otimes r} \to \bigwedge^r E$ splits when r is "small". It is convenient to record at this point the following list of semisimple q-Schur algebras. Here r is arbitrary once more. The result for $q = 1$ is given in [42].

(7) $S(n, r)$ *is semisimple if and only if either:*
(i) *q is not a root of unity or k has characteristic 0 and $q = 1$; or*
(ii) *q is a primitive lth root of unity with $l > 1$ and $r < l$; or*
(iii) *k has characteristic $p > 0$, $q = 1$ and $r < p$; or*
(iv) *$n = 2$, $l = 2$ and $r = 3$; or*
(v) *$n = 2$, $l = 1$, $p = 2$ and $r = 3$; or*
(vi) *$n = 1$ and l, p are arbitrary.*
In particular, in these cases, the natural map $E^{\otimes r} \to \bigwedge^\alpha E$ splits, for $\alpha \in \Lambda(n, r)$.

Proof If (i),(ii) or (iii) then $S(n, r)$ is semisimple by the argument of [36; Section 4, (7)]. Also, in cases (iv),(v) one gets $L(\lambda) = \Delta(\lambda) = \nabla(\lambda)$, for $\lambda = (3, 0), (2, 1)$, using the Steinberg tensor product theorem, and again one gets that $S(n, r)$ is semisimple by the argument of [36; Section 4, (7)]. If $n = 1$ then $S(n, r)$ is 1-dimensional, and hence semisimple.

Now suppose that $S(n, r)$ is semisimple. Then we have $L(\lambda) = \nabla(\lambda)$ for all $\lambda \in \Lambda^+(n, r)$. Suppose that $n \geq 3$. Suppose $l > 1$ and $r \geq l$. We write $r = r_{-1} + ls$, with $0 \le r_{-1} < l$. By the Steinberg tensor product theorem we have $S^r E = L(r\epsilon_1) \cong L(r_{-1}\epsilon_1) \otimes \bar{L}(s\epsilon_1)^F$. Moreover $L(r_{-1}\epsilon_1)$ is a submodule of $S^{r-1}E$ so that every weight of $S^r E$ is a weight of $S^{r-1}E \otimes \bar{L}(s\epsilon_1)^F$. In particular $(ls - 1, r_{-1}, 1, 0, \ldots, 0)$ may be expressed in the form $\alpha + l\beta$, with $\alpha, \beta \in X$ and α a weight of $S^{r-1}E$. But then we have $|\alpha| > r_{-1}$, a contradiction. Hence $n \le 2$. If $l = 1$ and $r \ge p$ we get $n \le 2$ by the same argument but using the ordinary Steinberg tensor product theorem. It remains to consider the case $n = 2$. Suppose $l > 1$ and $r \ge l$. We write $r = r_{-1} + ls$ as above. Again $S^r E$ embeds in $S^{r-1}E \otimes L(s\epsilon_1)^F$ and the weight $(ls - 1, r_{-1} + 1)$ has the from $\alpha + l\beta$, with $|\alpha| = r_{-1}$. This can only happen if $r_{-1} = l - 1$, i.e. $r \equiv -1 \pmod{l}$. But applying the same argument to $L(r - 1, 1) = \nabla(r - 1, 1)$ we get that $r - 2 \equiv -1 \pmod{l}$. Hence $2 \equiv 0$ mod l so $l = 2$ and $r = 1 + 2s$ is odd. Suppose that r is at least 5. If $s = 2m$ is even we get

$$S^r E = L(r\epsilon_1) \cong L(\epsilon_1) \otimes \bar{L}(s\epsilon_1)^F \cong L(\epsilon_1) \otimes (\bar{L}(m\epsilon_1)^{\bar{F}})^F$$

and comparing dimensions gives $r + 1 \le 2(m + 1)$, i.e. $1 + 4m \le 2m + 2$, a contradiction. Hence s is even. Applying this argument to $L(r - 1, 1)$ we get that $s - 1$ is also even, which is impossible. Hence r is at least 2, is odd and is less than 5. Thus $r = 3$. Similarly, if $l = 1$, k has characteristic $p > 0$ and $r \ge p$ we get $p = 2$ and $r = 3$ by the same argument, using the ordinary Steinberg tensor product theorem.

We now embark on the converse of (1).

(8)   For $a_1, \ldots, a_m, b \ge 0$ with $a_1 + \cdots + a_m + b \le n$ there is a G-module isomorphism $\psi : \bigwedge^{a_1} E \otimes \cdots \otimes \bigwedge^{a_m} E \otimes \bigwedge^b E \to \bigwedge^b E \otimes \bigwedge^{a_1} E \otimes \cdots \otimes \bigwedge^{a_m} E$ such that $\psi(x_1 \otimes \cdots \otimes x_m \otimes e_1 e_2 \ldots e_b)$ is a scalar multiple of $e_1 e_2 \ldots e_b \otimes x_1 \otimes \cdots \otimes x_m$, for all $x_i \in \bigwedge^{a_i} E$, $1 \le i \le m$.

Proof   In the case $m = 1$ we can take the inverse of the map ϕ of (6). Now assume $m > 1$. Let $y = e_1 e_2 \ldots e_b$. We have an isomorphism $\psi_1 : \bigwedge^{a_m} E \otimes \bigwedge^b E \to \bigwedge^b E \otimes \bigwedge^{a_m} E$ taking $x \otimes y$ to a multiple of $y \otimes x$, for $x \in \bigwedge^{a_m} E$. We put $\psi_1' = (\mathrm{id} \otimes \psi_1) : \bigwedge^{a_1} E \otimes \cdots \otimes \bigwedge^{a_{m-1}} E \otimes \bigwedge^{a_m} E \otimes \bigwedge^b E \to \bigwedge^{a_1} E \otimes \cdots \otimes \bigwedge^{a_{m-1}} E \otimes \bigwedge^b E \otimes \bigwedge^{a_m} E$. We can assume inductively that there is an isomorphism $\psi_2 : \bigwedge^{a_1} E \otimes \cdots \otimes \bigwedge^{a_{m-1}} E \otimes \bigwedge^b E \to \bigwedge^b E \otimes \bigwedge^{a_1} E \otimes \cdots \otimes \bigwedge^{a_{m-1}} E$ such that $\psi_2(x_1 \otimes \cdots \otimes x_{m-1} \otimes y)$ is a multiple of $y \otimes x_1 \otimes \cdots \otimes x_{m-1}$, for $x_i \in \bigwedge^{a_i} E$, $1 \le i \le m - 1$. Putting $\psi_2' = \psi_2 \otimes \mathrm{id} : \bigwedge^{a_1} E \otimes \cdots \otimes \bigwedge^{a_{m-1}} E \otimes \bigwedge^b E \otimes \bigwedge^{a_m} E \to \bigwedge^b E \otimes \bigwedge^{a_1} E \otimes \cdots \otimes \bigwedge^{a_{m-1}} E \otimes \bigwedge^{a_m} E$ we have that the map $\psi = \psi_2' \circ \psi_1' : \bigwedge^{a_1} E \otimes \cdots \otimes \bigwedge^{a_m} E \otimes \bigwedge^b E \to \bigwedge^b E \otimes \bigwedge^{a_1} E \otimes \cdots \otimes \bigwedge^{a_m} E$ has the required property.

We are now ready to prove the main result of this section.

(9)   For $\lambda \in \Lambda^+(n,r)$ we have $L(\lambda)^\omega \neq 0$ if and only if $\lambda \in \Lambda^+(n,r)_{\text{col}}$.

Proof   If $L(\lambda)^\omega \neq 0$ then $\lambda \in \Lambda^+(n,r)_{\text{col}}$, by (1).

The reverse is equivalent, by (4), to the statement that $T(\lambda)$ is a component of $E^{\otimes r}$ for $\lambda \in \Lambda^+(n,r)_{\text{row}}$, and this is what we shall prove. So let $\lambda \in \Lambda^+(n,r)_{\text{row}}$. Let $\lambda' = \alpha = (\alpha_1, \ldots, \alpha_n)$. Then $T(\lambda)$ is isomorphic to an indecomposable component of $\bigwedge^\alpha E$ containing the highest weight vector. If $S(n,r)$ is semisimple then we get that $T(\lambda)$ is a component of $E^{\otimes r}$ by (7). Thus we can, and do, assume either $l > 1$ or $l = 1$ and k has characteristic $p > 0$.

We shall produce a G-endomorphism θ of $\bigwedge^\alpha E$ which is non-zero on the highest weight space $(\bigwedge^\alpha E)^\lambda$ and factors through $E^{\otimes r}$. For the rest of this proof we shall write simply $\bigwedge^b$ for $\bigwedge^b E$, $b \geq 0$, and $\bigwedge^\beta$ for $\bigwedge^\beta E$, for $\beta = (\beta_1, \ldots, \beta_h)$, $\beta_1, \ldots, \beta_h \geq 0$. Let $\bar\alpha = (\alpha_2, \ldots, \alpha_m)$ and let $s = r - \alpha_1$. We can assume, inductively, that there is an endomorphism $\phi : \bigwedge^{\bar\alpha} \to \bigwedge^{\bar\alpha}$ which is non-zero on the highest weight space and which factors through $E^{\otimes s}$. For a subset X of $[1,n]$ we write $\hat{e}_X$ for the element $e_{x_1} \ldots e_{x_t}$ of $\bigwedge^t$, where $X = \{x_1, \ldots, x_t\}$ with $x_1 < \cdots < x_t$. Suppose that $a_1, \ldots, a_t \geq 0$ and $a = a_1 + \cdots + a_t$. By iterating the map of Lemma 1.2.3 we get an injective G-module homomorphism $\chi : \bigwedge^a \to \bigwedge^{a_1} \otimes \bigwedge^{a_2} \otimes \cdots \otimes \bigwedge^{a_t}$ given by $\chi(e_{i_1} \ldots e_{i_a}) = \sum_{A \in \mathcal{A}} c_A \hat{e}_{A_1} \otimes \cdots \otimes \hat{e}_{A_t}$, where $\mathcal{A}$ is the set of sequences $A = (A_1, \ldots, A_t)$ of subsets of $[1,n]$ such that $\{i_1, \ldots, i_a\}$ is the disjoint union of $A_1, \ldots, A_t$ and where each c_A is ± 1. Taking $a_1 = \alpha_1 - \alpha_2$, $a_2 = \alpha_2 - \alpha_3, \ldots,$ $a_{m-1} = \alpha_{m-1} - \alpha_m$, $a_m = \alpha_m$, we get a G-module homomorphism $\psi = \chi \otimes \phi :$ $\bigwedge^\alpha \to \bigwedge^{\alpha_1 - \alpha_2} \otimes \cdots \otimes \bigwedge^{\alpha_{m-1} - \alpha_m} \otimes \bigwedge^{\alpha_m} \otimes \bigwedge^{\alpha_2} \otimes \bigwedge^{\alpha_3} \otimes \cdots \otimes \bigwedge^{\alpha_m}$.

We put $y_i = e_1 \ldots e_{\alpha_i} \in \bigwedge^{\alpha_i}$, for $1 \leq i \leq m$. Now by (8) we have a G-module isomorphism $f : \bigwedge^{\alpha_2 - \alpha_3} \otimes \cdots \otimes \bigwedge^{\alpha_m} \otimes \bigwedge^{\alpha_2} \to \bigwedge^{\alpha_2} \otimes \bigwedge^{\alpha_2 - \alpha_3} \otimes \cdots \otimes \bigwedge^{\alpha_m}$ taking $x_2 \otimes \cdots \otimes x_m \otimes y_2$ to a multiple of $y_2 \otimes x_2 \otimes \cdots \otimes x_m$, for $x_i \in \bigwedge^{\alpha_i - \alpha_{i+1}}$, $1 \leq i < m$, and $x_m \in \bigwedge^{\alpha_m}$. Hence we have the isomorphism $g_1 = (\text{id} \otimes f \otimes \text{id}) : \bigwedge^{\alpha_1 - \alpha_2} \otimes \bigwedge^{\alpha_2 - \alpha_3} \otimes \cdots \otimes \bigwedge^{\alpha_m} \otimes \bigwedge^{\alpha_2} \otimes \bigwedge^{\alpha_3} \otimes \cdots \otimes \bigwedge^{\alpha_m} \to$ $\bigwedge^{\alpha_1 - \alpha_2} \otimes \bigwedge^{\alpha_2} \otimes \bigwedge^{\alpha_2 - \alpha_3} \otimes \cdots \otimes \bigwedge^{\alpha_m} \otimes \bigwedge^{\alpha_3} \otimes \cdots \otimes \bigwedge^{\alpha_m}$ taking $x_1 \otimes x_2 \otimes \cdots \otimes x_m \otimes y_2 \otimes y_3 \otimes \cdots \otimes y_m$ to a multiple of $x_1 \otimes y_2 \otimes x_2 \otimes \cdots \otimes x_m \otimes y_3 \otimes \cdots \otimes y_m$, for $x_i \in \bigwedge^{\alpha_i - \alpha_{i+1}}$, for $1 \leq i < m$ and $x_m \in \bigwedge^{\alpha_m}$.

Similarly, we have an isomorphism $g_2 : \bigwedge^{\alpha_1 - \alpha_2} \otimes \bigwedge^{\alpha_2} \otimes \bigwedge^{\alpha_2 - \alpha_3} \otimes$ $\bigwedge^{\alpha_3 - \alpha_4} \otimes \cdots \otimes \bigwedge^{\alpha_m} \otimes \bigwedge^{\alpha_3} \otimes \cdots \otimes \bigwedge^{\alpha_m} \to \bigwedge^{\alpha_1 - \alpha_2} \otimes \bigwedge^{\alpha_2} \otimes \bigwedge^{\alpha_2 - \alpha_3} \otimes \bigwedge^{\alpha_3} \otimes$ $\bigwedge^{\alpha_3 - \alpha_4} \otimes \cdots \otimes \bigwedge^{\alpha_m} \otimes \bigwedge^{\alpha_3} \otimes \cdots \otimes \bigwedge^{\alpha_m}$ taking $x_1 \otimes y_2 \otimes x_2 \otimes x_3 \otimes \cdots \otimes x_m \otimes$ $y_3 \otimes \cdots \otimes y_m$ to a multiple of $x_1 \otimes y_2 \otimes x_2 \otimes y_3 \otimes x_3 \otimes \cdots \otimes x_m \otimes y_3 \otimes \cdots \otimes y_m$, for $x_i \in \bigwedge^{\alpha_i - \alpha_{i+1}}$, for $1 \leq i < m$ and $x_m \in \bigwedge^{\alpha_m}$.

Continuing in this way we obtain isomorphisms $g_3, g_4, \ldots$ and the composite $g = g_{m-1} \circ \cdots \circ g_1 : \bigwedge^{\alpha_1 - \alpha_2} \otimes \bigwedge^{\alpha_2 - \alpha_3} \otimes \cdots \otimes \bigwedge^{\alpha_m} \otimes \bigwedge^{\alpha_2} \otimes \bigwedge^{\alpha_3} \otimes \cdots \otimes$ $\bigwedge^{\alpha_m} \to \bigwedge^{\alpha_1 - \alpha_2} \otimes \bigwedge^{\alpha_2} \otimes \bigwedge^{\alpha_2 - \alpha_3} \otimes \bigwedge^{\alpha_3} \otimes \cdots \otimes \bigwedge^{\alpha_{m-1} - \alpha_m} \otimes \bigwedge^{\alpha_m} \otimes \bigwedge^{\alpha_m}$ takes $x_1 \otimes x_2 \otimes \cdots \otimes x_m \otimes y_2 \otimes y_3 \otimes \cdots \otimes y_m$ to a multiple of $x_1 \otimes y_2 \otimes x_2 \otimes$ $y_3 \otimes \cdots \otimes x_{m-1} \otimes y_m \otimes x_m$, for $x_i \in \bigwedge^{\alpha_i - \alpha_{i+1}}$, for $1 \leq i < m$, and $x_m \in \bigwedge^{\alpha_m}$. Let $h_i : \bigwedge^{\alpha_i - \alpha_{i+1}} \otimes \bigwedge^{\alpha_{i+1}} \to \bigwedge^{\alpha_i}$ be the multiplication map, for $1 \leq i < m$,

and let $h = (h_1 \otimes \cdots \otimes h_{m-1} \otimes \mathrm{id}) : \bigwedge^{\alpha_1 - \alpha_2} \otimes \bigwedge^{\alpha_2} \otimes \bigwedge^{\alpha_2 - \alpha_3} \otimes \bigwedge^{\alpha_3} \otimes \cdots \otimes$
$\bigwedge^{\alpha_{m-1} - \alpha_m} \otimes \bigwedge^{\alpha_m} \otimes \bigwedge^{\alpha_m} \to \bigwedge^{\alpha}$.

We claim that the composite map $\theta = h \circ g \circ \psi : \bigwedge^\alpha \to \bigwedge^\alpha$ is a G-module homomorphism which factors through $E^{\otimes r}$ and acts as non-zero scalar multiplication on $y_1 \otimes \cdots \otimes y_m$. We first observe that θ does indeed factor through $E^{\otimes r}$. It suffices to show that ψ factors through $E^{\otimes r}$. Moreover $\psi = \chi \otimes \phi$ and, by assumption, ϕ factors through $E^{\otimes (r - \alpha_1)}$, so it suffices to show that $\chi : \bigwedge^{\alpha_1} \to \bigwedge^{\alpha_1 - \alpha_2} \otimes \bigwedge^{\alpha_2 - \alpha_3} \otimes \cdots \otimes \bigwedge^{\alpha_m}$ factors through $E^{\otimes \alpha_1}$. Since $\lambda \in \Lambda^+(n, r)_{\mathrm{row}}$ we have $l > 1$ and $0 \le \alpha_1 - \alpha_2, \ldots, \alpha_{m-1} - \alpha_m, \alpha_m < l$ or $l = 1$ and k has characteristic $p > 0$. By (7) we have homomorphisms $\zeta_i : \bigwedge^{\alpha_i - \alpha_{i+1}} \to E^{\otimes (\alpha_i - \alpha_{i+1})}$ and $\zeta_m : \bigwedge^{\alpha_m} \to E^{\otimes \alpha_m}$ such that $\zeta_i \circ \eta_i = \mathrm{id}$, where the maps $\eta_i : E^{\otimes (\alpha_i - \alpha_{i+1})} \to \bigwedge^{\alpha_i - \alpha_{i+1}}$ and $\eta_m : E^{\otimes \alpha_m} \to \bigwedge^{\alpha_m}$ are the natural maps, $1 \le i < m$. Thus we have $\chi = \eta \circ \zeta$, where $\zeta = (\zeta_1 \otimes \cdots \otimes \zeta_m) : \bigwedge^{\alpha_1 - \alpha_2} \otimes \cdots \otimes \bigwedge^{\alpha_{m-1} - \alpha_m} \otimes \bigwedge^{\alpha_m} \to E^{\otimes \alpha_1}$ and $\eta = (\eta_1 \otimes \cdots \otimes \eta_m) : E^{\otimes \alpha_1} \to \bigwedge^{\alpha_1 - \alpha_2} \otimes \cdots \otimes \bigwedge^{\alpha_{m-1} - \alpha_m} \otimes \bigwedge^{\alpha_m}$

We now track the effect of $\theta = h \circ g \circ \psi$ on $y_1 \otimes y_2 \otimes \cdots \otimes y_m$. We have $\psi(y_1 \otimes y_2 \otimes \cdots \otimes y_m) = \sum_{A \in \mathcal{A}} c_A \hat{e}_{A_1} \otimes \cdots \otimes \hat{e}_{A_m} \otimes y_2 \otimes \cdots \otimes y_m$, where $\mathcal{A}$ is the set of sequences $A = (A_1, \ldots, A_m)$ of sets such that $[1, \alpha_1]$ is the disjoint union of $A_1, \ldots, A_m$, and each c_A is ± 1. Applying g we get $\sum_{A \in \mathcal{A}} b_A \hat{e}_{A_1} \otimes y_2 \otimes \cdots \otimes \hat{e}_{A_2} \otimes y_3 \otimes \cdots \otimes \hat{e}_{A_{m-1}} \otimes y_m \otimes \hat{e}_{A_m}$, where each b_A is a non-zero scalar. Thus we obtain $\theta(y_1 \otimes \cdots \otimes y_m) = \sum_{A \in \mathcal{A}} b_A \hat{e}_{A_1} y_2 \otimes \cdots \otimes \hat{e}_{A_{m-1}} y_m \otimes \hat{e}_{A_m}$. However, A_1 is some subset of $[1, \alpha_1]$ of size $\alpha_1 - \alpha_2$ and $y_2 = e_1 e_2 \ldots e_{\alpha_2}$ so that $\hat{e}_{A_1} y_2$, when written as a product of e_i's, contains a repeated entry and hence is 0 unless $A_1 = [\alpha_2 + 1, \alpha_1]$. Thus for any non-zero term $b_A \hat{e}_{A_1} y_2 \otimes \cdots \otimes \hat{e}_{A_{m-1}} y_m \otimes \hat{e}_{A_m}$ we have $A_1 = [\alpha_2 + 1, \alpha_1]$. Hence, in a non-zero term, we have $A_2 \subseteq [1, \alpha_2]$ (since $[1, \alpha_1]$ is the disjoint union of the A_i's) and the condition $\hat{e}_{A_2} y_3 \ne 0$ now gives $A_2 = [\alpha_3 + 1, \alpha_2]$. Carrying on this way we get that, in any non-zero term, we have $A_i = [\alpha_{i+1}, \alpha_i]$, for $1 \le i \le m - 1$, and since $[1, \alpha_1]$ is the disjoint union of $A_1, \ldots, A_m$, we must also have $A_m = [1, \alpha_m]$. For this choice of $A_1, \ldots, A_m$ we have $\hat{e}_{A_i} y_{i+1} = \pm y_i$, for $1 \le i < m$, and $\hat{e}_{A_m} = y_m$. Thus we have $\theta(y_1 \otimes \cdots \otimes y_m) = \pm b y_1 \otimes y_2 \otimes \cdots \otimes y_m$, where $b = b_A$, for $A = (A_1, \ldots, A_m)$ as above. This proves the claim.

Now $T(\lambda)$ occurs exactly once as a component of $\bigwedge^\alpha$, by 3.3(1). Let $i : T(\lambda) \to \bigwedge^\alpha$ and $j : \bigwedge^\alpha \to T(\lambda)$ be G-module maps such that $j \circ i$ is the identity map on $T(\lambda)$. Then $\kappa = j \circ \theta \circ i$ is an endomorphism of $T(\lambda)$ which is non-zero on $T(\lambda)^\lambda$ and factors through $E^{\otimes r}$. Since $T(\lambda)$ is absolutely indecomposable $\mathrm{End}_G(T(\lambda))$ is local and κ is an isomorphism. Hence the identity map $T(\lambda) \to T(\lambda)$ factors through $E^{\otimes r}$ and so $T(\lambda)$ occurs as a component of $E^{\otimes r}$.

(10)   (i) *For* $\lambda \in \Lambda^+(n, r)_{\mathrm{row}}$ *we have* $T(\lambda)^\circ \cong T(\lambda)$ *and* $I(\lambda')^\circ \cong I(\lambda')$.
(ii) *There is a bijection* $i : \Lambda^+(n, r)_{\mathrm{col}} \to \Lambda^+(n, r)_{\mathrm{row}}$ *such that* $I(\lambda) \cong T(i(\lambda))$ *(for all* $\lambda \in \Lambda^+(n, r)_{\mathrm{col}}$*).*

(iii) For $\lambda \in \Lambda^+(n,r)_{\text{col}}$ the injective module $I(\lambda)$ has simple head $L(\lambda)$ and the tilting module $T(i(\lambda))$ has simple head $L(\lambda)$.

(iv) For $\lambda \in \Lambda^+(n,r)_{\text{col}}$ the module $\nabla(i(\lambda))$ has simple head $L(\lambda)$ and the module $\Delta(i(\lambda))$ has simple socle $L(\lambda)$.

Proof Since $T(\lambda)$ is a component of $E^{\otimes r}$ and $(E^{\otimes r})^{\circ} \cong E^{\otimes r}$ we have that $T(\lambda)^{\circ}$ is component of $E^{\otimes r}$. Hence $T(\lambda)^{\circ}$ is a tilting module $T(\mu)$ for some μ. Since tilting modules are classified by highest weight and ch $J^{\circ} = $ ch J, for any finite dimensional $S(n,r)$-module J, we have $\mu = \lambda$, i.e. $T(\lambda)^{\circ} \cong T(\lambda)$. Since the modules $I(\lambda)$, with $\lambda \in \Lambda^+(n,r)_{\text{col}}$, and the modules $T(\lambda)$, with $\lambda \in \Lambda^+(n,r)_{\text{row}}$, are exactly the indecomposable components of $E^{\otimes r}$ we must have, for $\lambda \in \Lambda^+(n,r)_{\text{col}}$, that $I(\lambda)$ is isomorphic to $T(\mu)$, for some $\mu \in \Lambda^+(n,r)_{\text{row}}$, and therefore $I(\lambda)^{\circ} \cong I(\lambda)$. This completes the proof of (i) and also proves (ii). Part (iii) follows from (ii). Part (iv) follows from (iii) since $\nabla(i(\lambda))$ is a homomorphic image of $T(i(\lambda))$, for $\lambda \in \Lambda^+(n,r)_{\text{col}}$.

4.4 Connections with the Hecke algebra

In the first part of this section we apply the Schur functor

$$f : \text{mod}(S(n,r)) \to \text{Hec}(r)$$

to various results already obtained for the q-Schur algebra to obtain results for the Hecke algebra. In particular we obtain the parametrization of Young modules and signed Young modules (many treatments of the parametrization are available in the classical case, see e.g. Section 3 of [33] and the references given there). Applying the Schur functor to filtration results on injective modules and tilting modules for $S(n,r)$ gives, as in the classical case (see [32; Section 2] for Young modules), the existence of Specht filtrations for Young modules and signed Young modules (compare with the paper of Martin, [66]). Applying the Schur functor to the irreducible $S(n,r)$-module and using 4.3(9) gives the parametrization of irreducible Hec(r)-modules due to Dipper and James, [20; Section 6]. Our approach is close to the treatment of the classical case given in [51; Chapter 6]. We give the relationship between filtration multiplicities for tilting modules of $S(n,r)$ and decomposition numbers for the Hecke algebra and give the q-version of a result of Erdmann, [47; (2.3) Proposition].

In the second part of the section we form a graded ring whose degree d component is the Grothendieck group of finite dimensional Hec(d)-modules. We give a description of this ring by generators and relations. This is related to recent work of Erdmann, [48].

We assume that $r \leq n$. Throughout this section e denotes the idempotent involved in the definition of f. We write S short for $S(n,r)$ and H or $H(r)$ short for Hec(r).

Definitions Let $\lambda \in \Lambda^+(n,r)$. We write $Y(\lambda)$ for $fI(\lambda)$ (where $I(\lambda)$ is the $S(n,r)$-module injective envelope of $L(\lambda)$) and call $Y(\lambda)$ the *Young module* labelled by λ. We write $Y_s(\lambda)$ for $fT(\lambda)$ and call $Y_s(\lambda)$ the *signed Young module* labelled by λ.

We summarize the main properties of Young modules and signed Young modules. For the first assertion see also [20].

(1)   (i) *Suppose* $\alpha, \beta \in \Lambda(n,r)$ *have the same content. Then we have* $H(r) \otimes_{H(\alpha)} k \cong H(r) \otimes_{H(\beta)} k$ *and* $H(r) \otimes_{H(\alpha)} k_s \cong H(r) \otimes_{H(\beta)} k_s$.
(ii) *For finite dimensional S-modules I_1, I_2, with I_1 injective and $I_2 \in \mathcal{F}(\nabla)$, the natural map* $\mathrm{Hom}_S(I_1, I_2) \to \mathrm{Hom}_H(fI_1, fI_2)$ *is an isomorphism and for finite dimensional S-modules T_1, T_2, with $T_1 \in \mathcal{F}(\nabla)$ and T_2 a tilting module, the natural map* $\mathrm{Hom}_S(T_1, T_2) \to \mathrm{Hom}_H(fT_1, fT_2)$ *is an isomorphism.*
(iii) *The modules $\{Y(\lambda) \mid \lambda \in \Lambda^+(n,r)\}$ are pairwise non-isomorphic and are precisely (up to isomorphism) the indecomposable summands of the modules* $H(r) \otimes_{H(\alpha)} k$, $\alpha \in \Lambda(n,r)$.
(iv) *The modules $\{Y_s(\lambda) \mid \lambda \in \Lambda^+(n,r)\}$ are pairwise non-isomorphic and are precisely (up to isomorphism) the indecomposable summands of the modules* $H(r) \otimes_{H(\alpha)} k_s$, $\alpha \in \Lambda(n,r)$.
(v) *For* $\lambda \in \Lambda^+(n,r)$, $\mu \in \Lambda(n,r)$ *we have*

$$(H(r) \otimes_{H(\mu)} k \mid Y(\lambda)) = (S^\mu E \mid I(\lambda)) = \dim L(\lambda)^\mu.$$

In particular $(H(r) \otimes_{H(\mu)} k \mid Y(\lambda))$ *is 1 for* $\lambda = \mu$ *and is 0 for* $\lambda \not\trianglerighteq \mu$.
(vi) *For* $\lambda \in \Lambda^+(n,r)$, $\mu \in \Lambda(n,r)$ *we have* $(H(r) \otimes_{H(\mu)} k_s \mid Y_s(\lambda)) = (\bigwedge^\mu E \mid T(\lambda)) = \dim L(\lambda')^\mu$. *In particular* $(H(r) \otimes_{H(\mu)} k_s \mid Y(\lambda))$ *is 1 for* $\lambda' = \mu$ *and is 0 for* $\lambda' \not\trianglerighteq \mu$.

Proof   (i) To prove $H(r) \otimes_{H(\alpha)} k \cong H(r) \otimes_{H(\beta)} k$ it suffices, by 2.1(20)(i), to note that $S^\alpha E \cong S^\beta E$, which is true since $S^\alpha E$ and $S^\beta E$ are injective modules with the same character. To prove $H(r) \otimes_{H(\alpha)} k_s \cong H(r) \otimes_{H(\beta)} k_s$ it suffices, by 2.1(20)(ii), to note that $\bigwedge^\alpha E \cong \bigwedge^\beta E$, which is true since $\bigwedge^\alpha E$ and $\bigwedge^\beta E$ are tilting modules with the same character.
(ii) Since every injective module is a direct sum of $I(\lambda)$'s it suffices to show that $\mathrm{Hom}_S(I(\lambda), I_2) \to \mathrm{Hom}_H(fI(\lambda), fI_2)$ is an isomorphism. Since $I(\lambda)$ is a direct summand of $S^\lambda E$ it suffices to note that $\mathrm{Hom}_S(S^\lambda E, I_2) \to \mathrm{Hom}_H(fS^\lambda E, fI_2)$ is an isomorphism and this is the case by 2.1(16)(ii).
    Similarly it suffices to observe that

$$\mathrm{Hom}_S(T_1, \textstyle\bigwedge^\lambda E) \to \mathrm{Hom}_H(fT_1, f\bigwedge^\lambda E)$$

is an isomorphism, for $\lambda \in \Lambda^+(n,r)$, and this is true by 2.1(16)(iii).
(iii) For $\lambda \in \Lambda^+(n,r)$, $\mathrm{End}_S(I(\lambda))$ is a local algebra since $I(\lambda)$ is indecomposable. Hence $\mathrm{End}_H(fI(\lambda))$ is local by (ii) and so $Y(\lambda) = fI(\lambda)$ is indecomposable. Let $\lambda, \mu \in \Lambda^+(n,r)$ and suppose that $\phi : fI(\lambda) \to fI(\mu)$ is an

H-module isomorphism with inverse ψ, say. Then ϕ is the restriction of some $\bar{\phi} \in \mathrm{Hom}_S(I(\lambda), I(\mu))$ and ψ is the restriction of some $\tilde{\psi} \in \mathrm{Hom}_S(I(\mu), I(\lambda))$, by (ii). Then $\tilde{\psi} \circ \bar{\phi} \in \mathrm{End}_S(I(\lambda))$ is the identity on $fI(\lambda)$ and hence $\tilde{\psi} \circ \bar{\phi} = \mathrm{id}_{I(\lambda)}$, by (ii). Similarly $\tilde{\psi} \circ \bar{\phi} = \mathrm{id}_{I(\mu)}$ so that $I(\lambda) \cong I(\mu)$ and $\lambda = \mu$. Hence the modules $\{Y(\lambda) \mid \lambda \in \Lambda^+(n,r)\}$ are pairwise non-isomorphic. Let $\lambda \in \Lambda^+(n,r)$. The multiplicity of $I(\lambda)$ as a component of $S^\lambda E$ is $\dim \mathrm{Hom}_S(L(\lambda), S^\lambda E) = 1$. Thus $Y(\lambda) = fI(\lambda)$ occurs as a component of $fS^\lambda E \cong H(r) \otimes_{H(\lambda)} k$. For $\alpha \in \Lambda(n,r)$ the module $S^\alpha E$ is injective and hence a direct sum of modules $I(\mu)$, $\mu \in \Lambda^+(n,r)$ and thus $H(r) \otimes_{H(\alpha)} k \cong fS^\alpha E$ is a direct sum of modules $Y(\mu)$, $\mu \in \Lambda^+(n,r)$. This completes the proof of (iii).
(iv) Similar to (iii).
(v) This follows from (i),(iii) and the fact that the multiplicity of $I(\lambda)$ as a component of $S^\alpha E$ (for $\lambda \in \Lambda^+(n,r)$, $\alpha \in \Lambda(n,r)$) is $\dim L(\lambda)^\alpha$.
(vi) This follows from (iv) and 4.3(3).

(2) *(i) The modules $\{fL(\lambda) \mid \lambda \in \Lambda^+(n,r)_{\mathrm{col}}\}$ form a complete set of inequivalent irreducible representations of $H(r)$.*
(ii) The modules $Y(\lambda)$, $\lambda \in \Lambda^+(n,r)_{\mathrm{col}}$, are precisely the projective indecomposable $H(r)$-modules and the modules $Y_s(\lambda)$, $\lambda \in \Lambda^+(n,r)_{\mathrm{row}}$, are precisely the projective indecomposable $H(r)$-modules. We have $Y(\lambda) \cong Y_s(i(\lambda))$, for $\lambda \in \Lambda^+(n,r)_{\mathrm{col}}$, where i is the bijection of 4.3(10)(ii).

Proof (i) This is true by 4.3(9) and the Appendix, A1(4).
(ii) This is true by A1(4)(v) and 4.3(4) and (10).

Definition For $\lambda \in \Lambda^+(n,r)$ we call the Hec(r)-module $\mathrm{Sp}(\lambda) = f\nabla(\lambda)$ the *Specht module* labelled by λ.

(3) *Let $\lambda \in \Lambda^+(n,r)$.*
(i) There exist $\mu^1, \ldots, \mu^u \in \Lambda^+(n,r)$ and an H-module filtration $0 = Y^0 < Y^1 < \cdots < Y^u = Y(\lambda)$ such that $\mu_i \not> \mu_j$ for $1 \le i < j \le u$, such that $Y^i/Y^{i-1} \cong \mathrm{Sp}(\mu^i)$ for $1 \le i \le u$ and such that for each $\mu \in \Lambda^+(n,r)$ we have $|\{i \in [1,u] \mid \mu^i = \mu\}| = [\mu : \lambda]$.
(ii) There exist $\mu^1, \ldots, \mu^v \in \Lambda^+(n,r)$ and an H-module filtration $0 = Y_s^0 < Y_s^1 < \cdots < Y_s^v = Y_s(\lambda)$ such that $\mu_i \not> \mu_j$ for $1 \le i < j \le v$, such that $Y^i/Y^{i-1} \cong \mathrm{Sp}(\mu^i)$ for $1 \le i \le v$ and such that for each $\mu \in \Lambda^+(n,r)$ we have $|\{i \in [1,s] \mid \mu^i = \mu\}| = [\mu' : \lambda']$.

Proof These results are obtained by applying the Schur functor from the corresponding results for $I(\lambda)$ and $T(\lambda)$ (see [36; Section 4, (6)] and Proposition 4.1.5(ii) above) and 2.1(13).

(4) *Suppose q is a root of unity.*

(i) For $\lambda \in \Lambda^+(n,r)_{\text{row}}$ the head $\operatorname{hd}\operatorname{Sp}(\lambda)$ of the Specht module $\operatorname{Sp}(\lambda)$ is simple.

(ii) The modules $\operatorname{hd}\operatorname{Sp}(\lambda)$, $\lambda \in \Lambda^+(n,r)_{\text{col}}$, form a complete set of inequivalent irreducible H-modules and we have $fL(\lambda) \cong \operatorname{hd}\operatorname{Sp}(i(\lambda))$, for $\lambda \in \Lambda^+(n,r)_{\text{col}}$, where $i : \Lambda^+(n,r)_{\text{col}} \to \Lambda^+(n,r)_{\text{row}}$ is the bijection of 4.3(10)(ii).

Proof (i) Let $\lambda \in \Lambda^+(n,r)_{\text{row}}$. We have a filtration $0 = Y_0 < Y^1 < \cdots < Y^v = Y_s(\lambda)$ as in (3)(ii). For $1 \le i \le v$ we have $[\mu^i : \lambda'] \ne 0$ and hence $\mu^i \le \lambda$. Thus we must have $\mu^v = \lambda$. Hence $\operatorname{Sp}(\lambda)$ is an epimorphic image of $Y_s(\lambda)$. Now $Y_s(\lambda)$ is a projective indecomposable H-module and hence has a simple head. Thus $\operatorname{hd}\operatorname{Sp}(\lambda) = \operatorname{hd}Y_s(\lambda)$ is simple.

(ii) Since $\{Y_s(\lambda) \mid \lambda \in \Lambda^+(n,r)_{\text{row}}\}$ is a full set of projective indecomposable modules, $\{\operatorname{hd}Y_s(\lambda) \mid \lambda \in \Lambda^+(n,r)_{\text{row}}\}$ is a full set of irreducible H-modules, i.e. $\{\operatorname{hd}\operatorname{Sp}(\lambda) \mid \lambda \in \Lambda^+(n,r)_{\text{row}}\}$ is a full set of simple H-modules. For $\lambda \in \Lambda^+(n,r)_{\text{col}}$ we have an epimorphism $\nabla(i(\lambda)) \to L(\lambda)$, by 4.3(10)(iii), and hence we have an epimorphism $\operatorname{Sp}(i(\lambda)) = f\nabla(i(\lambda)) \to fL(\lambda)$. Hence we have $fL(\lambda) \cong \operatorname{hd}\operatorname{Sp}(i(\lambda))$, as required.

Definition We write $D(\lambda)$ for the irreducible H-module $\operatorname{hd}\nabla(\lambda)$, for $\lambda \in \Lambda^+(n,r)_{\text{row}}$.

For the $q = 1$ case of the following see [45; 4.5 Lemma].

(5) Suppose q is a root of unity. For $\lambda \in \Lambda^+(n,r)_{\text{row}}$ and $\mu \in \Lambda^+(n,r)$ we have $(T(\lambda) : \nabla(\mu)) = [\operatorname{Sp}(\mu) : D(\lambda)]$.

Proof We write $\lambda = i(\nu)$, for $\nu \in \Lambda^+(n,r)_{\text{col}}$ (where i is as in 4.3(10)(ii). We have $(T(i(\nu)) : \nabla(\mu)) = [I(\nu) : \nabla(\mu)]$, by 4.3(10)(ii), which is $[\nabla(\mu) : L(\nu)]$, by [36; Section 4, (6)]. However, by applying the Schur functor to a composition series of $\nabla(\mu)$, one sees that $[\nabla(\mu) : L(\nu)] = [f\nabla(\mu) : fL(\nu)]$, which is $[\operatorname{Sp}(\mu) : D(i(\nu))]$, by (4)(ii).

We can now read off the Hecke algebra decomposition multiplicities for 2-part partitions from 3.4(3) and 4.2(6). We assume for definiteness that $l > 1$ and k has characteristic $p > 0$. (A derivation along similar lines in the classical case $l = 1$ is given in [45; Section 6].) As James has pointed out, the result may be obtained by methods similar to those used in [56; Section 20] to determine the decomposition numbers for 2-part partitions for the finite general linear groups.

Recall, from Section 3.4, that if a has (l,p) decomposition $a = a_{-1} + l\sum_{i=0}^{m-1} p^i a_i$ and J is a subset of $[-1, m-1]$ then a_J denotes the sum of the terms indexed by J, i.e. a_J denotes $a_{-1} + \sum_{-1 \ne i \in J} lp^i a_i$ if $-1 \in J$, and denotes $\sum_{i \in J} lp^i a_i$ if $-1 \notin J$.

(6) Let $\lambda = (\lambda_1, \lambda_2)$, $\mu = (\mu_1, \mu_2)$ be 2-part partitions of r and assume $\mu \in \Lambda^+(n, r)_{\mathrm{row}}$. If $\mu_1 - \mu_2 \le l-1$ then $[\mathrm{Sp}(\lambda) : D(\mu)] = \delta_{\lambda\mu}$. If $\mu_1 - \mu_2 \ge l-1$ and $\mu_1 - \mu_2 = lp^m - 1 + a + lp^m b$ with $a \le lp^m - 1$ and $0 \le b < p$ then for $\lambda \in X^+$ we have

$$(\mathrm{Sp}(\lambda) : D(\mu)) = \begin{cases} 1, & \text{if } \lambda = (\mu_1 - a_J, \mu_2 + a_J) \text{ for some } J \subseteq [-1, m-1]; \\ 0, & \text{otherwise.} \end{cases}$$

Combining (5) and 3.3(9) we get (where t is as in 3.3(9)):

(7) Suppose that $q \ne 1$ is a root of unity. For $\lambda, \mu \in \Lambda^+(n, r)$ we have $[\mathrm{Sp}(t(\mu')) : \mathrm{Sp}(t(\lambda'))] = [\bar{\Delta}(\mu) : \bar{L}(\lambda)]$.

Here $\bar{\Delta}(\mu)$ denotes the Weyl module of highest weight μ for the ordinary general linear group $\bar{G}$ of degree n. This is the formal q-analogue of Erdmann's result, [47; (2.3) Proposition], proved for ordinary general linear groups and symmetric groups. The result in the classical case has the consequence that each decomposition number for a general linear group is also a decomposition number for a symmetric group. However, in our case the decomposition number on the left of the equation is for a Hecke algebra and on the right is for an ordinary general linear group, so we have no such q-analogue of this punch-line.

We now consider an involution $\# : \Lambda^+(n, r)_{\mathrm{row}} \to \Lambda^+(n, r)_{\mathrm{row}}$ which is closely related to the bijection $\iota : \Lambda^+(n, r)_{\mathrm{col}} \to \Lambda^+(n, r)_{\mathrm{row}}$. From the standard description of $H(r)$ by generators and relations one sees that there is an involutory algebra automorphism $\# : H(r) \to H(r)$ given on the generators by $\#(T_{s_a}) = -T_{s_a} + (q - 1)1$, for $1 \le a < r$. We also write $\#(h)$ as $h^\#$, for $h \in H(r)$. For an $H(r)$-module U affording the representation $\pi : H(r) \to \mathrm{End}_k(U)$ we write $U^\#$ for the vector space U regarded as an $H(r)$-module via the representation $\pi \circ \#$. Putting $\phi^\# = \phi$, for a morphism $\phi : U \to U'$ of finite dimensional $H(r)$-modules, we obtain an isomorphism from the category of finite dimensional $H(r)$-modules to itself.

From the definitions, we have:

(8) $\# \circ J = J \circ \#$ and therefore $(U^\circ)^\# \cong (U^\#)^\circ$, for $U \in \mathrm{mod}(H(r))$.

We shall also need:

(9) for $\alpha = (\alpha_1, \ldots, \alpha_m) \in \Lambda^+(n, r)$ we have $x(\alpha)^\# = (-1)^{n(\alpha)} y(\alpha)$, where $n(\alpha)$ is the number of parts of α of size 2.

Proof It is enough to consider the case $\alpha = (r)$. Since $H(r)$ is a Frobenius algebra, there is a unique submodule of $H(r)$ isomorphic to the trivial module. It follows that $kx(r) = \{h \in H(r) \mid T_{s_a} h = qh, \text{ for all } 1 \le a < r\}$, and similarly we have $ky(r) = \{h \in H(r) \mid T_{s_a} h = -h, \text{ for all } 1 \le a < r\}$. Now for $1 \le a < r$ we have

$$
\begin{aligned}
T_{s_a} x(r)^{\#} &= (-T_{s_a}^{\#} + (q-1)1)x(r)^{\#} \\
&= -(T_{s_a} x(r))^{\#} + (q-1)x(r)^{\#} \\
&= -x(r)^{\#}
\end{aligned}
$$

so that $x(r)^{\#} \in ky(r)$ and $x(r)^{\#} = cy(r)$ for some $c \in k$. Comparing coefficients of T_{w_0} in $x(r)^{\#}$ and $cy(r)$ we get $c = (-1)^{\binom{r}{2}}$, which gives the result.

(10)   For $\alpha = (\alpha_1, \ldots, \alpha_m) \in \Lambda^+(n, r)$ we have

$$
(H(r) \otimes_{H(\alpha)} k)^{\#} \cong H(r) \otimes_{H(\alpha)} k_s.
$$

Proof   We have $H(r) \otimes_{H(\alpha)} k \cong Hx(\alpha)$ and $H(r) \otimes_{H(\alpha)} k_s \cong Hy(\alpha)$ by 2.1(20). The restriction of $\# : H(r) \to H(r)$ defines an $H(r)$-module isomorphism $(Hx(\alpha))^{\#} \to Hy(\alpha)$.

We also have the following, from Proposition 4.1.7 and the remarks before Proposition 4.1.4.

(11)   For $\alpha = (\alpha_1, \ldots, \alpha_m) \in \Lambda^+(n, r)$ we have

$$
(H(r) \otimes_{H(\alpha)} k_s)^{\circ} \cong H(r) \otimes_{H(\alpha)} k_s.
$$

Now from (8) and (10) we obtain:

(12) $(H(r) \otimes_{H(\alpha)} k)^{\circ} \cong H(r) \otimes_{H(\alpha)} k$, for $\alpha \in \Lambda(n, r)$.

Let S be a finite dimensional k-algebra. We write $\mathrm{Grot}(S)$ for the Grothendieck group of the category of finite dimensional left S-modules. For a finite dimensional left S-module X we write $[X]$ for the corresponding element of $\mathrm{Grot}(S)$. Note that since the functor $f : \mathrm{mod}(S(n, r)) \to \mathrm{mod}(H(r))$ is exact, it defines a homomorphism $[f] : \mathrm{Grot}(S(n, r)) \to \mathrm{Grot}(H(r))$ satisfying $[f]([X]) = [fX]$, for a finite dimensional $S(n, r)$-module X. Similarly we have defined a homomorphism $\mathrm{Grot}(H(r)) \to \mathrm{Grot}(H(r))$ taking $[U]$ to $[U^{\#}]$, for a finite dimensional $H(r)$-module U.

(13)   For $\lambda \in \Lambda^+(n, r)$ we have $[\mathrm{Sp}(\lambda)^{\#}] = [\mathrm{Sp}(\lambda')]$.

Remark   We give a more precise version of this in the next section, see Proposition 4.5.9.

Proof We write $\bar{\omega}$ for the standard involution of symmetric function theory (denoted ω in [**63**]). We identify the formal character of a G-module which is polynomial of degree r with a symmetric function of degree r. We have $\bar{\omega}(\text{ch} \bigwedge^{\mu} E) = \text{ch} \, S^{\mu} E$, from the definition of $\bar{\omega}$, and $\bar{\omega}(\chi(\mu)) = \chi(\mu')$ for all $\mu \in \Lambda^{+}(n, r)$, by [**63**; I, (3.8)]. We have $\text{ch} \bigwedge^{\lambda'} E = \chi(\lambda) + \sum_{\mu} a_{\mu} \chi(\mu)$, for non-negative integers a_{μ}, $\mu \in \Lambda^{+}(n, r)$, with a_{μ} zero unless $\mu > \lambda$. Applying $\bar{\omega}$, we also have $\text{ch} \, S^{\lambda'} E = \chi(\lambda') + \sum_{\mu} a_{\mu} \chi(\mu')$. Hence we have the formulas

$$[\bigwedge^{\lambda'} E] = [\nabla(\lambda)] + \sum_{\mu} a_{\mu}[\nabla(\mu)]$$

and

$$[S^{\lambda'} E] = [\nabla(\lambda')] + \sum_{\mu} a_{\mu}[\nabla(\mu')]$$

in $\text{Grot}(S(n, r))$. Applying $[f]$ and using 2.1(20), and (1) we obtain the formulas

$$[H(r) \otimes_{H(\lambda')} k_s] = [\text{Sp}(\lambda)] + \sum_{\mu} a_{\mu}[\text{Sp}(\mu)]$$

and

$$[H(r) \otimes_{H(\lambda')} k] = [\text{Sp}(\lambda')] + \sum_{\mu} a_{\mu}[\text{Sp}(\mu')]$$

in $\text{Grot}(H(r))$. Applying the homomorphism $\text{Grot}(H(r)) \to \text{Grot}(H(r))$ defined by the automorphism $\# : H(r) \to H(r)$ to the first formula and invoking (11) we obtain

$$[H(r) \otimes_{H(\lambda')} k] = [\text{Sp}(\lambda)^{\#}] + \sum_{\mu} a_{\mu}[\text{Sp}(\mu^{\#})].$$

Assuming inductively that $[\text{Sp}(\mu)^{\#}] = [\text{Sp}(\mu')]$ for all $\mu > \lambda$ and comparing the above displayed formula with the previous one we obtain $[\text{Sp}(\lambda)^{\#}] = [\text{Sp}(\lambda')]$, as required.

We now give the promised relationship between $\#$ and ι. We define an involution $\lambda \to \lambda^{\#}$ on $\Lambda^{+}(n, r)_{\text{row}}$ by $D(\lambda)^{\#} = D(\lambda^{\#})$.

(14)   For $\lambda \in \Lambda^{+}(n, r)_{\text{row}}$ we have $\lambda^{\#} = \iota(\lambda')$.

Proof We have

$$1 = (T(\iota(\lambda)) : \nabla(\lambda'))$$
$$= [\text{Sp}(\lambda') : D(\iota(\lambda))] = [\text{Sp}(\lambda) : D(\iota(\lambda)^{\#})]$$
$$= (T(\iota(\lambda)^{\#}) : \nabla(\lambda))$$

and hence $\lambda \leq \iota(\lambda')^{\#}$. Thus we have a bijection

$$\phi : \Lambda^+(n,r)_{\text{row}} \to \Lambda^+(n,r)_{\text{row}},$$

$\phi(\lambda) = \iota(\lambda')^{\#}$, such that $\phi(\lambda) \geq \lambda$ for all λ. It follows that ϕ is the identity, i.e. $\lambda^{\#} = \iota(\lambda')$, for all $\lambda \in \Lambda^+(n,r)_{\text{row}}$.

We consider the direct sum of the Grothendieck groups of finite dimensional $H(r)$-modules, for $r \geq 0$. We make a graded commutative ring with identity $R = \bigoplus_{d \geq 0} R^d$, where $R^d = \text{Grot}(H(d))$ for $d > 0$ and where R^0 is freely generated, as an abelian group, by 1_R. The multiplication $R^a \times R^b \to R^{a+b}$ is given by $[X] \cdot [Y] = [\text{Ind}_{H(a,b)}^{H(a+b)}(X \otimes Y)]$, for $X \in \text{mod}(H(a))$, $Y \in \text{mod}(H(b))$ (and $a, b > 0$). We leave it to the reader to check that multiplication is well-defined, associative and commutative (cf. the case $q = 1$ in [63; I,Section 7]). This ring is also discussed, in the classical case $q = 1$, in [39; Section 4.7, Remark 3] and in [48]. We shall give a description of R by generators and relations.

Let Λ denote the ring of symmetric functions in infinitely many variables $x_1, x_2, \ldots$, as in [63]. Thus $\Lambda = \bigoplus_{d \geq 0} \Lambda^d$ is a graded ring, freely generated by the complete symmetric functions $h_d \in \Lambda_d$, $d > 0$. For $\lambda = (\lambda_1, \lambda_2, \ldots)$ we put $h_\lambda = h_{\lambda_1} h_{\lambda_2} \ldots$ (as in [63]). The complete symmetric functions h_λ form a $\mathbb{Z}$-basis of Λ, consisting of homogeneous elements, as do the monomial symmetric functions m_λ (as λ varies over partitions). We define a ring homomorphism $\theta : \Lambda \to R$ by putting $\theta(h_d) = [k] \in R^d$, for $d \geq 0$. Then $\theta = \bigoplus_{d \geq 0} \theta^d$ is a homomorphism of graded rings and $\text{Ker}(\theta) = \bigoplus_{d \geq 0} \text{Ker}(\theta^d)$ is a homogeneous ideal of Λ.

Let $n \geq d$. We identify a character of an $S(n,d)$-module with an element of Λ^d, in the usual way (see Section 4.3). Thus we have $\theta(\text{ch } S^\lambda E) = \theta(h_\lambda) = [H(d) \otimes_{H(\lambda)} k] = [fS^\lambda E]$, by 2.1(20), for any $\lambda \in \Lambda^+(n,d)$, where $f : \text{mod}(S(n,d)) \to \text{mod}(H(d))$ denotes the Schur functor. However, the elements $h_\lambda = \text{ch } S^\lambda E$, $\lambda \in \Lambda^+(n,d)$, generate the additive group Λ^d and so we must have $\theta(\text{ch } X) = [fX]$, for every $X \in \text{mod}(S(n,d))$. Thus, from (2)(ii), we have $\theta(\Lambda^d) = \text{Grot}(H(d))$ and $\theta : \Lambda \to R$ is surjective. If q is not a root of unity, or $q = 1$ and k has characteristic 0, then the Schur functor $f : \text{mod}(S(n,d)) \to \text{mod}(H(d))$ (for $n \geq d$) is an equivalence of categories, Section 2.1 Remark (iii), and it follows that $\theta : \Lambda \to R$ is an isomorphism (cf. [63; I, Section 7], where the case $q = 1$, $k = \mathbb{C}$ is discussed). So we assume for the rest of this section that q is a primitive lth root of unity and either that $l > 1$ or that $l = 1$ and k has characteristic $p > 0$.

For an integer t, we denote by $\psi^t : \Lambda \to \Lambda$ the corresponding Adams operation. Thus ψ^t is the ring homomorphism such that $\psi^t(m_\lambda) = m_{t\lambda}$, for a partition λ. We now take t to be l if $l > 1$ and take t to be p if $l = 1$ and k has characteristic $p > 0$. The map ψ^t is essentially the map induced on characters by the appropriate Frobenius morphism. Suppose first that $l > 0$. We fix a degree $d > 0$ and choose $n \geq ld$. We have the quantum general linear group G of degree n, the "classical" general linear group scheme $\bar{G}$ of degree n and the quantum Frobenius morphism $F : G \to \bar{G}$. For $\lambda \in \Lambda^+(n, d)$ we have the simple $\bar{G}$-module $\bar{L}(\lambda)$ and hence the simple $S(n, ld)$-module $\bar{L}(\lambda)^F \cong L(l\lambda)$. Thus the character of $L(l\lambda)$ is $\psi^t(\mathrm{ch}\,\bar{L}(\lambda))$. Since $fL(l\lambda) = 0$, by 4.3(1), we have $\theta(\mathrm{ch}\,L(l\lambda)) = 0$. Moreover Λ^d is spanned by the characters $\mathrm{ch}\,\bar{L}(\lambda)$, $\lambda \in \Lambda^+(n, d)$, so that $\theta(\psi^t(\Lambda^d)) = 0$. Similarly, in the case $l = 1$, one may argue as above using the ordinary Frobenius morphism $\bar{G} \to \bar{G}$ so that we have $\theta(\psi^t(\Lambda^d)) = 0$ in that case too.

Let $I \leq \mathrm{Ker}(\theta)$ be the ideal of Λ generated by $\psi^t(\Lambda^d)$, for all $d > 0$. Note that $I = \bigoplus_{d \geq 0} I^d$ is a homogeneous ideal of Λ. A partition λ may be expressed in the form $\alpha + t\beta$, for unique partitions α, β with α restricted. Moreover, we have $m_\alpha m_{t\beta} \in m_{\alpha + t\beta} + \sum_{\mu < \alpha + t\beta} \mathbb{Z} m_\mu$. Hence the elements $m_\alpha m_{t\beta}$ (α, β partitions and α restricted) form a $\mathbb{Z}$-basis of Λ, consisting of homogeneous elements. Thus Λ^d / I^d is a free abelian group of rank equal to the number of restricted partitions of d, for $d > 0$. Thus θ^d induces a surjective homomorphism $\bar{\theta}^d : \Lambda^d / I^d \to \mathrm{Grot}(H(d))$ and, since Λ^d / I^d and $\mathrm{Grot}(H(d))$ are free abelian groups of the same rank, $\bar{\theta}^d$ is an isomorphism and hence $I^d = \mathrm{Ker}(\theta)^d$, for $d > 0$, and $I = \mathrm{Ker}(\theta)$.

Remark The characteristic map of symmetric function theory is an isomorphism between the graded commutative ring whose dth component is the group $X(\mathrm{Sym}(d))$ of generalized characters of $\mathrm{Sym}(d)$ and Λ^d, see [63; I, Section 7]. Under the isomorphism $X(\mathrm{Sym}(d)) \to \Lambda^d$, the subgroup $I^d = \mathrm{Ker}(\theta)^d$ corresponds to the additive group of generalized characters which vanish of all r-regular elements of $\mathrm{Sym}(d)$, where we say that $g \in \mathrm{Sym}(d)$ is r-regular if no cycle length of g is divisible by r (cf. [48]). We leave the details to the interested reader.

Now Λ is free on the complete symmetric functions h_d, $d \geq 1$, so that, to describe $R = \Lambda / I$ by generators and relations, it is enough to give an explicit formula for $\psi^t(h_d)$ in terms of the generators h_r. Let $\zeta \in \mathbb{C}$ be a primitive tth root of unity. We have the formula

$$\psi^t(h_d) = \sum_{\lambda \in \Lambda^+(t, td)} m_\lambda(1, \zeta, \ldots, \zeta^{t-1}) h_\lambda \qquad (*)$$

due to Nuttall, [70].

There are many instances of a formal similarity between the representation theory of the symmetric groups over a field of characteristic $p > 0$ and that of the Hecke algebras over a field of characteristic 0 at a pth root of unity, for example in the works of Dipper and James [20]. From $(*)$, and the fact that $I = \mathrm{Ker}(\theta)$, proved above, we have another example of this phenomenon, as follows.

(15)   Let p be a prime. Let k_1 be a field of characteristic $p > 0$ and let q be a primitive pth root of unity in a field k_2. Then the rings $\bigoplus_{d \geq 0} \mathrm{Grot}(k_1 \mathrm{Sym}(d))$ and $\bigoplus_{d \geq 0} \mathrm{Grot}(H(d))$ are isomorphic (as graded rings), where $H(d)$ denotes the Hecke algebra of degree d over k_2 constructed via q.

It remains to give the promised explicit description of $\psi^t(h_d)$, for $d > 0$. Let S be the set of sequences of positive integers of length at most t, including the empty sequence which we denote by 0. For $0 \neq \alpha = (\alpha_1, \ldots, \alpha_u) \in S$, we define the "unnormalized monomial symmetric polynomial" $\tilde{m}_\alpha = \sum_{(i_1, \ldots, i_u)} x_{i_1}^{\alpha_1} \ldots x_{i_u}^{\alpha_u}$, where the sum runs over sequences $(i_1, \ldots, i_u)$ of distinct elements of $[1, t]$. We put $\tilde{m}_0 = 1$. Our aim is to give an explicit formula for $f(\alpha) = \tilde{m}_\alpha(1, \zeta, \ldots, \zeta^{t-1})$.

Let $0 \neq \alpha = (\alpha_1, \ldots, \alpha_u) \in S$. Note that if t divides a then we have $x^a = 1$, for $x \in \{1, \zeta, \ldots, \zeta^{t-1}\}$, from which we get:

(i) $f(\alpha_1, \ldots, \alpha_u) = (t - u + 1)f(\alpha_1, \ldots, \alpha_{u-1})$, if α_u is a multiple of t.

Note also that if t does not divide a then $\sum_{i=1}^{t} \zeta^{(i-1)a} = 0$, from which we get:

(ii) $f(\alpha_1, \ldots, \alpha_u) = -f(\alpha_1 + \alpha_u, \alpha_2, \ldots, \alpha_{u-1}) - f(\alpha_1, \alpha_2 + \alpha_u, \ldots, \alpha_{u-1})$
$\cdots - f(\alpha_1, \ldots, \alpha_{u-1} + \alpha_u)$
$= -\sum_{i=1}^{u-1} f(\alpha(i))$
if α_u is not a multiple of t, where $\alpha(i) = (\alpha_1, \ldots, \alpha_{i-1}, \alpha_i + \alpha_u, \alpha_{i+1}, \ldots, \alpha_{u-1})$, for $1 \leq i \leq u - 1$.

We denote the length of $\alpha \in S$ by $u(\alpha)$. For $0 \neq \alpha = (\alpha_1, \ldots, \alpha_u) \in S$ we consider a set $\mathcal{S}(\alpha)$ of associated partitions. Precisely, we denote by $\mathcal{S}(\alpha)$ the set of all sequences $A = (A_1, \ldots, A_r)$ of non-empty subsets of $[1, u]$ such that $[1, u]$ is the disjoint union of $A_1, \ldots, A_r$, and such that $\sum_{i \in A_j} \alpha_i$ is divisible by t, for $1 \leq j \leq r$. We denote the number of components of $A \in \mathcal{S}(\alpha)$ by $r(A)$. For $A \in \mathcal{S}(\alpha)$ we set

$$n_\alpha(A) = \frac{1}{r(A)!}(-1)^{r(A)+u(\alpha)}t^{r(A)}(|A_1| - 1)! \ldots (|A_{r(A)}| - 1)!$$

We claim that

$$\tilde{m}_\alpha(1, \zeta, \ldots, \zeta^{t-1}) = \sum_{A \in \mathcal{S}(\alpha)} n_\alpha(A) \qquad (\dagger)$$

that is, $f(\alpha) = g(\alpha)$, where

$$g(\alpha) = \sum_{A \in \mathcal{S}(\alpha)} n_\alpha(A)$$

for $0 \neq \alpha \in S$. We set $g(0) = 1$. Clearly, $f : S \to \mathbf{Z}$ is determined by the properties (i),(ii), and the condition $f(0) = 1$, so it suffices to show that $g : S \to \mathbf{Z}$ also has these properties.

Consider property (i). Let $\alpha = (\alpha_1, \ldots, \alpha_u)$ and suppose that α_u is divisible by t. Let $\mathcal{S}(\alpha)'$ denote the set of $A \in \mathcal{S}(\alpha)$ such that $\{u\}$ occurs as some component of A and let $\mathcal{S}(\alpha)''$ denote the complement of $\mathcal{S}(\alpha)'$ in $\mathcal{S}(\alpha)$. Note that $\mathcal{S}(\alpha)'$ is the disjoint union of the sets $\mathcal{S}(\alpha, j)'$, $j \geq 1$, where $\mathcal{S}(\alpha, j)'$ is the set of $A = (A_1, \ldots, A_r) \in \mathcal{S}(\alpha)$ such that $A_j = \{u\}$. Thus we have

$$g(\alpha) = \sum_{j \geq 1} \sum_{A \in \mathcal{S}(\alpha, j)'} n_\alpha(A) + \sum_{A \in \mathcal{S}(\alpha)''} n_\alpha(A).$$

Let $\beta = (\alpha_1, \ldots, \alpha_{u-1})$. For $B = (B_1, \ldots, B_r) \in \mathcal{S}(\beta)$ we define $B'(j) = (B_1, \ldots, B_{j-1}, \{u\}, B_j, B_{j+1}, \ldots, B_r)$, for $1 \leq j \leq r + 1$. Then we have

$$\sum_{A \in \mathcal{S}(\alpha)'} n_\alpha(A) = \sum_{B \in \mathcal{S}(\beta)} \sum_{j=1}^{r(B)+1} n_\alpha(B'(j)).$$

Now we have

$$n_\alpha(B'(j)) = \frac{1}{(r(B) + 1)!} (-1)^{u(\beta)+1+r(B)+1} t^{r(B)+1} (|B_1| - 1)! \ldots (|B_{r(B)}| - 1)!$$

$$= \frac{t}{r(B) + 1} n_\beta(B)$$

for $B \in \mathcal{S}(\beta)$, $1 \leq j \leq r(B) + 1$. Hence we have $\sum_{A \in \mathcal{S}(\alpha)'} n_\alpha(A) = t g(\beta)$.

Now consider $\sum_{A \in \mathcal{S}(\alpha)''} n_\alpha(A)$. For $B = (B_1, \ldots, B_r) \in \mathcal{S}(\beta)$ we define $B''(j) = (B_1, \ldots, B_{j-1}, B_j \bigcup \{u\}, B_{j+1}, \ldots, B_r)$, for $1 \leq j \leq r$. We get

$$\sum_{A \in \mathcal{S}(\alpha)''} n_\alpha(A) = \sum_{B \in \mathcal{S}(\beta)} \sum_{j=1}^{r(B)} n_\alpha(B''(j)).$$

Now we have

$$n_\alpha(B''(j)) = \frac{1}{r(B)!} (-1)^{u(\beta)+1+r(B)} t^{r(B)}$$

$$\times (|B_1| - 1)! \ldots (|B_{j-1}| - 1)! |B_j|! (|B_{j+1}| - 1)! \ldots (|B_{r(B)}| - 1)!$$

$$= -|B_j| n_\beta(B)$$

for $B \in \mathcal{S}(\beta)$, $1 \leq j \leq r(B)$. Thus we have

$$\sum_{j=1}^{r(B)} n_\alpha(B''(j)) = -(u - 1) n_\beta(B).$$

and hence

$$\sum_{A \in \mathcal{S}(\alpha)''} n_\alpha(A) = -(u-1)g(\beta).$$

Thus we have $g(\alpha) = (t - u + 1)g(\beta)$.

We now consider property (ii). So let $\alpha = (\alpha_1, \ldots, \alpha_u) \in S$ with α_u not divisible by t. Let $X \subseteq [1, u-1] \times \mathbb{N} \times \mathcal{S}(\alpha)$ be the subset consisting of the triples (i, j, A) such that i and u belong to the component A_j of $A = (A_1, A_2, \ldots, A_r)$. Consider the sum

$$g'(\alpha) = \sum_{(i,j,A) \in X} \frac{n_\alpha(A)}{|A_j| - 1}.$$

We have

$$g'(\alpha) = \sum_{j \in \mathbb{N}} \sum_{(i,A) \in X_j} \frac{n_\alpha(A)}{|A_j| - 1}$$

where X_j is the subset of $[1, u-1] \times \mathcal{S}(\alpha)$ consisting of the pairs (i, A) such that $(i, j, A) \in X$. Now for given j and A such that A_j contains u, there are $|A_j| - 1$ values of $i \in [1, u-1]$ such that $(i, A) \in X_j$. Hence we have

$$g'(\alpha) = \sum_{j \in \mathbb{N}} \sum_{A \in \mathcal{S}(\alpha,j)} n_\alpha(A) = \sum_{A \in \mathcal{S}(\alpha)} n_\alpha(A)$$

$$= g(\alpha)$$

where $\mathcal{S}(\alpha, j) = \{A \in \mathcal{S}(\alpha) \mid u \in A_j\}$.

We now break the sum up another way. We have

$$g(\alpha) = g'(\alpha) = \sum_{i=1}^{u-1} \sum_{(j,A) \in X^i} \frac{n_\alpha(A)}{|A_j| - 1}$$

where $X^i = \{(j, A) \in \mathbb{N} \times \mathcal{S}(\alpha) \mid (i, j, A) \in X\}$, for $1 \le i \le u - 1$. We write Y^i for the subset of $\mathbb{N} \times \mathcal{S}(\alpha(i))$ consisting of those pairs (j, B) such that $i \in B_j$. Note that if $(j, B) \in Y^i$, then $(j, B(i)) \in X^i$, where $B(i) = (B_1, \ldots, B_{j-1}, B_j \bigcup\{u\}, B_{j+1}, \ldots, B_r)$, and moreover, the assignment $(j, B) \mapsto (j, B(i))$ determines a bijection from Y^i to X^i. Hence we have

$$g(\alpha) = \sum_{i=1}^{u-1} \sum_{(j,B) \in Y^i} \frac{n_\alpha(B(i))}{|B_j|}.$$

Now we have

$$n_\alpha(B(i)) = \frac{1}{r(B)!}(-1)^{r(B)+u(\alpha(i))+1}$$

$$\times (|B_1| - 1)! \ldots (|B_{j-1}| - 1)! |B_j|!(|B_{j+1}| - 1)! \ldots (|B_{r(B)}| - 1)!$$

$$= -|B_j|n_{\alpha(i)}(B)$$

for $(j, B) \in Y^i$. Thus we get

$$
\begin{aligned}
g(\alpha) &= -\sum_{i=1}^{u-1} \sum_{(j,B)\in Y^i} n_{\alpha(i)}(B) \\
&= -\sum_{i=1}^{u-1} g(\alpha(i))
\end{aligned}
$$

as required.

We have shown that $g : S \to \mathbf{Z}$ has the properties (i),(ii) enjoyed by $f : S \to \mathbf{Z}$. Therefore $f = g$ and (†) holds for all $0 \neq \alpha \in S$.

We can rewrite (†) so that the summands do not involve denominators as follows. We write $\bar{\mathcal{S}}(\alpha)$ for the set of "normalized" partitions in $\mathcal{S}(\alpha)$, that is, the set of $A = (A_1, \ldots, A_r) \in \mathcal{S}(\alpha)$ such that 1 occurs in A_1, such that the minimal element of $A \backslash A_1$ occurs in A_2, and so on. We define $\bar{n}_\alpha(A) = (-1)^{r(A)+u(\alpha)}(|A_1| - 1)! \ldots (|A_{r(A)}| - 1)!$, for $A \in \bar{\mathcal{S}}(\alpha)$. Note that $\bar{n}_\alpha(A)$ is the sum of all terms $n_\alpha(A')$, in which A' is obtained by permuting the components of A. Thus we have

$$
\tilde{m}_\alpha(1, \zeta, \ldots, \zeta^{t-1}) = \sum_{A \in \bar{\mathcal{S}}(\alpha)} \bar{n}_\alpha(A).
$$

Finally, note that, for $\lambda = (1^{y_1} 2^{y_2} \ldots)$, we have

$$
m_\lambda(1, \zeta, \ldots, \zeta^{t-1}) = \frac{1}{y_1! y_2! \ldots} \tilde{m}_\lambda(1, \zeta, \ldots, \zeta^{t-1})
$$

so that identifying Λ/I with R, via the map induced by θ, we get the following.

(16) *R is the commutative ring given by generators h_d, $d \geq 1$ and relations*

$$
\sum_{\lambda=(1^{y_1} 2^{y_2} \ldots)} \frac{1}{y_1! y_2! \ldots} \sum_{A \in \bar{\mathcal{S}}(\lambda)} \bar{n}_\lambda(A) h_1^{y_1} h_2^{y_2} \ldots = 0
$$

for $d > 1$, where the sum is over partitions of td having at most t parts. In this presentation h_d corresponds to $[k] \in \mathrm{Grot}(H(d))$, $d > 0$.

4.5 Some identifications

We identify the module $\nabla(\lambda)$ with the module of bideterminants, as in [51], and we identify the Specht module $\mathrm{Sp}(\lambda)$ with the Dipper–James Specht module, as in [20]. We first describe $\nabla(\lambda)$ by bideterminants.

Recall that, for $\lambda \in X(n)$, we have a 1-dimensional B-module k_λ with weight λ. Let $\tau_\lambda : k_\lambda \to k_\lambda \otimes k[B]$ be the structure map and let $a_\lambda \in k[B]$ be the element given by $\tau_\lambda(x) = x \otimes a_\lambda$, for all $x \in k_\lambda$. By definition, for $\lambda \in X^+(n)$, the module $\nabla(\lambda) = \mathrm{Ind}_B^G k_\lambda$ is the space of elements $x \otimes f \in k_\lambda \otimes k[G]$ such that $(\tau_\lambda \otimes \mathrm{id})(x \otimes f) = (\mathrm{id} \otimes (\pi \otimes \mathrm{id})\delta)(x \otimes f)$, where $\pi : k[G] \to k[B]$ is the restriction map. We identify $\nabla(\lambda)$ with the G-submodule of $k[G]$ consisting of the elements f such that $a_\lambda \otimes f = (\pi \otimes \mathrm{id})\delta(f)$.

Proposition 4.5.1 *Let M be a finite dimensional G-module with high weight $\lambda \in X^+(n)$ and $\dim M^\lambda = 1$. Let $m_1, \cdots, m_r$ be a basis of weight vectors with $m_1 \in M^\lambda$, and let $f_{ij}, 1 \le i, j \le r$, be the corresponding coefficient functions.*
(i) We have $f_{1i} \in \nabla(\lambda)$, for $1 \le i \le r$.
(ii) If M has a good filtration then $\nabla(\lambda)$ is spanned by the elements $f_{1i}, 1 \le i \le r$.

Proof (i) Let $N = \sum_{\mu \ne \lambda} M^\mu = km_2 + \ldots + km_r$. Since λ is a high weight of M, we have that N is a B-submodule of M. Since $M/N \cong k_\lambda$ we have $a_\lambda = \pi(f_{11})$. To show that $f_{1i} \in \nabla(\lambda)$, we must verify that $a_\lambda \otimes f_{1i} = (\pi \otimes \mathrm{id})\delta(f_{1i})$, i.e.

$$\pi(f_{11}) \otimes f_{1i} = \sum_j \pi(f_{1j}) \otimes \pi(f_{ji}) \tag{*}$$

for $1 \le i \le r$. However, for $2 \le j \le r$, we have $\tau_N(m_j) = \sum_i m_i \otimes \pi(f_{ij}) \in N \otimes k[B]$ and hence $\pi(f_{1j}) = 0$. This gives (*).
(ii) Since M has high weight λ, occurring with multiplicity 1, and has a good filtration, there is a G-submodule M', say, such that $M/M' \cong \nabla(\lambda)$ and λ is not a weight of M'. The linear map $\phi : M \to \nabla(\lambda)$, defined by $\phi(m_i) = f_{1i}, 1 \le i \le r$, is a G-module homomorphism. Since $\varepsilon(f_{11}) = 1$, the map ϕ is not zero. Since $L(\lambda) \le \nabla(\lambda)$, and has weight λ, we have $\phi(M') \cap L(\lambda) = 0$. But $L(\lambda)$ is the G-socle of $\nabla(\lambda)$ so that $\phi(M') = 0$ and so ϕ induces a non-zero G-module homomorphism $\bar\phi : M/M' \to \nabla(\lambda)$. Now $M/M' \cong \nabla(\lambda)$ and $\mathrm{End}_G(\nabla(\lambda)) = k$, so that any non-zero homomorphism $M/M' \to \nabla(\lambda)$ is an isomorphism. In particular $\bar\phi$ is an isomorphism and ϕ is surjective, i.e. $\nabla(\lambda)$ is spanned by the coefficient functions $f_{1i}, 1 \le i \le r$.

Proposition 4.5.2 *Let $\lambda \in \Lambda^+(n,r)$ and let $\lambda' = \mu = (\mu_1, \mu_2, \ldots)$. Let S denote the μ-tableau with first row entries $\mu_1, \ldots, 2, 1$ (from left to right), second row entries $\mu_2, \ldots, 2, 1$ and so on. Then $\nabla(\lambda)$ has k-basis $\{(S : T) \mid T \in \mathrm{AStan}(\mu)\}$.*

Proof The module $\bigwedge^\mu E$ has a good filtration and has highest weight λ, which occurs with multiplicity 1. The module $\bigwedge^\mu E$ has basis of weight vectors $\hat{e}_T$, $T \in \mathrm{Tab}_1(\mu)$, in the notation of Section 1.2. Furthermore we have $\hat{e}_S \in (\bigwedge^\mu E)^\lambda$, so that, by Proposition 4.5.1 and Lemma 1.3.1, $\nabla(\lambda)$ is spanned by $\{(S : T) \mid T \in \mathrm{Tab}_1(\mu)\}$. By Theorem 1.3.4, the elements

$(S : T)$, $T \in \text{AStan}(\mu)$, are linearly independent and hence the dimension of the space $\nabla(\lambda)'$ spanned by all such bideterminants has dimension equal of the number of antistandard μ-tableaux. This is equal to the number of standard λ-tableaux. Moreover, the dimension of $\nabla(\lambda)$ is independent of q (e.g. by Weyl's character formula) and is equal to the number of standard λ-tableaux when $q = 1$ (e.g. by [51; (4.5a)]). Hence we have $\nabla(\lambda)' = \nabla(\lambda)$ and $\{(S : T) \mid T \in \text{AStan}(\mu)\}$ is a k-basis.

For the rest of this section we assume $r \leq n$. To identify the Specht module $\text{Sp}(\lambda)$ with that of Dipper and James we give another realization of $\nabla(\lambda)$, this time as the image of a homomorphism between an exterior power and a symmetric power. For $\alpha \in \Lambda(n,r)$, we write $\bar{\alpha}$ for the partition obtained by writing the parts of α in descending order.

Proposition 4.5.3 Let $\alpha, \beta \in \Lambda(n,r)$ with $\lambda = \bar{\alpha}$ and $\mu = \bar{\beta}$.
(i) If $\text{Hom}_G(\bigwedge^{\alpha} E, S^{\lambda} E) \neq 0$ then we have $\lambda' \geq \mu$ and if $\lambda' = \mu$ then we have $\text{Hom}_G(\bigwedge^{\alpha} E, S^{\beta} E) = k$.
(ii) The image of any non-zero homomorphism from $\bigwedge^{\mu'} E \to S^{\mu} E$ is isomorphic to $\nabla(\mu)$.

Proof By 3.3 Remark (i) we have $\bigwedge^{\alpha} E \cong \bigwedge^{\lambda} E$ and $S^{\beta} E \cong S^{\mu} E$ so that we may assume $\alpha = \lambda$ and $\beta = \mu$. By 2.1(8), we have $\text{Hom}_G(\bigwedge^{\lambda} E, S^{\mu} E) \cong (\bigwedge^{\lambda} E)^{\mu}$. Now both assertions of (i) follow from the fact that $\bigwedge^{\lambda} E$ has unique highest weight λ' and this occurs with multiplicity 1.

Let $\phi : \bigwedge^{\mu'} E \to S^{\mu} E$ be a non-zero homomorphism. Since $\bigwedge^{\mu'} E \in \mathcal{F}(\nabla)$ and has highest weight μ occurring with multiplicity 1, there is a submodule M of $\bigwedge^{\mu'} E$ such that $\bigwedge^{\mu'} E/M$ is isomorphic to $\nabla(\mu)$ and all weights of M are less than μ. We now get $\text{Hom}_G(M, S^{\mu} E) = 0$, by 2.1(8), and hence $\phi(M) = 0$. Thus ϕ induces a non-zero homomorphism $\bar{\phi} : \nabla(\mu) \to S^{\mu} E$ and the image of $\bar{\phi}$ is the image of ϕ. If $\bar{\phi}(L(\mu)) = 0$ then $\bar{\phi}$ induces a non-zero map $\nabla(\mu)/L(\mu) \to S^{\mu} E$. But all weights of $\nabla(\mu)/L(\mu)$ are less than μ so that another application of 2.1(8) gives $\text{Hom}_G(\nabla(\mu)/L(\mu), S^{\mu} E) = 0$. Thus we must have $\bar{\phi}(L(\mu)) \neq 0$ and, since $L(\mu)$ is the socle of $\nabla(\mu)$, we get that $\bar{\phi}$ is injective. Thus the image of $\bar{\phi}$, and therefore also of ϕ, is isomorphic to $\nabla(\mu)$, as required.

To continue we shall need the following generality.

Proposition 4.5.4 Let $X \in \mathcal{F}(\nabla)$ be polynomial of degree r. We have $X = S(n,r) \cdot X^{\omega}$, where $\omega = (1,1,\ldots,1) \in \Lambda^{+}(n,r)$.

Proof We first take $X = Se$, where $S = S(n,r)$ and $e = \xi_{\omega}$. We have $S \cdot X^{\omega} = SeSe = Se = X$. Now suppose that $X = \nabla(\lambda)$ for some $\lambda \in \Lambda^{+}(n,r)$.

We have $Se \cong E^{\otimes r}$ by 2.1(7) and we have an S-module epimorphism $\theta :$ $E^{\otimes r} \to \nabla(\lambda)$. Hence we have $S \cdot \nabla(\lambda)^\omega = S \cdot \theta((E^{\otimes r})^\omega) = \theta(S \cdot (E^{\otimes r})^\omega) = \theta(E^{\otimes r}) = X$. Now suppose that X has a submodule $Y \in \mathcal{F}(\nabla)$ with $X/Y \cong \nabla(\lambda)$ for some $\lambda \in \Lambda^+(n,r)$ and that $Y = S \cdot Y^\omega$. From the above we get $X/Y = S \cdot (X/Y)^\omega = S \cdot (X^\omega + Y)$. Hence we have $X = S \cdot X^\omega + Y = S \cdot X^\omega + S \cdot Y^\omega = S \cdot X^\omega$ and so the result follows for arbitrary $X \in \mathcal{F}(\nabla)$ by induction on filtration length.

We now execute a base change argument. Let $R = k[t, t^{-1}]$, where t is an indeterminant. Recall that any $S(n,r)_R$-module X has its weight space decomposition $X = \bigoplus_{\alpha \in \Lambda(n,r)} X^\alpha$, where $X^\alpha = \xi_\alpha X$, $\alpha \in \Lambda(n,r)$.

Proposition 4.5.5 *Suppose X is a finitely generated $S(n,r)_R$-module such that, for every homomorphism of R into a field, $R \to k'$, the $S(n,r)_{k'}$-module $X_{k'} = k' \otimes_R X$ has a good filtration. Then we have $X = S(n,r)_R \cdot X^\omega$.*

Remark The base change condition on X is equivalent to the condition that it has a good filtration (suitably defined) as an $S(n,r)_R$-module; see [66].

Proof Let $Y = S(n,r)_R \cdot X^\omega$. Let $R \to k'$ be a homomorphism into a field. The image of $k' \otimes_R Y \to k' \otimes_R X$ is $S(n,r)_{k'} \cdot X_{k'}^\omega$, and this is X_k by the hypothesis. Thus we have $Y = X$, as in the last paragraph of the proof of Theorem 1.3.4.

We fix $\lambda = (\lambda_1, \lambda_2, \ldots) \in \Lambda^+(n,r)$. Let $T_1 : [\lambda] \to [1,r]$ be the tableau whose first row is $1, \ldots, \lambda_1$, whose second row is $\lambda_1+1, \ldots, \lambda_1+\lambda_2$, and so on, and let $T_2 : [\lambda] \to [1,r]$ be the tableau whose first *column* is $1, \ldots, \mu_1$, whose second column is $\mu_1 + 2, \ldots, \mu_1 + \mu_2$, and so on, where $\mu = (\mu_1, \mu_2, \ldots)$ is the transpose λ' of λ. We have a uniquely determined element $w_\lambda \in \mathrm{Sym}(r)$ such that $T_2 = w_\lambda \circ T_1$.

By [20; proof of 4.1 Lemma], we have a unique $(\mathrm{Sym}(\lambda'), \mathrm{Sym}(\lambda))$ double coset in $\mathrm{Sym}(r)$, $D = \mathrm{Sym}(\lambda')d\mathrm{Sym}(\lambda)$ such that $d^{-1}\mathrm{Sym}(\lambda')d \cap \mathrm{Sym}(\lambda) = \{1\}$. By [20; 1.6 Lemma (iii)], each element $g \in \mathrm{Sym}(\lambda')d\mathrm{Sym}(\lambda)$ has a unique expression $g = udv$ with $u \in \mathrm{Sym}(\lambda')$, $v \in \mathrm{Sym}(\lambda)$ and, if d is chosen to have minimal length in this double coset, then we have $l(udv) = l(u) + l(d) + l(v)$. Thus we have $y(\lambda')T_d x(\lambda) = \sum_{g \in D} c_g T_g$, where $c_g = (-q)^{N-l(u)}$ for $g \in D$ with $g = udv$, $u \in \mathrm{Sym}(\lambda')$, $v \in \mathrm{Sym}(\lambda)$. In particular we have $y(\lambda')T_d x(\lambda) \neq 0$. Moreover, if $g = udv \in D$ (with $u \in \mathrm{Sym}(\lambda')$ and $v \in \mathrm{Sym}(\lambda)$) then we have $y(\lambda')T_g x(\lambda) = (-q)^{l(u)} y(\lambda')T_d x(\lambda) \neq 0$. Furthermore we have $D = \mathrm{Sym}(\lambda')T_{w_\lambda}\mathrm{Sym}(\lambda)$, by [20; proof of 4.1 Lemma], so we have:

Proposition 4.5.6 $y(\lambda')T_{w_\lambda} x(\lambda) \neq 0$.

We append k' to our usual notation when we wish to emphasize that the construction is with respect to the general linear quantum group over the field k'. Let $K = k(t)$ and take, as above, $R = k[t, t^{-1}]$ (where t is an indeterminant). Let $\lambda' = \mu = (\mu_1, \ldots, \mu_m)$ (with $\mu_m > 0$) and let α denote the element of $\Lambda(n, r)$ obtained by reversing the order of the parts of λ', i.e. $\alpha = (\mu_m, \ldots, \mu_1)$.

By Proposition 4.5.6, right multiplication by $T_{w_{\lambda'}, K} x(\lambda)_K$ defines a non-zero $H(r)_K$-module $H(r)_K y(\lambda')_K \to H(r)_K x(\lambda)$. By 2.1(20), we have the isomorphism $H(r)_K \otimes_{H(\alpha)_K} K_s \to H(r)_K y(\lambda')_K$ taking $1 \otimes 1$ to $y(\lambda')_K$ and the isomorphism $H(r)_K \otimes_{H(\lambda)_K} k \to H(r)_K x(\lambda)_K$ taking $1 \otimes 1$ to $x(\lambda)_K$. Hence we have a non-zero homomorphism $H(r)_K \otimes_{H(\alpha)_K} K_s \to H(r)_K \otimes_{H(\lambda)_K} K$ taking $1 \otimes 1$ to $y(\lambda')_K T_{w_{\lambda'}, K} \otimes 1$. In view of the isomorphisms $H(r)_K \otimes_{H(\alpha)} K_s \to f_K \bigwedge^{\lambda'} E_K$ and $H(r)_K \otimes_{H(\lambda)} K \to f_K S^\lambda E_K$, of 2.1(20), we have a non-zero $H(r)_K$-module homomorphism $f_K \bigwedge^\alpha E_K \to f_K S^\lambda E_K$ taking $\hat{e}_{v,K}$ to $T_{w_{\lambda'}, K} \bar{e}_{u,K}$. Now by 2.1 Remark (iii), this $H(r)_K$-module homomorphism extends to an $S(n, r)_K$-module homomorphism $\bigwedge^\alpha E_K \to S^\lambda E_K$. Thus we have an $S(n, r)_K$-module $\phi : \bigwedge^\alpha E_K \to S^\lambda E_K$ such that $\phi(\hat{e}_{v,K}) = T_{w_{\lambda'}, K} \bar{e}_{u,K}$.

We write E_R for $Re_{1,K} + \cdots + Re_{n,K}$. We write $H(r)_R$ for the R-subalgebra of $H(r)_K$ generated by T_{s_a}, $a \in [1, n-1]$, and write $H(\alpha)_R$ for the R-subalgebra generated by T_{s_a}, $a \in J(\alpha)$, where $J(\alpha)$ is as in 2.1. We write R_s for R viewed as an $H(r)_R$-submodule of K_s. We write $\bigwedge^\alpha E_R$ for the R-submodule of $\bigwedge^\alpha E_K$ spanned by the elements $\hat{e}_{i,K}$, $i \in I(n, r)$. Similarly we write $S^\lambda E_R$ for the R-submodule of $S^\lambda E_K$ spanned by the elements $\bar{e}_{i,R}$, $i \in I(n, r)$. It is easy to check that $\bigwedge^\alpha E_R$ is an $S(n, r)_R$-submodule of $\bigwedge^\alpha E_K$ and that $S^\lambda E_R$ is an $S(n, r)_R$-submodule of $S^\lambda E_K$. Furthermore, it is easy to check that one has, for any homomorphism $R \to k'$ into a field, an $S(n, r)_{k'}$-module isomorphism $k' \otimes_R \bigwedge^\alpha E_R \to \bigwedge^\alpha E_{k'}$ taking $1 \otimes \hat{e}_{i,K}$ to $\hat{e}_{i,k'}$ and an $S(n, r)_{k'}$-module isomorphism $k' \otimes_R S^\lambda E_R \to S^\lambda E_{k'}$ taking $1 \otimes \bar{e}_{i,K}$ to $\hat{e}_{i,k'}$, for $i \in I(n, r)$.

We leave it to the reader to check that the isomorphism $H(r)_K \otimes_{H(\alpha)_K} K_s \to f_K \bigwedge^\alpha E_K$ restricts to an isomorphism $H(r)_R \otimes_{H(\alpha)_R} R_s \to (\bigwedge^\alpha E_R)^\omega$. In particular we have $(\bigwedge^\alpha E_R)^\omega = H(r)_R \hat{e}_v$. Now $\bigwedge^\alpha E_{k'}$ has a good $S(n, r)_{k'}$-module filtration for every homomorphism $R \to k'$ and every field k' so that, by Proposition 4.5.4, we have $\bigwedge^\alpha E_R = S(n, r)_R \cdot H(r)_R \hat{e}_v = S(n, r)_R \hat{e}_v$. Thus we have $\phi(\bigwedge^\alpha E_R) \leq S(n, r)_R y(\lambda')_K T_{w_{\lambda'}, K} \hat{e}_{u,K} \leq S^\lambda E_R$. Thus, by base change, we obtain an $S(n, r)$-module homomorphism $\theta : \bigwedge^\alpha E \to S^\lambda E$ such that $\theta(\hat{e}_v) = y(\lambda') T_{w_{\lambda'}} \bar{e}_u$. Now we have $\text{Im}(\theta) = \theta(S(n, r)\hat{e}_v) = S(n, r) y(\lambda') T_{w_{\lambda'}} \bar{e}_u$. Thus from Proposition 4.5.3 we have the following.

Proposition 4.5.7 $\nabla(\lambda)$ is isomorphic to the submodule of $S^\lambda E$ generated by $y(\lambda') T_{w_{\lambda'}} \bar{e}_u$.

Applying the Schur functor and identifying $f S^\lambda E$ with $H x(\lambda)$, as in 2.1(20), we obtain our promised identification of $\text{Sp}(\lambda)$ with the Dipper–James definition of the Specht module.

Proposition 4.5.8 $\text{Sp}(\lambda) \cong H(r)y(\lambda')T_{w_{\lambda'}}x(\lambda)$.

Remarks (i) This shows incidentally that the module $\text{Sp}(\lambda)$ is independent of the choice of $n \geq r$.
(ii) The presence of $T_{w_{\lambda'}}$ in the above rather than T_{w_λ}, which occurs in [20], may be accounted for by the fact that our Specht module is a left $H(r)$-module and that of Dipper and James is a right $H(r)$-module.

Examples Taking $\lambda = (r)$ we get $\text{Sp}(r) \leq Hx(r) = kx(r)$ so that $\text{Sp}(r) \cong k$. Similarly we get $\text{Sp}(1^r) \cong k_s$.

We now give the promised improvement on 4.4(13).

Proposition 4.5.9 For $\lambda \in \Lambda^+(n,r)$ we have $\text{Sp}(\lambda')^\# \cong \text{Sp}(\lambda)$.

Proof Note that if $S(n,r)$ (and hence $H(r)$) is semisimple then this is true by 4.4,(13). Let $\lambda' = \mu = (\mu_1, \ldots, \mu_m)$ and let $\alpha = (\alpha_m, \ldots, \alpha_2, \alpha_1)$. We have a surjective homomorphism $\phi : \bigwedge^\alpha E \to \nabla(\lambda)$ and so a surjection $f\phi : f\bigwedge^\alpha E \to f\nabla(\lambda)$ with kernel N, say. We have a non-singular contravariant form on $\bigwedge^\alpha E$, inducing one on $f\bigwedge^\alpha E$ (see the final paragraph of Section 4.1). Thus we get $N^\perp \cong \text{Sp}(\lambda)^\circ$. On the other hand, since $\text{Sp}(\lambda')$ is isomorphic to the left ideal $H(r)y(\lambda)T_{w_\lambda}x(\lambda')$, by Proposition 4.5.8, we get $\text{Sp}(\lambda')^\# \cong \#(H(r)y(\lambda)T_{w_\lambda}x(\lambda')) = H(r)x(\lambda')T_{w_\lambda}^\# y(\lambda)$ so that, in view of 2.1(20)(ii), we have

$$\text{Sp}(\lambda')^\# \cong H(r)x(\lambda)T_{w_\lambda}^\# \hat{e}_v \leq f\bigwedge^\alpha E. \tag{†}$$

Thus it suffices to prove that $H(r)x(\lambda)T_{w_\lambda}^\# \hat{e}_v = N^\perp$. Indeed it suffices to prove that

$$x(\lambda)T_{w_\lambda}^\# \hat{e}_v \in N^\perp$$

for then we get $\text{Sp}(\lambda')^\# \cong H(r)x(\lambda)T_{w_\lambda}^\# \hat{e}_v \leq N^\perp \cong \text{Sp}(\lambda)^\circ$ and, since $\dim \text{Sp}(\lambda')^\# = \dim \text{Sp}(\lambda)^\circ$ (e.g. by 4.4(13)), we must have equality.

Note that if $S(n,r)$ is semisimple then $\bigwedge^\alpha E$ contains a unique submodule isomorphic to $L(\lambda)$ and $f\bigwedge^\alpha E$ contains a unique submodule isomorphic to $\text{Sp}(\lambda)$, and this is isomorphic to $\text{Sp}(\lambda')^\#$, by 4.4(13). Thus, by (†) we get $H(r)x(\lambda)T_{w_\lambda}^\# \hat{e}_v = N^\perp \cong \text{Sp}(\lambda)^\circ \cong \text{Sp}(\lambda)$. In particular we have $x(\lambda)T_{w_\lambda}^\# \hat{e}_v \in N^\perp$, in this case.

We now append k' to our notation to indicate that constructions are made with respect to the quantum general linear group over the field k'. Let $K = k(t)$, where t is an indeterminant, and let $R = k[t]$. Let $\phi_K : \bigwedge^\alpha E_K \to S^\lambda E_K$ be the G_K-homomorphism such that $\phi_K(\hat{e}_{v,K}) = T_{w_{\lambda'},K}\bar{e}_{u,K}$ (see the proof of Proposition 4.5.6). Then ϕ_K restricts to an $S(n,r)_R$-module homomorphism $\phi_R : \bigwedge^\alpha E_R \to S^\lambda E_R$ (where the notation is as in the proof of Proposition 4.5.6). Let N_R denote the intersection of the kernel of ϕ_R with $f\bigwedge^\alpha E_K \cap \bigwedge^\alpha E_R$. Note that the form on $\bigwedge^\alpha E_R$ is obtained by base change from the form on $\bigwedge^\alpha E_K$ and that $N_k = k \otimes_R N_R$. Since $x(\lambda)_K T^{\#}_{w_\lambda,K} \hat{e}_{v,K} \in N_K^\perp$ we get $x(\lambda)_k T^{\#}_{w_\lambda,k} \hat{e}_{v,k} \in N_k^\perp$ by base change, as required.

4.6 Standard Levi subalgebras of Schur algebras

Let $\nu = (\nu_1,\ldots,\nu_m)$ be a composition of n. We put $Z_1(\nu) = [1,\nu_1]$, $Z_2(\nu) = [\nu_1, \nu_1 + \nu_2]$, ..., $Z_m(\nu) = [\nu_1 + \cdots + \nu_{m-1} + 1, n]$. We write $D(\nu, r)$ for the set of $(i,j) \in I(n,r) \times I(n,r)$ such that, for all $1 \le a \le r$ and $1 \le b \le m$, we have $i_a \in Z_b(\nu)$ if and only if $j_a \in Z_b(\nu)$. We define the ν-*type* of $i \in I(n,r)$ to be $\rho = (\rho_1,\ldots,\rho_m) \in \Lambda(m,r)$, where $\rho_b = |i^{-1}Z_b(\nu)|$, for $1 \le b \le m$. We define $D(\nu,\rho)$ to be the set of elements $(i,j) \in D(\nu,r)$ such that i (and hence j) is of ν-type ρ.

Let $\lambda = (\lambda_1, \lambda_2,\ldots) \in \Lambda(n,r)$. We say that $j \in I(n,r)$ is λ-*decreasing* if $j_{\lambda_1} \ge \cdots \ge j_2 \ge j_1$, $j_{\lambda_1+\lambda_2} \ge \cdots \ge j_{\lambda_1+1}$, and so on. We write $Q(n,r)$ for the set of all $(i,j) \in I(n,r) \times I(n,r)$ such that i is weakly increasing and j is λ-decreasing, where λ is the content of i. We recall, from Section 2.1, that $A(n,r)$ has k-basis $\{c_{ij} \mid (i,j) \in Q(n,r)\}$ and hence $S(n,r)$ has the dual basis $\{\xi_{ij} \mid (i,j) \in Q(n,r)\}$.

We have the quantum submonoid $M(\nu)$ of $M(n)$, as defined in [36; Section 2]. Let $A(\nu) = k[M(\nu)]$ and let $R(\nu)$ be the kernel of restriction $k[M(n)] \to k[M(\nu)]$ (i.e. $R(\nu)$ is the defining ideal of $M(\nu)$ in $M(n)$). For $i,j \in I(n,r)$ we write $\bar{c}_{ij}$ for the restriction of c_{ij} to $M(\nu)$. Now $R(\nu) = \bigoplus_{r=0}^\infty R(\nu,r)$ is a graded biideal and coideal so that $A(\nu)$ inherits a grading $A(\nu) = \bigoplus_{r=0}^\infty A(\nu,r)$. Furthermore, we have the natural isomorphism $M(\nu) \to M(\nu_1) \times \cdots \times M(\nu_m)$ so that, by transport of structure, we obtain a multigrading $A(\nu) = \bigoplus_{\rho \in \mathbb{N}_0^m} A(\nu,\rho)$, with $A(\nu,r) = \bigoplus_{\rho \in \Lambda(m,r)} A(\nu,\rho)$, for $r \ge 0$. We write $\phi_{\nu\rho}$ for the coalgebra isomorphism $A(\nu_1,\rho_1) \otimes \cdots \otimes A(\nu_m,\rho_m) \to A(\nu,\rho)$, obtained by restricting the isomorphism $k[M(\nu_1) \times \cdots \times M(\nu_m)] \to k[M(\nu)]$ to degree $\rho = (\rho_1,\ldots,\rho_m) \in \mathbb{N}_0^m$. We have

$$\dim A(\nu,\rho) = \prod_{i=1}^m \dim A(\nu_i,\rho_i) = \prod_{i=1}^m |Q(\nu_i,\rho_i)|$$

and so

$$\dim A(\nu,r) = \sum_{\rho \in \Lambda(m,r)} \dim A(\nu,\rho) = \sum_{\rho \in \Lambda(m,r)} \prod_{i=1}^m |Q(\nu_i,\rho_i)|$$

for $\rho = (\rho_1,\ldots,\rho_m) \in \mathbb{N}_0^m$.

We set $Q(\nu, \rho) = Q(n, r) \cap D(\nu, \rho)$ and note that $|Q(\nu, \rho)| = \prod_{i=1}^{m} |Q(\mu, \rho)|$. Suppose that $(i, j) \in Q(n, r) \backslash \bigcup_{\rho \in \Lambda(m, r)} Q(\nu, \rho)$. Then, for some $1 \leq a \leq r$ and $1 \leq b \leq m$, we have $i_a \in Z_b(\nu)$ and $j_a \notin Z_b(\nu)$ so that $c_{i_a j_a} \in R(\nu)$ and hence $c_{ij} \in R(\nu)$. By a dimension count we therefore get:

(1) $\{c_{ij} \mid (i, j) \in Q(n, r) \backslash \bigcup_{\rho \in \Lambda(m, r)} Q(\nu, \rho)\}$ is a k-basis of $R(\nu)$ and $\{\bar{c}_{ij} \mid (i, j) \in \bigcup_{\rho \in \Lambda(m, r)} Q(\nu, \rho)\}$ is a k-basis of $A(\nu, r)$.

Similarly we get:

(2) For $\rho \in \Lambda(m, r)$ we have that $\{\bar{c}_{ij} \mid (i, j) \in Q(\nu, \rho)\}$ is a k-basis of $A(\nu, \rho)$.

We define the dual algebras $S(\nu, r) = A(\nu, r)^*$ and $S(\nu, \rho) = A(\nu, \rho)^*$, for $r \in \mathbb{N}_0$, $\rho \in \Lambda(m, r)$. The dual of restriction $A(n, r) \to A(\nu, r)$ gives an injective algebra map $S(\nu, r) \to S(n, r)$ by which we identify $S(\nu, r)$ with a subalgebra of $S(n, r)$. Dualizing the coalgebra decomposition $A(\nu, r) = \bigoplus_{\rho \in \Lambda(m, r)} A(\nu, \rho)$ gives an algebra decomposition $S(\nu, r) = \bigoplus_{\rho \in \Lambda(m, r)} S(\nu, \rho)$. We call the algebras $S(\nu, r)$ and $S(\nu, \rho)$ *Levi subalgebras* of $S(n, r)$, for $r \in \mathbb{N}_0$, $\rho \in \Lambda(m, r)$. From (1) and (2) we get:

(3) $\{\xi_{ij} \mid (i, j) \in Q(\nu, \rho)\}$ is a k-basis of $S(\nu, \rho)$.

Writing $1 = \sum_{\rho \in \Lambda(m, r)} \eta_{\nu \rho}$, as an orthogonal sum of central idempotents, according to the algebra decomposition $S(\nu, r) = \bigoplus_{\rho \in \Lambda(m, r)} S(\nu, \rho)$, it is not difficult to convince oneself of the following.

(4) For $\rho \in \Lambda(m, r)$ we have $\eta_{\nu \rho} = \sum_\alpha \xi_\alpha$, where the sum is over all $\xi_\alpha = \xi_{ii}$ with i of ν-type ρ.

We have an exact functor $d_{\nu \rho} : \mathrm{mod}(S(\nu, r)) \to \mathrm{mod}(S(\nu, \rho))$ as follows. For $U \in \mathrm{mod}(S(\nu, r))$ we set $d_{\nu \rho} U = \eta_{\nu \rho} U$ and for a morphism $\theta : U \to U'$ of $S(\nu, r)$-modules we set $d_{\nu \rho} \theta : \eta_{\nu \rho} U \to \eta_{\nu \rho} U'$ to be the restriction of θ.

We now take $r = n$. Recall that $e = \xi_\omega$, where $\omega = (1, \ldots, 1)$.

(5) $eS(\nu, \nu)e = H(\nu)$.

Proof Note that $eS(\nu, \nu)e = eS(n, n)e \cap S(\nu, \nu)$. For $\pi \in \mathrm{Sym}(\nu)$ we have $\xi_{u, u\pi} \in eS(\nu, \nu)e$, by (3). Hence $H(\nu)$, the k-span of $\{\xi_{u, u\pi} \mid \pi \in \mathrm{Sym}(\nu)\}$, is contained in $eS(\nu, \nu)e$. Now let $(i, j) \in Q(\nu, \nu)$ and suppose that $\xi_{uu}\xi_{ij}\xi_{uu} \neq 0$. Then we have $i \sim u$, by 2.1(4). Since i is weakly increasing we have $i = u$. Also, $\xi_{ij}\xi_{uu} \neq 0$ gives $j \sim u$. Hence $j = u\pi$, for some $\pi \in \mathrm{Sym}(r)$, and the condition $(u, u\pi) \in Q(\nu, \nu)$ gives $\pi \in \mathrm{Sym}(\nu)$.

Thus we have an exact functor $f_\nu : \mathrm{mod}(S(\nu,\nu)) \to \mathrm{mod}(H(\nu))$, defined on objects by $f_\nu U = eU$, and defined on morphisms by restriction.

We need to keep track of the degree of the general linear group over which we are working so we now write $\nabla(n,\lambda)$ for the $G(n)$-module $\nabla(\lambda)$ if we wish to emphasize the role of n. For $\rho = (\rho_1,\ldots,\rho_m) \in \mathbb{N}_0^m$ let $\Lambda^+(\nu,\rho) = \Lambda^+(\nu_1,\rho_1) \times \cdots \times \Lambda^+(\nu_m,\rho_m)$. For $\lambda = (\lambda(1),\ldots,\lambda(m)) \in \Lambda^+(\nu,\rho)$ we have the module $\nabla_\Sigma(\lambda)$ for $G(\nu)$ (where Σ is the subset of Π corresponding to ν, as in Section 4.2). Regarding the $G(\nu_1) \times \cdots \times G(\nu_m)$-module $\nabla(\nu_1,\lambda(1)) \otimes \cdots \otimes \nabla(\nu_m,\lambda(m))$ as a $G(\nu)$-module via the natural isomorphism $G(\nu) \to G(\nu_1) \times \cdots \times G(\nu_m)$ we have $\nabla_\Sigma(\lambda) \cong \nabla(\nu_1,\lambda(1)) \otimes \cdots \otimes \nabla(\nu_m,\lambda(m))$.

Now $\nabla_\Sigma(\lambda)$ is a polynomial $G(\nu)$-module of degree r and any polynomial $G(\nu)$-module U of degree r decomposes as a direct sum $U = \bigoplus_{\rho \in \Lambda(m,r)} \eta_{\nu\rho} U$. Since $\nabla_\Sigma(\lambda)$ is indecomposable we must have $\eta_{\nu\rho}\nabla_\Sigma(\lambda) \neq 0$ for precisely one $\rho \in \Lambda^+(m,r)$. We also write $\lambda \in \Lambda(n,r)$ for the result of concatenating $\lambda(1), \lambda(2), \ldots, \lambda(m)$. We have $\dim \nabla(\lambda)^\lambda = 1$ so that $\xi_\lambda \nabla(\lambda)^\lambda \neq 0$ and hence $\eta_{\nu\rho}\nabla(\lambda)^\lambda \neq 0$, where $\rho = (\rho_1,\ldots,\rho_m)$, with $\rho_i = |\lambda(i)|$, for $1 \leq i \leq m$. Hence we have:

(6) $\eta_{\nu\rho}\nabla_\Sigma(\lambda) = \begin{cases} \nabla_\Sigma(\lambda), & \text{if } \rho = (|\lambda(1)|,\ldots,|\lambda(m)|); \\ 0, & \text{otherwise} \end{cases}$

for $\rho \in \Lambda(m,r)$.

We express the character $\chi(\lambda)$ as a sum of characters of $\nabla_\Sigma(\mu)$'s, i.e.

$$\chi(\lambda) = \sum_{\rho \in \Lambda(m,r)} \sum_{\mu \in \Lambda^+(\nu,\rho)} a_\mu \chi_\Sigma(\mu).$$

The coefficient a_μ is the multiplicity of $\nabla_\Sigma(\mu)$ as a composition factor of $\nabla(\lambda)|_{G(\nu)}$ in the (very) classical case $k = \mathbb{C}$ and $q = 1$.

(7) $\nabla(\lambda)$ has an $S(\nu,r)$-module filtration $0 = \nabla_0 < \nabla_1 < \cdots < \nabla_h = \nabla(\lambda)$, where $\nabla_i/\nabla_{i-1} \cong \nabla_\Sigma(\mu^i)$ and $|\{i \in [1,h] \mid \mu^i = \mu\}| = a_\mu$, for $\mu \in \Lambda^+(\nu,r)$. Moreover, we may arrange the order such that $\mu^i <_\Sigma \mu^j$ implies $i < j$.

Applying $\eta_{\nu\rho}$, for $\rho \in \Lambda(m,r)$, we get the following.

(8) $\eta_{\nu\rho}\nabla(\lambda)$ has an $S(\nu,\rho)$-module filtration $0 = \nabla_0' < \nabla_1' < \cdots < \nabla_g' = \eta_{\nu\rho}\nabla(\lambda)$, where $\nabla_i'/\nabla_{i-1}' \cong \nabla_\Sigma(\tau^i)$, with $\tau^i \in \Lambda^+(\nu,\rho)$, for $1 \leq g \leq g$, and $|\{i \in [1,g] \mid \tau^i = \tau\}| = a_\tau$, for $\tau \in \Lambda^+(\nu,\rho)$. Moreover, we may arrange the order such that $\tau^i <_\Sigma \tau^j$ implies $i < j$.

We now specialize to the case $n = r$ and $\rho = \nu$. For $\mu = (\mu(1),\ldots,\mu(m)) \in \Lambda^+(\nu,\nu)$ we set $\mathrm{Sp}_\Sigma(\mu) = e\nabla_\Sigma(\mu)$, which is naturally an $H(\nu)$-module, as $H(\nu) = eS(\nu,\nu)e$. Our identification of $\nabla_\Sigma(\mu)$ with $\nabla(\nu_1,\mu(1)) \otimes \cdots \otimes$

$\nabla(\nu_m, \mu(m))$ identifies $\nabla_\Sigma(\mu)^\omega$ with $\nabla(\nu_1, \mu(1))^{\omega_1} \otimes \cdots \otimes \nabla(\nu_m, \mu(m))^{\omega_m}$ (where $\omega_i = (1, 1, \ldots, 1) \in \Lambda(\nu_i, \nu_i)$ and $|\mu(i)| = \nu_i$, for $1 \le i \le m$) so we have:

(9) $\operatorname{Sp}_\Sigma(\mu) \cong \operatorname{Sp}(\mu(1)) \otimes \cdots \otimes \operatorname{Sp}(\mu(m))$, as $H(\nu) = H(\nu_1) \otimes \cdots \otimes H(\nu_m)$-modules.

Moreover, from (8) we get:

(10) $\operatorname{Sp}(\lambda) = e\nabla(\lambda)$ has an $H(\nu)$-module filtration $0 = \operatorname{Sp}_0 < \operatorname{Sp}_1 < \cdots < \operatorname{Sp}_g = \operatorname{Sp}(\lambda)$ with $\operatorname{Sp}_i / \operatorname{Sp}_{i-1} \cong \operatorname{Sp}_\Sigma(\tau^i)$ and $|\{i \in [1, g] \mid \tau^i = \tau\}| = a_\tau$, for $\rho \in \Lambda^+(\nu, \rho)$. Moreover, we may arrange the order such that $\tau^i <_\Sigma \tau^j$ implies $i < j$.

We conclude with a result which will be needed in Section 4.7.

(11) Let $X \in \mathcal{F}(\nabla)$.
(i) For any $Y \in \operatorname{mod}(S(n, r))$ the natural map $\operatorname{Hom}_G(X, Y) \to \operatorname{Hom}_H(fX, fY)$ is injective.
(ii) Assume that $1 + q \ne 0$. For $\mu \in \Lambda(n, r)$, the natural map $\operatorname{Hom}_G(X, S^\mu E) \to \operatorname{Hom}_H(fX, fS^\mu E)$ is an isomorphism.

Proof (i) Let $\theta \in \operatorname{Hom}_G(X, Y)$. If $\theta(fX) = 0$ then $\theta = 0$ by Proposition 4.5.4.
(ii) We have $\dim \operatorname{Hom}_G(X, S^\mu E) = \dim X^\mu$ so, by part (i), it suffices to show that $\dim \operatorname{Hom}_H(fX, fS^\mu E) \le \dim X^\mu$. Thus, by left exactness of $\operatorname{Hom}_H(f-, S^\mu E)$, it suffices to prove this in the case $X = \nabla(\lambda)$, for $\lambda \in \Lambda^+(n, r)$.
We prove this first in the case $\mu = (r)$. Assume $\operatorname{Hom}_H(f\nabla(\lambda), fS^\mu E) \ne 0$. Let α denote the element of $\Lambda^+(n, r)$ obtained by writing the entries in λ' in reverse order. Thus we have a surjection $\bigwedge^\alpha E \twoheadrightarrow \nabla(\lambda)$, by Proposition 4.5.3(ii). We have $Hx(r) = k$ so that if $\operatorname{Hom}_H(\operatorname{Sp}(\lambda), Hx(r)) \ne 0$ then $\operatorname{Hom}_H(f\bigwedge^\alpha E, k) \ne 0$. Moreover we have $f\bigwedge^\alpha E \cong H \otimes_{H(\lambda')} k_s$, by 2.1(20)(ii). Thus we get $\operatorname{Hom}_{H(\lambda')}(k_s, k) \ne 0$, by Frobenius reciprocity, i.e. k_s and k are isomorphic as $H(\lambda')$-modules. In particular, if a is any entry of λ' then $k_s \cong k$ as $H(a)$-modules. Since $q + 1 \ne 0$, this can only happen if $a = 1$. Thus $\lambda' = (1^r)$ and $\lambda = (r)$. For $\lambda = (r)$ we have $\operatorname{Hom}_H(\operatorname{Sp}(\lambda), Hx(r)) = \operatorname{Hom}_H(k, k) = k$ and we get $\dim \operatorname{Hom}_H(\operatorname{Sp}(\lambda), Hx(r)) = \dim \nabla(\lambda)^{(r)}$ in all cases.
Now let λ, μ be arbitrary. Note that, for any $U \in \operatorname{mod}(H)$, we have

$$\operatorname{Hom}_H(U, Hx(\mu)) \cong \operatorname{Hom}_H((Hx(\mu))^\circ, U^\circ)$$
$$\cong \operatorname{Hom}_H(Hx(\mu), U^\circ) \cong \operatorname{Hom}_{H(\mu)}(k, U^\circ)$$
$$\cong \operatorname{Hom}_{H(\mu)}(U, k)$$

by 4.4(11) and Frobenius reciprocity. Thus we have $\text{Hom}_H(\text{Sp}(\lambda), Hx(\mu)) \cong \text{Hom}_{H(\mu)}(\text{Sp}(\lambda), k)$. Moreover, by (10), we have an $H(\mu)$-module filtration $0 = \text{Sp}_0 < \text{Sp}_1 < \cdots < \text{Sp}_g = \text{Sp}(\lambda)$ with $\text{Sp}_i/\text{Sp}_{i-1} \cong \text{Sp}_\Sigma(\tau^i)$ and $|\{i \in [1, g] \mid \tau^i = \tau\}| = a_\tau$, where Σ is the subset of Π corresponding to μ and the numbers a_τ are defined by the equation

$$\chi(\lambda) = \sum_{\rho \in \Lambda(m,r)} \sum_{\tau \in \Lambda^+(\mu,\rho)} a_\tau \chi_\Sigma(\tau).$$

Thus we get, by left exactness,

$$\dim \text{Hom}_H(\text{Sp}(\lambda), Hx(\mu)) \leq \sum_\tau a_\tau \dim \text{Hom}_{H(\mu)}(\text{Sp}_\Sigma(\tau), k).$$

and, by the case already considered, and (9), this is a_μ. Moreover, in the generic case (q an indeterminate) this gives $\dim \text{Hom}_H(\text{Sp}(\lambda), Hx(\mu)) = a_\mu$, by complete reducibility. Furthermore, in the generic case, the Schur functor is an equivalence of categories and hence $a_\mu = \dim \text{Hom}_G(\nabla(\lambda), S^\mu E) = \dim \nabla(\lambda)^\mu$. Thus, in general, we have
$\dim \text{Hom}_G(\nabla(\lambda), S^\mu E) \geq \dim \text{Hom}_H(\text{Sp}(\lambda), Hx(\mu))$, completing the proof.

4.7 Quotients of Hecke algebras

Let $n \geq r$ and let $\pi \subseteq \Lambda^+(n,r)$ be a cosaturated set of dominant weights consisting of row regular partitions. Here cosaturated means that the complement $\pi^c = \Lambda^+(n,r)\backslash\pi$ is saturated, in the sense that whenever we have $\lambda \in \pi^c$ and $\mu \in \Lambda^+(n,r)$ is such that $\mu \leq \lambda$ then we also have $\mu \in \pi^c$. Associated to π we have an idempotent $\xi \in S(n,r)$. The algebra $\xi S(n,r)\xi$ is quasihereditary and we shall show that the Ringel dual $(\xi S(n,r)\xi)'$ is equivalent to a certain quotient of the Hecke algebra $H(r)$. This generalizes a result of Erdmann, [45], which applies when $q = 1$ and $\pi = \Lambda^+(m,r)$ (for $m \leq n$) consists of row regular partitions.

We first observe that $S(n,r)$ has a theory of weights, in the sense of the Appendix, Definition A3.8. We define $\theta : \Lambda(n,r) \to S(n,r)$ by $\theta(\alpha) = \xi_\alpha$. Note that $1 = \sum_{\alpha \in \Lambda(n,r)} \theta(\alpha) = \sum_{\alpha \in \Lambda(n,r)} \xi_\alpha$ is an decomposition of 1 as an orthogonal sum of non-zero idempotents, and condition (i) for a theory of weights is satisfied. For $\alpha \in \Lambda(n,r)$ and $\lambda \in \Lambda^+(n,r)$ we have $\dim \text{Hom}_{S(n,r)}(S(n,r)\xi_\alpha, L(\lambda)) = \dim L(\lambda)^\alpha$. Thus, writing

$$S(n,r)\xi_\alpha = \bigoplus_{\alpha \in \Lambda^+(n,r)} P(\lambda)^{(d_\lambda)}$$

(where $P(\lambda)$ is the projective cover of $L(\lambda)$) we have $d_\lambda = \dim L(\lambda)^\alpha$. In particular we have $d_{\bar{\alpha}} = 1$ (where $\bar{\alpha}$ is the dominant weight conjugate obtained by writing the parts of α in descending order) and if $d_\mu \neq 0$, for $\mu \in \Lambda^+(n,r)$, then $L(\mu)^\alpha \neq 0$ and hence $\mu \geq \bar{\alpha}$. This verifies condition (ii) and shows that the element α^+ of condition (ii) is $\bar{\alpha}$, for $\alpha \in \Lambda(n,r)$. Certainly the map $\Lambda(n,r) \to \Lambda^+(n,r)$, taking α to $\alpha^+ = \bar{\alpha}$, is onto, so that property (iii) is satisfied and $\theta : \Lambda(n,r) \to S(n,r)$ is a theory of weights. (Indeed this is the motivating example for the general definition.)

Let π be a cosaturated subset of $\Lambda^+(n,r)$ and let $\Gamma \subseteq \Lambda$ be such that $\Gamma^+ = \pi$ (where $\Gamma^+ = \{\alpha^+ \mid \alpha \in \Gamma\}$). We put $\xi = \xi_\Gamma$, i.e. $\xi_\Gamma = \sum_{\alpha \in \Gamma} \xi_\alpha$. Assume that $n \geq r$ and let $e = \xi_\omega$. We put $S = S(n,r)$ and $S_\xi = \xi S \xi$. We have $Se \cong E^{\otimes r}$. In particular Se is a tilting module. Hence, by the Appendix, Lemma A3.9, the restriction map $\rho : \operatorname{End}_S(Se) \to \operatorname{End}_{S_\xi}(\xi Se)$ is surjective. We calculate the dimension of the kernel of ρ. For $\lambda \in \Lambda^+(n,r)$ we have $(Se : \nabla(\lambda)) = \dim \operatorname{Hom}_S(Se, \nabla(\lambda)) = \dim \nabla(\lambda)^\omega = \dim \operatorname{Sp}(\lambda)$, by Proposition A2.2(ii). Thus we have $\operatorname{ch} E^{\otimes r} = \sum_{\lambda \in \Lambda^+(n,r)} \dim \operatorname{Sp}(\lambda) \operatorname{ch} \nabla(\lambda)$. Since we have $\operatorname{ch} \nabla(\lambda) = \operatorname{ch} \Delta(\lambda)$ we also have $\operatorname{ch} E^{\otimes r} = \sum_{\lambda \in \Lambda^+(n,r)} \dim \operatorname{Sp}(\lambda) \operatorname{ch} \Delta(\lambda)$ and so $(Se : \Delta(\lambda)) = \dim \operatorname{Sp}(\lambda)$, for $\lambda \in \Lambda^+(n,r)$. Hence, by Proposition A2.2(ii), we have $\dim \operatorname{End}_S(Se) = \sum_{\lambda \in \Lambda^+} (\dim \operatorname{Sp}(\lambda))^2$. It follows from Proposition A3.11(ii) (and Proposition A2.2(ii)) that $\dim \operatorname{End}_{S_\xi}(\xi Se) = \sum_{\lambda \in \pi} \dim (\operatorname{Sp}(\lambda))^2$. Hence we have:

(1) $\rho : \operatorname{End}_S(Se) \to \operatorname{End}_{S_\xi}(\xi Se)$ is surjective and $\operatorname{Ker}(\rho)$ has dimension $\sum_{\lambda \in \pi^c} \dim \operatorname{Sp}(\lambda)^2$, where $\pi^c = \Lambda^+ \backslash \pi$.

Note we have an isomorphism $(eSe)^{\mathrm{op}} \to \operatorname{End}_S(Se)$ and so, combining this with the above, we have the surjective algebra homomorphism $\rho' : (eSe)^{\mathrm{op}} \to \operatorname{End}_{S_\xi}(\xi Se)$, given by $\rho'(x)(y) = yx$, for $x \in eSe$, $y \in \xi Se$. Since $Se \cong E^{\otimes r}$ we have $Se \in \mathcal{F}(\nabla)$ and hence $O_{\pi^c}(Se) \in \mathcal{F}_{\pi^c}(\nabla)$ and it follows from Proposition A3.11(i), that $\xi O_{\pi^c}(Se) = 0$. Hence $eO_{\pi^c}(Se) \leq \operatorname{Ker}(\rho')$. However, $O_{\pi^c}(Se)$ has a filtration with sections $\nabla(\lambda)$, $\lambda \in \pi^c$, by Lemma A3.1(ii), with $\nabla(\lambda)$ occurring $\dim \nabla(\lambda)^\omega = \dim \operatorname{Sp}(\lambda)$ times. Thus we have:

(2) $eO_{\pi^c}(Se)$ has dimension $\sum_{\lambda \in \pi^c} (\dim \operatorname{Sp}(\lambda))^2$ and is precisely the kernel of ρ'. Hence ρ' induces an isomorphism $(eSe)^{\mathrm{op}}/eO_{\pi^c}(Se) \to \operatorname{End}_{S_\xi}(\xi Se)$.

We put $I(\pi^c) = eO_{\pi^c}(Se)$. Now $eSe = H(r)$ and we have the antiautomorphism $J : H(r) \to H(r)$ taking $T_{s_a} \to T_{s_a}$, for $1 \leq a < r$. Combining this with the above we obtain an isomorphism $\rho'' : H(r)/I(\pi^c) \to \operatorname{End}_{S_\xi}(\xi Se)$.

There is another description of $I(\pi^c)$, which we now give. Let $K = k(t)$ (where t is an indeterminate). Let $S_K = S(n,r)_K$ be the Schur algebra over K and $S(n,r)_R$ the Schur algebra over $R = k[t, t^{-1}]$. Let $e_R, \xi_{\Gamma,R} \in S(n,r)_R$ be the idempotents as above but over R. We let $I(\pi^c)_K$ be the corresponding ideal over K and $I(\pi^c)_R = I(\pi^c)_K \cap H(r)_R$. Then we have $I(\pi^c)_R \leq \operatorname{Ker}(\rho'')$ and by base change we obtain $k \otimes I(\pi^c)_R \leq \operatorname{Ker}(\rho'')$. By dimensions we obtain:

(3) $I(\pi^c) = k \otimes_R I(\pi^c)_R$.

Note that, from the semisimplicity of $S(n,r)_K$, we get that $I(\pi^c)_K$ is the direct sum of all submodules of $H(r)_K$ which are isomorphic to $\mathrm{Sp}(\lambda)_K$ for some $\lambda \in \pi^c$.

Now assume that π consists of row regular partitions. We define $T = Se \oplus (\bigoplus_{\lambda \in \pi^c} T(\lambda))$. By 4.3(4), (9) and (10) we have that $T(\lambda)$ occurs as a component of Se, for all $\lambda \in \pi$. Thus T is a full tilting module. Hence we have $S'(\pi) \cong \mathrm{End}_{S_\xi}(\xi T)^{\mathrm{op}} = \mathrm{End}_{S_\xi}(\xi Se)^{\mathrm{op}}$, by Proposition A4.9, and so:

(4) $\quad S'(\pi) \cong S_\xi^{\mathrm{op}} \cong \mathrm{End}_{S_\xi}(\xi Se)^{\mathrm{op}}$ *is isomorphic to* $H(r)/I(\pi^c)$.

In particular $H(r)/I(\pi^c)$ is a quasihereditary algebra.

It was claimed in [45] that the category $\mathcal{F}_\pi(\mathrm{Sp})$ of finite dimensional $H(r)$-modules which have a filtration by the modules of the form $\mathrm{Sp}(\lambda)$, $\lambda \in \pi$ (for $q = 1$, $\pi = \Lambda^+(m,r)$, $m \le n$), is equivalent to the category of $S(n,r)$-modules which have a filtration with terms $\Delta(\lambda)$, $\lambda \in \pi$. However, as noticed by Cline, Parshall and Scott, [12], it is not shown in [45] that such an $H(r)$-module must be annihilated by $I(\pi)$. This problem is rectified in [12; (3.8.3)] (for the case $q = 1$, $\pi = \Lambda^+(n,r)$). We found the following general argument (which is anyway quite different) before seeing [12], but after being informed of the difficulty in [45].

(5) *Suppose that* $1 + q \ne 0$. *Let* π *be a cosaturated subset of* $\Lambda^+(n,r)$ *consisting of row regular weights. Then* $I(\pi)$ *acts trivially on every* $H(r)$-*module which admits a filtration with sections of the form* $\mathrm{Sp}(\lambda)$, $\lambda \in \pi$.

Proof Let $H = H(r)$, $I = I(\pi)$ and $A = H/I$. We claim that $\mathrm{Ext}_H^1(A,Y) = 0$ for all $Y \in \mathrm{mod}(H)$ which are filtered by $\mathrm{Sp}(\mu)$'s, with $\mu \in \pi$. It suffices to prove that $\mathrm{Ext}_H^1(A, \mathrm{Sp}(\mu)) = 0$, for $\mu \in \pi$. We have an exact sequence

$$0 \to \mathrm{Hom}_H(A, \mathrm{Sp}(\mu)) \to \mathrm{Hom}_H(H, \mathrm{Sp}(\mu)) \to \mathrm{Hom}_H(I, \mathrm{Sp}(\mu))$$
$$\to \mathrm{Ext}_H^1(A, \mathrm{Sp}(\mu)) \to 0.$$

Now $\mathrm{Hom}_H(A, \mathrm{Sp}(\mu)) = \mathrm{Hom}_A(A, \mathrm{Sp}(\mu))$ (as $\mathrm{Sp}(\mu)$ is an A-module) so $\dim \mathrm{Hom}_H(A, \mathrm{Sp}(\mu)) = \dim \mathrm{Sp}(\mu)$ and of course $\dim \mathrm{Hom}_H(H, \mathrm{Sp}(\mu)) = \dim \mathrm{Sp}(\mu)$ so that $\mathrm{Hom}_H(I, \mathrm{Sp}(\mu)) \cong \mathrm{Ext}_H^1(A, \mathrm{Sp}(\mu))$. To establish the claim it therefore suffices to show that $\mathrm{Hom}_H(I, \mathrm{Sp}(\mu)) = 0$, for all $\mu \in \pi$. Since $\mathrm{Sp}(\mu)$ embeds in $Hx(\mu)$, it therefore suffices to prove that $\mathrm{Hom}_H(I, Hx(\mu)) = 0$. Since I is filtered by modules $\mathrm{Sp}(\lambda)$, with $\lambda \in \pi^c$, it suffices to note that $\mathrm{Hom}_H(\mathrm{Sp}(\lambda), Hx(\mu)) = 0$, for $\lambda \in \pi^c$, $\mu \in \pi$. By 4.6(10), we have $\mathrm{Hom}_H(\mathrm{Sp}(\lambda), Hx(\mu)) \cong \mathrm{Hom}_G(\nabla(\lambda), S^\mu E) \cong \nabla(\lambda)^\mu$. If this is non-zero then one must have $\mu \le \lambda$ and since π^c is saturated, $\mu \in \pi^c$, which is impossible since $\mu \in \pi$. This completes the proof of the claim.

Now suppose that $X, Y \in \mathrm{mod}(A)$ admit filtrations by $\mathrm{Sp}(\lambda)$'s, with $\lambda \in \pi$. Let $F \in \mathrm{mod}(A)$ be a free module mapping onto X with kernel N, say. We get exact sequences

$$0 \to \mathrm{Hom}_A(X,Y) \to \mathrm{Hom}_A(F,Y) \to \mathrm{Hom}_A(N,Y) \to \mathrm{Ext}_A^1(X,Y) \to 0$$

and

$$0 \to \text{Hom}_H(X,Y) \to \text{Hom}_H(F,Y) \to \text{Hom}_H(N,Y) \to \text{Ext}_H^1(X,Y) \to 0$$

since $\text{Ext}_H^1(F,Y) = 0$, by the claim just proved. But $\text{Hom}_H(X,Y) = \text{Hom}_A(X,Y)$ (for all A-modules X,Y). Hence we have dim $\text{Ext}_A^1(X,Y) = $ dim $\text{Hom}_H(X,Y)$ so the natural (injective) map $\text{Ext}_A^1(X,Y) \to \text{Ext}_H^1(X,Y)$, on equivalence classes of extensions, is an isomorphism. In particular, every H-module extension of Y by X arises via an A-module extension of Y by X. By induction on the filtration length one therefore concludes that every H-module which admits a filtration by $\text{Sp}(\lambda)$'s, with $\lambda \in \pi$, arises from an A-module (i.e. I acts as zero on every such module).

Now the quasihereditary algebra S_ξ has costandard modules $\xi \nabla(\lambda)$, $\lambda \in \pi$, by Proposition A3.11(ii). Hence, by Theorem A4.7, the quasihereditary algebra $\text{End}_{S_\xi}(\xi Se)$ has standard modules $\Delta'(\lambda) = \text{Hom}_{S_\xi}(\xi Se, \xi \nabla(\lambda))$, $\lambda \in \pi$. Hence the isomorphism $H(r)/I(\pi^c) \to \text{End}_{S_\xi}(\xi Se)^{\text{op}}$ gives $H(r)/I(\pi)$ the structure of a quasihereditary algebra with standard modules

$$\text{Hom}_{S_\xi}(\xi Se, \xi \nabla(\lambda)).$$

Now by Proposition A3.13, the natural map

$$\text{Hom}_S(Se, \nabla(\lambda)) \to \text{Hom}_{S_\xi}(\xi Se, \xi \nabla(\lambda))$$

is a linear isomorphism. Clearly this commutes with the action of $H(r)$ so we get that $H(r)/I(\pi)$ has standard modules $\text{Hom}_S(Se, \nabla(\lambda)) \cong \text{Sp}(\lambda)$, for $\lambda \in \pi$.

Thus we get that the category of $H(r)/I(\pi)$-modules with a filtration by $\text{Sp}(\lambda)$, $\lambda \in \pi$, is equivalent to the category of $S'(\pi)$-modules with a filtration by $\Delta'(\lambda)$, $\lambda \in \pi$. By Proposition A3.3, this is equivalent to the category of modules for S' which have a filtration by modules $\Delta'(\lambda)$, $\lambda \in \pi$, and it follows from Proposition A4.8(i), that this is equivalent to the category of S-modules with a filtration by $\nabla(\lambda)$, $\lambda \in \pi$. Summarizing and using (5), we have:

(6)   *Let π be a cosaturated subset of $\Lambda^+(n,r)$ consisting of row regular partitions. The category of $H(r)/I(\pi)$-modules which have a filtration with sections from $\{\text{Sp}(\lambda) \mid \lambda \in \pi\}$ is equivalent to the category of $S(n,r)$-modules which have a filtration with sections from $\{\nabla(\lambda) \mid \lambda \in \pi\}$. If $1 + q \neq 0$, these categories are equivalent to the category of $H(r)$-modules which have a filtration with sections from $\{\text{Sp}(\lambda) \mid \lambda \in \pi\}$.*

Let $m \leq n$ and suppose that $\Lambda^+(m,r)$ consists of row regular partitions. We take $\Gamma = \Lambda(m,r)$. We write E_m for the natural left $G(m)$-module, and we view this as a $G(m)$-submodule of E. We identify S_ξ with $S(m,r)$ via the natural isomorphism given at the beginning of Section 2.2. Now $\xi S e \cong \xi E^{\otimes r} = E_m^{\otimes r}$ and the isomorphism $H(r)/I(\pi) \to \mathrm{End}_{S_\xi}(\xi S e)^{\mathrm{op}}$ of (4) gives an isomorphism $H(r)/I(\pi) \to S(m,r)'$, the Ringel dual of $S(m,r)$. Thus $H(r)/I(\pi)$ is quasihereditary and, as above, we find that the standard modules are $\mathrm{Sp}(\lambda)$, $\lambda \in \Lambda^+(n,r)$, and obtain the following.

(7) *Let $m \leq n$ and suppose that $\pi = \Lambda^+(m,r)$ consists of row regular partitions. Then $H(r)/I(\pi)$ is the Ringel dual of $S(m,r)$. Furthermore, the category of $H(r)/I(\pi)$-modules which have a filtration with sections from $\{\mathrm{Sp}(\lambda) \mid \lambda \in \Lambda^+(m,r)\}$ is equivalent to the category of $S(m,r)$-modules which have a filtration with sections from $\{\nabla(\lambda) \mid \lambda \in \Lambda^+(m,r)\}$. If $1+q \neq 0$, these categories are equivalent to the category of $H(r)$-modules which have a filtration with sections from $\{\mathrm{Sp}(\lambda) \mid \lambda \in \Lambda^+(m,r)\}$.*

This is the q-analogue of Erdmann, [45; 4.4 Theorem].

4.8 The global dimension of the Schur algebras for $r \leq n$

We calculate the global dimension of the Schur algebra $S(n,r)$, for $r \leq n$. Our result generalizes a recent result of Totaro, [76], who determined this in the classical case, $q = 1$, and our arguments are based very firmly on his work. If q is not a root of unity then $S(n,r)$ is semisimple. We assume from now on that q is a primitive lth root of unity and that $r \leq n$. Let the characteristic of k be $p \geq 0$. If $p = 0$ and r has (l,p) expansion $r = r_{-1} + lr'$ we define $d(r) = r_{-1} + r'$ and if $p > 0$ and r has (l,p) expansion $r = r_{-1} + lr_0 + lpr_1 + lp^2 r_2 + \cdots$ we define $d(r) = r_{-1} + r_0 + r_1 + \cdots$ (the sum of the digits). We show that the global dimension of $S(n,r)$ is precisely $2(r - d(r))$.

We begin by introducing the divided powers modules. Recall that we have the pairing $V^{\otimes r} \times E^{\otimes r} \to k$. Let $\lambda = (\lambda_1, \lambda_2, \ldots)$ be a composition of r and let X^λ denote the kernel of the natural map $V^{\otimes r} \to S^\lambda V$. We define the divided powers module $D^\lambda E = (X^\lambda)^\perp = \{y \in E^{\otimes r} \mid (X^\lambda, y) = 0\}$. Then we have the induced form $S^\lambda V \times D^\lambda E \to k$. From the definitions we have

$$X^\lambda = X^{\lambda_1} \otimes V^{\otimes \lambda_2} \otimes \cdots + V^{\otimes \lambda_1} \otimes X^{\lambda_2} \otimes \cdots + V^{\otimes \lambda_1} \otimes V^{\otimes \lambda_2} \otimes X^{\lambda_3} \otimes \cdots + \cdots$$

so that

$$(X^\lambda)^\perp = (X^{\lambda_1})^\perp \otimes (X^{\lambda_2})^\perp \otimes \cdots = D^{\lambda_1} \otimes D^{\lambda_2} \otimes \cdots.$$

Thus we have:

(1) $D^\lambda E = D^{\lambda_1} E \otimes D^{\lambda_2} E \otimes \cdots.$

The right divided powers modules $D^\lambda V$, $\lambda \in \Lambda(n,r)$, may be defined analogously (though these are not used in the sequel). Dualizing the isomorphism $A(n,r) \cong \bigoplus_{\lambda \in \Lambda(n,r)} S^\lambda V$, 2.1(1)(ii), of right G-modules, and using Lemma 1.1.2(iii), we obtain a left $S(n,r)$-module decomposition:

(2) $S(n,r) \cong \bigoplus_{\lambda \in \Lambda(n,r)} D^\lambda E$.

We write glob(S) for the global dimension of an algebra S. For $X \in$ mod($S(n,r)$) we write inj(X) for the injective dimension and proj(X) for the projective dimension of X. Let $N = $ glob(S), where $S = S(n,r)$, and let $X, Y \in$ mod(S) be such that $\text{Ext}_S^N(X,Y) \neq 0$. We have $Y \cong F/R$ for some free module $F \in$ mod(S) and submodule R. We have an exact sequence $\text{Ext}_S^N(X,F) \to \text{Ext}_S^N(X,Y) \to \text{Ext}_S^{N+1}(X,R)$. Since $\text{Ext}_S^N(X,Y) \neq 0$ and $\text{Ext}_S^{N+1}(X,R) = 0$ we have $\text{Ext}_S^N(X,F) \neq 0$. Thus the global dimension of S is the injective dimension of F, which is the injective dimension of the left regular module S. Thus, from (2), we get:

(3)   glob($S(n,r)$) = max{inj($D^\lambda E$) $| \lambda \in \Lambda(n,r)$}.

This prompts the following observation.

(4)   *Suppose that $r, s \in \mathbb{N}_0$, $r+s \leq n$, $X \in$ mod($S(n,r)$), $Y \in$ mod($S(n,s)$). Then we have* inj($X \otimes Y$) $\leq$ inj(X) + inj(Y) *(resp.* proj($X \otimes Y$) $\leq$ proj(X) + proj(Y)). *In particular if X and Y are injective (resp. projective) then $X \otimes Y$ is injective (resp. projective).*

Proof We suppose first that X and Y and injective and prove that $X \otimes Y$ is injective. We may assume that X and Y are indecomposable. By 2.1(1)(ii), we have that X is isomorphic to a component of $S^\lambda E$ and Y is isomorphic to a summand of $S^\mu E$ for some $\lambda = (\lambda_1, \ldots, \lambda_r) \in \Lambda(n,r)$ and some $\mu = (\mu_1, \ldots, \mu_s) \in \Lambda(n,s)$. But then $X \otimes Y$ is isomorphic to a direct summand of $S^\tau E$, where $\tau = (\lambda_1, \ldots, \lambda_r, \mu_1, \ldots, \mu_s) \in \Lambda(n,r+s)$ and this is injective, by a further application of 2.1(1)(ii). Hence $X \otimes Y$ is injective. In general one gets that inj($X \otimes Y$) $\leq$ inj(X) + inj(Y), either by dimension shifting, or by tensoring an injective resolution of X with an injective resolution of Y to get one of $X \otimes Y$, in the usual way.

The arguments for projective dimension are similar, using (2) to get that $X \otimes Y$ is projective if both X and Y are projective.

Now from (3) we get:

(5)   *For $r \leq n$ we have* glob($S(n,r)$) $\leq$ max{$\sum_i$ inj($D^{\lambda_i} E$) $| \lambda = (\lambda_1, \lambda_2, \ldots) \in \Lambda^+(n,r)$}.

We shall also need to construct resolutions for the divided powers modules. We could do this by arguing as in [1], to produce projective resolutions, but (for the sake of variety) our treatment is based on the Koszul resolution, which we now produce. We write A_n for the algebra given by generators $x_1, \ldots, x_n$ subject to the relations $x_j x_i = q x_i x_j$, for $i < j$. We write simply k for $A_n/(x_1, \ldots, x_n)$. We identify A_{n-1} with a subalgebra of A_n and note that $A_n = A_n x_n \oplus A_{n-1}$ (for $n > 1$). We write W_n for the n-dimensional vector space on basis $y_1, \ldots, y_n$ and, as usual, write $\bigwedge W_n$ for the exterior algebra $\bigoplus_{j \geq 0} \bigwedge^j W_n$. We identify $\bigwedge W_{n-1}$ with a subalgebra of $\bigwedge W_n$ (for $n > 1$). For $1 \leq r \leq n$ we define $\phi_n^r : A_n \otimes \bigwedge^r W_n \to A_n \otimes \bigwedge^{r-1} W_n$ to be the linear map such that

$$\phi_n^r(f \otimes y_{i_1} \ldots y_{i_r}) = f x_{i_1} \otimes y_{i_2} \ldots y_{i_r} - q f x_{i_2} \otimes y_{i_1} y_{i_3} \ldots y_{i_r} + \cdots$$
$$\cdots + (-q)^{r-1} f x_{i_r} \otimes y_{i_1} \ldots y_{i_{r-1}}$$

for $f \in A_n$ and $n \geq i_1 > \cdots > i_r \geq 1$. We claim that:

(6) $0 \to A_n \otimes \bigwedge^n W_n \to \cdots \to A_n \otimes \bigwedge^r W_n \to A_n \otimes \bigwedge^{r-1} W_n \to \cdots \to A_n \otimes W_n \to A_n \to k \to 0$ *is exact.*

We call this the q-Koszul resolution of the A_n-module k. It is easy to check that it is a complex. Also, exactness at A_n is clear and we leave it to the reader to check that $A_n \otimes \bigwedge^n W_n \to A_n \otimes \bigwedge^{n-1} W_n$ is injective. To prove that $\operatorname{Ker}(\phi_n^r) \leq \operatorname{Im}(\phi_n^{r+1})$, for $n > r \geq 1$, we argue by induction on n. For $n = 1$ there is nothing to prove. We now suppose that $n > 1$ and that the result holds for $n - 1 > r \geq 1$. We note that the restriction of ϕ_n^r maps $A_{n-1} \otimes \bigwedge^r W_{n-1}$ into $A_{n-1} \otimes \bigwedge^{r-1} W_{n-1}$ via ϕ_{n-1}^r. Note also that we have a k-space decomposition

$$A_n \otimes \bigwedge^r W_n = A_{n-1} \otimes \bigwedge^r W_{n-1} \oplus C_n^r$$

where $C_n^r = A_n x_n \otimes \bigwedge^r W_{n-1} + A_n \otimes y_n \bigwedge^{r-1} W_{n-1}$. Let $\pi : A_n \otimes \bigwedge^r W_n \to A_{n-1} \otimes \bigwedge^r W_{n-1}$ be the projection.

We must show that if $\phi_n^r(F) = 0$, where $F = \sum_{i \in I_1(n,r)} f_i \otimes y_i$, with $f_i \in A_n$ and $y_i = y_{i_1} \ldots y_{i_r}$, for $i = (i_1, \ldots, i_r) \in I_1(n, r)$, then $F \in \operatorname{Im}(\phi_n^{r+1})$. Let $X = I_1(n, r) \backslash I_1(n-1, r)$. For each $i \in I_1(n-1, r)$ we write $f_i = f_i' x_n + f_i''$, with $f_i' \in A_n$, $f_i'' \in A_{n-1}$. We have

$$\phi_n^r(F) = \phi_n^r\left(\sum_{i \in I_1(n,r)} f_i' x_n \otimes y_i\right) + \phi_n^r\left(\sum_{i \in I_1(n-1,r)} f_i'' \otimes y_i\right) + \phi_n^r\left(\sum_{i \in X} f_i \otimes y_i\right) = 0.$$

Note that $\phi_n^r(f_i' x_n \otimes y_i) \in C_n^r$, for $i \in I_1(n-1, r)$, and $\phi_n^r(f_i \otimes y_i) \in C_n^r$, for $i \in X$. Thus, applying π, we get $\phi_n^r(\sum_{i \in I_1(n,r)} f_i'' \otimes y_i) = 0$, i.e.

$$\phi_{n-1}^r\left(\sum_{i \in I_1(n-1,r)} f_i'' \otimes y_i\right) = 0.$$

If $r = n - 1$ we get $\sum_{i \in I_1(n,r)} f_i'' \otimes y_i = 0$ and if $r < n - 1$ we get, by the inductive hypothesis,

$$\sum_{i \in I_1(n,r)} f_i'' \otimes y_i \in \mathrm{Im}(\phi_{n-1}^{r+1}) \leq \mathrm{Im}(\phi_n^{r+1}).$$

Thus F is congruent to $\sum_{i \in I_1(n-1,r)} f_i' x_n \otimes y_i + \sum_{i \in X} f_i \otimes y_i$, modulo $\mathrm{Im}(\phi_n^{r+1})$. We may therefore assume $F = \sum_{i \in I_1(n-1,r)} f_i' x_n \otimes y_i + \sum_{i \in X} f_i \otimes y_i$.

We put $U = \sum_{i \in X} A_n \otimes y_i = A_n \otimes y_n \bigwedge^{r-1} W_{n-1}$. Now, for $i = (i_1, \ldots, i_r) \in I(n-1,r)$, we have

$$\phi_n^{r+1}(f_i' \otimes y_n y_i)$$
$$= f_i' x_n \otimes y_{i_1} \cdots y_{i_r} - q f_i' x_{i_1} \otimes y_n y_{i_2} \cdots y_{i_r} + \cdots + (-q)^r f_i' x_{i_r} \otimes y_n y_{i_1} \cdots y_{i_{r-1}}$$
$$= f_i' \otimes y_i + G_i$$

for some $G_i \in U$. Thus we have that F is congruent to $\sum_{i \in I_1(n-1,r)} G_i + \sum_{i \in X} f_i \otimes y_i$ modulo $\mathrm{Im}(\phi_n^{r+1})$. Hence we can assume that $F \in U$. We have $F = \sum_{i \in I_1(n-1,r)} g_i \otimes y_n y_i$ for some $g_i \in A_n$. But the condition $\phi_n^r(F) = 0$ gives

$$\sum_{i=(i_1,\ldots,i_r) \in I_1(n-1,r)} (g_i x_n \otimes y_{i_1} \cdots y_{i_r} - q g_i x_{i_1} \otimes y_n y_{i_2} \cdots y_{i_r} + \cdots$$

$$+ (-q)^r g_i x_{i_r} \otimes y_n y_{i_1} \cdots y_{i_{r-1}}) = 0$$

and hence $g_i x_n = 0$ and therefore $g_i = 0$, for all $i \in I_1(n-1,r)$. Thus we have $F = 0 \in \mathrm{Im}(\phi_n^{r+1})$, as required.

We fix a degree r and, for $1 \leq a \leq r$, let $\theta_a : S^{r-a} V \otimes \bigwedge^a V \to S^{r+1-a} V \otimes \bigwedge^{a-1} V$ be the k-map given by

$$\theta_a(f \otimes v_{i_1} \wedge \cdots \wedge v_{i_a}) = f v_{i_1} \otimes v_{i_2} \wedge \cdots \wedge v_{i_a} - q f v_{i_2} \otimes v_{i_1} \wedge v_{i_3} \wedge \cdots \wedge v_{i_a}$$
$$+ \cdots + (-q)^{a-1} f v_{i_a} \otimes v_{i_1} \wedge \cdots \wedge v_{i_{a-1}}$$

for $f \in S^{a-1} V$, $n \geq i_1 > \cdots > i_a \geq 1$.

(7) $0 \to \bigwedge^r V \to V \otimes \bigwedge^{r-1} V \to \cdots \to S^{r-a} V \otimes \bigwedge^a V \to S^{r+1-a} V \otimes \bigwedge^{a-1} V \to \cdots \to S^{r-1} V \otimes V \to S^r V \to 0$ is an exact sequence of right G-modules.

Proof We have a natural grading on A_n such that x_i has degree 1, and an isomorphism of graded algebras $A_n \to S(V)$, taking x_i to v_i, for $1 \leq i \leq n$. Moreover, the linear isomorphism $W_n \to V$, taking y_i to v_i, $1 \leq i \leq n$, induces an isomorphism of k-spaces $\bigwedge^j W_n \to \bigwedge^j V$, for each $j \geq 0$. Thus we have natural isomorphisms $A_n \otimes \bigwedge^j W_n \to S(V) \otimes \bigwedge^j V$. We obtain, by transport of structure from (6), a resolution of graded $S(V)$-modules

$$0 \to S(V) \otimes \bigwedge^n V \to S(V) \otimes \bigwedge^{n-1} V \to \cdots \to S(V) \otimes V \to S(V) \to k \to 0$$

and the rth component of this is the sequence (7), which is therefore exact. It remains to show that each map $\theta_a : S^{r-a}V \otimes \bigwedge^a V \to S^{r+1-a}V \otimes \bigwedge^{a-1}V$ is a G-module map. Note that θ_a factorizes as $(m \otimes \mathrm{id}) \circ (\mathrm{id} \otimes \phi)$, where $m : S^{r-a}V \otimes V \to S^{r+1-a}V$ is multiplication and where $\phi : \bigwedge^a V \to V \otimes \bigwedge^{a-1}V$ is the linear map given by

$$\phi(v_{i_1} \wedge \cdots \wedge v_{i_a}) = v_{i_1} \otimes v_{i_2} \wedge \cdots \wedge v_{i_a} - q v_{i_2} \otimes v_{i_1} \wedge v_{i_3} \wedge \cdots \wedge v_{i_a}$$
$$+ \cdots + (-q)^{a-1} v_{i_a} \otimes v_{i_a} \otimes v_{i_1} \wedge \cdots \wedge v_{i_{a-1}}$$

for $i_1 > \cdots > i_a$. Thus it suffices to prove that ϕ is a G-module homomorphism. This may be easily checked by observing that ϕ may be obtained by dualizing multiplication $E \otimes \bigwedge^{a-1}E \to \bigwedge^a E$, via the natural pairing $V \otimes \bigwedge^{a-1}V \times E \otimes \bigwedge^{a-1}E \to k$.

Now for $a, b \in \mathsf{N}_0$ we have pairings $S^a V \times D^a E \to k$ and $\bigwedge^b V \times \bigwedge^b E \to k$ and hence the product pairing $S^a V \otimes \bigwedge^b V \times D^a E \otimes \bigwedge^b E \to k$. Dualizing (7) using these pairings:

(8)   We have an exact sequence of left G-modules $0 \to D^r E \to D^{r-1}E \otimes E \to \cdots \to D^{r+1-a}E \otimes \bigwedge^{a-1}E \to D^{r-a}E \otimes \bigwedge^a E \to \cdots \to E \otimes \bigwedge^{r-1}E \to \bigwedge^r E \to 0$.

By dimension shifting we have the following.

(9)   If $0 \to A \to X_0 \to X_1 \to \cdots \to X_m \to 0$ is an exact sequence of modules over a ring then we have $\mathrm{inj}(A) \le \max\{\mathrm{inj}(X_j) + j \mid 0 \le j \le m\}$.

Thus from (4), the usual Koszul resolution $0 \to \bigwedge^m E \to E \otimes \bigwedge^{m-1}E \to \cdots \to S^{m-1}E \otimes E \to S^m E \to 0$ (we leave it to the reader to check that the maps are G-homomorphisms) and induction we get:

(10) $\mathrm{inj}(\bigwedge^m E) \le m - 1$, for $1 \le m \le n$.

From (8), (10), (9) and (4) (and induction) we get:

(11) $\mathrm{inj}(D^m E) \le 2(m - 1)$, for $1 \le m \le n$.

We fix non-negative integers r and a and let $f(x) \in \mathsf{Z}[x]$ be the corresponding Gaussian polynomial (see e.g. [2; p. 33]). Thus the value of $f(x)$ at a prime power q is the number of a-dimensional subspaces in a vector space of dimension r, over the field of cardinality q. We claim that for $r \geq a$ we have

$$f(x) = \sum_{\sigma \in X(a,b)} x^{l(\sigma)}$$

where $b = r - a$ and $X(a,b)$, as in Section 1.2, is the set of $\sigma \in \mathrm{Sym}(r)$ such that $\sigma(i) < \sigma(j)$ whenever $1 \leq i < j \leq a$ or $a+1 \leq i < j \leq r$. It suffices to show that the polynomials agree at prime powers q. We have $f(q) = [\mathrm{GL}_r(q) : P]$, where P is the stabilizer of an a-dimensional subspace. We take P to be the set of $g = (g_{ij}) \in \mathrm{GL}_r(q)$ such that $g_{ij} = 0$ for $(i,j) \in [1,a] \times [a+1, a+b]$. Let $\Phi^+ = \{(i,j) \mid 1 \leq i < j \leq n\} = \{\alpha_1, \dots, \alpha_N\}$ (where $N = \binom{r}{2}$). For $\alpha \in \Phi^+$ let U_α denote the corresponding root subgroups of $\mathrm{GL}_r(q)$. For $\sigma \in \mathrm{Sym}(n)$ let $\Phi_\sigma^+ = \{(i,j) \in \Phi^+ \mid \sigma(i) > \sigma(j)\}$ and let $U_\sigma = U_{\alpha_{h_1}} \cdots U_{\alpha_{h_m}}$, where $\Phi_\sigma^+ = \{\alpha_{h_1}, \dots, \alpha_{h_m}\}$ and $h_1 < \cdots < h_m$. Then $|U_\sigma| = q^{l(\sigma)}$, for $\sigma \in \mathrm{Sym}(r)$. We identify $\sigma \in \mathrm{Sym}(n)$ with the corresponding permutation matrix. It follows from the Bruhat decomposition that $GL_n(q)$ is the disjoint union of the sets $U_\sigma \sigma P$, $\sigma \in X(a,b)$, and that the map $U_\sigma \times P \to U_\sigma \sigma P$ is bijective (for $\sigma \in X(a,b)$). This gives $[GL_n(q) : P] = \sum_{\sigma \in X(a,b)} q^{l(\sigma)}$, as required.

We now return to our usual understanding that q is a non-zero element of the field k. We write $\begin{bmatrix} r \\ a \end{bmatrix}$ for $\sum_{\sigma \in X(a,b)} q^{l(\sigma)}$. By [75; Lemma 2.1(i)], if q is a primitive lth root of unity and $r = r_{-1} + lr'$, $s = s_{-1} + ls'$, with $0 \leq r_{-1}, s_{-1} < l$ then we have

$$\begin{bmatrix} r \\ s \end{bmatrix} = \begin{bmatrix} r_{-1} \\ s_{-1} \end{bmatrix} \binom{r'}{s'}. \tag{$*$}$$

(12) **Proposition**   Let $a, b \geq 0$. Let $X = \bigwedge^a E \otimes \bigwedge^b E$ (resp. $\bigwedge^{a+b} V$, resp. $S^a E \otimes S^b E$, resp. $D^{a+b} V$, resp. $D^{a+b} E$, resp. $S^a V \otimes S^b V$) and $Y = \bigwedge^{a+b} E$ (resp. $\bigwedge^a V \otimes \bigwedge^b V$, resp. $D^a V \otimes D^b V$, resp. $S^{a+b} E$, resp. $D^a E \otimes D^b E$, resp. $S^{a+b} V$). Then $\dim \mathrm{Hom}_G(X, Y) = 1$ and the natural map $X \to Y$ is split if and only if $\begin{bmatrix} a+b \\ a \end{bmatrix} \neq 0$.

Proof   By using suitable pairings and Lemma 1.1.2, we are reduced to proving this for $X = \bigwedge^a E \otimes \bigwedge^b E$ (resp. $S^a E \otimes S^b E$, resp. $S^a V \otimes S^b V$) and $Y = \bigwedge^{a+b} E$ (resp. $S^{a+b} E$, resp. $S^{a+b} V$).

Let $r = a + b$. We first consider the case $X = \bigwedge^a E \otimes \bigwedge^b E$ and $Y = \bigwedge^r E$. Since $\bigwedge^r E \in \mathcal{F}(\Delta)$ and $\bigwedge^a E \otimes \bigwedge^b E \in \mathcal{F}(\nabla)$ we get, from the Appendix, Proposition A2.2(ii), $\dim \mathrm{Hom}_G(\bigwedge^r E, \bigwedge^a E \otimes \bigwedge^b E) = (\bigwedge^a E \otimes \bigwedge^b E : \bigwedge^{a+b} E)$. This the coefficient of s_{1^r} in $s_{1^a} s_{1^b}$ (where s_λ denotes the Schur symmetric function determined by the partition λ) expressed as a **Z**-linear combination of Schur symmetric functions, and this is 1 e.g. by [63; I,(5.17)].

By Lemma 1.2.3, the natural G-module map $\eta : \bigwedge^a E \otimes \bigwedge^b E \to \bigwedge^{a+b} E$ is split if and only $\phi \circ \eta \in \operatorname{End}_G(\bigwedge^r E)$ is a non-zero multiple of the identity, where $\phi : \bigwedge^r E \to \bigwedge^a E \otimes \bigwedge^b E$ is the linear map such that $\phi(\hat{e}_i) = \sum_{\sigma \in X(a,b)} \pi(e_{i\sigma})$, for $i \in I_1(n,r)$ (and where $\pi : E^{\otimes r} \to \bigwedge^a E \otimes \bigwedge^b E$ is the natural map). Moreover, since $\operatorname{End}_G(\bigwedge^r E) = k$, we have that η is split if and only if $\pi \circ \eta \neq 0$. Now for $i = (i_1, \ldots, i_r) \in I_1(n,r)$ we have

$$\pi \circ \eta(\hat{e}_i) = \sum_{\sigma \in X(a,b)} (-1)^{l(\sigma)} \hat{e}_{i\sigma}.$$

Moreover, it follows from the defining relations for $\bigwedge(E)$ that $\hat{e}_{i\sigma} = (-q)^{l(\sigma)} \hat{e}_i$, for $i \in I_1(n,r)$ and $\sigma \in \operatorname{Sym}(r)$. Thus we have

$$\pi \circ \eta(\hat{e}_i) = \left(\sum_{\sigma \in X(a,b)} q^{l(\sigma)} \right) \hat{e}_i = \begin{bmatrix} a+b \\ a \end{bmatrix} \hat{e}_i$$

for $i \in I_1(n,r)$. Thus η is split if and only if $\begin{bmatrix} a+b \\ a \end{bmatrix} \neq 0$.

We now consider the case $X = S^a E \otimes S^b E$ and $Y = S^r E$. By 2.1(8) we have $\dim \operatorname{Hom}_G(S^r E, S^a E \otimes S^b E) = \dim S^r E^{(a,b,0,\ldots)}$ and this is 1.

We now consider the natural map $\eta : S^a E \otimes S^b E \to S^r E$. We produce a non-zero G-homomorphism $S^r E \to S^a E \otimes S^b E$. Recall we have the isomorphism $\phi : S^a E \otimes S^b E \to {}^\mu A(n,r)$, of 2.1(1), where $\mu = (a,b)$. Let ψ be the inverse of ϕ and let $\gamma : S^r E \to S^a E \otimes S^b E$ be the composite $\psi \circ \beta$, where $\beta : S^r E \to {}^\mu A(n,r)$ is given by $\beta(\bar{e}_i) = \sum_{j \in \mu} c_{ji}$. Note that β is a G-map since we have $\beta = (\nu \otimes \operatorname{id})\tau$, where $\tau : S^r E \to S^r E \otimes A(n,r)$ is the structure map and $\nu : S^r E \to k$ is the linear map such that $\nu(e_1^a e_2^b) = 1$ and $\nu(S^r E^\alpha) = 0$ for $\mu \neq \alpha \in \Lambda(n,r)$. Note also that $\varepsilon\beta(e_1^a e_2^b) = \varepsilon(c_{ii}) = 1$, where $i = (1,1,\ldots,1,2,2,\ldots,2)$. Thus β, and hence γ, is a non-zero G-map. Thus, multiplication $\eta : S^a E \otimes S^b E \to S^r E$ splits if and only if $\eta \circ \gamma \in \operatorname{End}_G(S^r E)$ is an isomorphism, and since $\operatorname{End}_G(S^r E) = k$, this is if and only if $\eta \circ \gamma$ is non-zero. Thus η is split if and only if $\eta\gamma(e_1^r) \neq 0$, i.e. if and only if $\eta\psi(\sum_{j \in \mu} c_{ji}) \neq 0$, where now $i = (1,1,\ldots,1)$. Let $h \in I(n,r)$ and let $h' \in I(n,r)$ be the result of writing the components of h in ascending order. It is easy to check, from the defining relations, that $c_{h,i} = q^g c_{h',i}$, where g is the number of pairs (u,v), of elements of $[1,r]$, such that $u < v$ and $h_u > h_v$. Let $h = (1,1,\ldots,1,2,2,\ldots,2)$. Then each $j \in \mu$ may be written uniquely in the form $h\sigma$, for $\sigma \in X(a,b)$. We get $c_{h\sigma,i} = q^{l(\sigma)} c_{h,i}$, for $\sigma \in X(a,b)$. Moreover, we have $\psi(c_{hi}) = e_1^a \otimes e_1^b$ so we get $\eta\psi(\sum_{j \in \mu} c_{ji}) = (\sum_{\sigma \in X(a,b)} q^{l(\sigma)}) e_1^r$ and so η is split if and only if $\begin{bmatrix} a+b \\ a \end{bmatrix} \neq 0$.

The case $X = S^a V \otimes S^b V$ and $Y = S^{a+b} V$ is similar.

We assume that $l > 0$ and that k has characteristic $p > 0$, and leave it to the reader to make the simplifications in the following arguments needed to cover the remaining cases. Let $m \leq n$ and let $m = m_{-1} + lm_0 + lpm_1 + \cdots$ be the (l,p)-expansion. By repeated application of (12), using (*), we get that $\bigwedge^m E$ is a G-module summand of $E^{\otimes m_{-1}} \otimes (\bigwedge^l E)^{\otimes m_0} \otimes (\bigwedge^{lp} E)^{m_1} \otimes \cdots$ and that $S^m V$ is a G-module direct summand of $V^{\otimes m_{-1}} \otimes (S^l V)^{\otimes m_0} \otimes (S^{lp} V)^{\otimes m_1} \otimes \cdots$, and hence, by duality, that $D^m E$ is a G-module summand of $E^{\otimes m_{-1}} \otimes (D^l E)^{\otimes m_0} \otimes (D^{lp})^{\otimes m_1} \otimes \cdots$ Thus, from (4), (10) and (11) we obtain:

(13) $\mathrm{inj}(\bigwedge^m E) \leq m - d(m)$ and $\mathrm{inj}(D^m E) \leq 2(m - d(m))$.

It is easy to check that, for $\lambda = (\lambda_1, \lambda_2, \ldots) \in \Lambda(n,r)$ we have $d(\lambda_1) + d(\lambda_2) + \cdots \leq d(r)$. Hence, from (3), (5) and (13) we have:

(14) $\mathrm{glob}(S(n,r)) \leq 2(r - d(r))$, for $r \leq n$.

We now show that in fact we have equality.

(15)   *For $r \leq n$ we have* $\mathrm{glob}(D^r E) = 2(r - d(r))$.

Proof   By (14) we may assume $r \geq l$ and $r \geq p$ if $l = 1$. Moreover, by (14), it suffices to demonstrate the existence of some $A, B \in \mathrm{mod}(S(n,r))$ with $\mathrm{Ext}_G^{2(r-d(r))}(A, B) \neq 0$.

Let $r = r_{-1} + lr_0 + lpr_1 + \cdots$ be the (l,p) expansion of r. We take $A = S^\lambda E$ and $B = D^\lambda E$ where $\lambda = (1^{r_{-1}} l^{r_0} (lp)^{r_1} \ldots)$.

For each component m of $\lambda = (1^{r_{-1}} l^{r_0} (lp)^{r_1} \ldots)$, we have, by (8), the exact sequence $0 \to D^m E \to X_0 \to X_1 \to \cdots \to X_{m-1} \to 0$, where $X_j = D^{m-1-j} E \otimes \bigwedge^{j+1} E$. By (4), (10) and (11) we have $\mathrm{inj}(X_j) \leq 2(m-1) - j$, for $0 \leq j \leq m - 1$, and $\mathrm{inj}(X_j) \leq 2(m-1) - j - 2$ for $j < m - 1$. Tensoring together all such sequences we obtain an exact sequence $0 \to D^\lambda E \to Y_0 \to \cdots \to Y_N \to 0$, where $N = r - d(r)$ and $Y_N = \bigwedge^\lambda E$. Moreover, by (4), we have $\mathrm{inj}(Y_j) \leq 2N - j - 2$, for $j < N$. Thus we have $\mathrm{Ext}_G^{2N}(S^\lambda E, D^\lambda E) = \mathrm{Ext}_G^{2N}(S^\lambda, Y_N) \cong \mathrm{Ext}_G^N(S^\lambda E, \bigwedge^\lambda E)$.

Now by [1], we have, for a positive integer m, an exact sequence of vector spaces $0 \to K \to X_0 \to X_1 \to \cdots \to X_{m-1} \to 0$, where X_{m-1-a} is the direct sum of all tensor products $S^{j_1} E \otimes \cdots \otimes S^{j_a} E$ with $j_1, \ldots, j_a \geq 1$ and $j_1 + \cdots + j_a = m$, for $0 \leq a \leq m - 1$. Moreover $K \to X_0$ is inclusion and $X_{a-1} \to X_a$ is derived from multiplication in the symmetric algebra $S(E)$, for $1 \leq a < m$. Thus $0 \to K \to X_0 \to X_1 \to \cdots \to X_{m-1} \to 0$ is an exact sequence of G-modules. In the classical case, by [1], we have $K \cong \bigwedge^m E$, and so the character of K, in general, is $\mathrm{ch} \bigwedge^m E$. Since $\bigwedge^m E$ is an irreducible G-module, we must have $K \cong \bigwedge^m E$, in general. Hence, for each component m of $\lambda = (1^{r_{-1}} l^{r_0} (lp)^{r_1} \ldots)$ we have an exact sequence of G-modules $0 \to \bigwedge^m E \to X_0 \to \cdots \to X_{m-1} \to 0$. Tensoring together all such sequences we obtain an injective resolution $0 \to \bigwedge^\lambda E \to Y_0 \to$

$\cdots \to Y_N \to 0$, with $Y_N = S^\lambda E$. Hence we get that $\mathrm{Ext}_G^N(S^\lambda E, \bigwedge^\lambda E) \neq 0$ provided that $\mathrm{Hom}_G(S^\lambda E, Y_{N-1}) \to \mathrm{Hom}_G(S^\lambda E, Y_N) = \mathrm{Hom}_G(S^\lambda E, S^\lambda E)$ is not surjective, and this will certainly be so if $Y_{N-1} \to Y_N$ is non-split.

We shall prove (cf. [76]) that any homomorphism $S^\lambda E \to Y_{N-1}$ is zero on the $(r, 0, \ldots)$ weight space. Note that Y_{N-1} is a direct sum of modules of the form $S^\mu E$, where μ is obtained from $\lambda = (1^{r-1} l^{r_0}(lp)^{r_1} \ldots)$ by replacing one of the components lp^r by the pair $(lp^r - 1, 1)$. Thus it suffices to show that, for all such μ, every G-module homomorphism $S^\lambda E \to S^\mu E$ is 0 on the $(r, 0, \ldots)$ weight space. One may do this by giving a q-analogue of the argument of Totaro, [76]. For the sake of variety, we give a deduction based on the Steinberg tensor product theorem. Suppose, for a contradiction, that we have a G-module homomorphism $S^\lambda E \to S^\mu E$ which is non-zero on the $(r, 0, \ldots)$ weight space. Now $S^r E$ is a G-module direct summand of $S^\lambda E$ (see the proof of (13)) so there must be a G-module homomorphism $S^r E \to S^\mu E$ which is non-zero on the G-socle $L = L(r, 0, \ldots)$ of $S^r E$. Thus we get $\dim \mathrm{Hom}_G(L, S^\mu E) \neq 0$ and therefore, by 2.1(8), μ is a weight of L. By the Steinberg tensor product theorem, 3.2(5), we have $L \cong L(r_{-1}, 0, \ldots) \otimes \bar{L}(s, 0, \ldots)^F$, where $r = r_{-1} + ls$. Thus μ has the form $\alpha + l\beta$, for suitable weights α of $L(r_{-1}, 0, \ldots)$ and β of $\bar{L}(s, 0, \ldots)$. Now, in μ, we have 1 and $lp^r - 1$ occurring as entries in consecutive positions. Restricting to these positions we get $(1, lp^r - 1) = \gamma + l\nu$, for $\gamma = (a, b)$, $\nu = (c, d)$ and $a + b < l$. This is clearly impossible. Thus we have that $L(r, 0, \ldots)^\mu = 0$, that any homomorphism $Y_N \to Y_{N-1}$ is 0 on the $(r, 0, \ldots)$ weight space of Y_N and that $Y_{N-1} \to Y_N$ is non-split, as required.

Remark Though we knew at the outset that $S(n, r)$ has finite global dimension, this follows from the resolutions given in this section. We have $\mathrm{inj}(S^\lambda E) = 0 < \infty$, for all $\lambda \in \Lambda(n, r)$. We have $\mathrm{inj}(\bigwedge^m E) < \infty$, by (6), (4) and induction. Hence we have $\mathrm{inj}(\bigwedge^\alpha E) < \infty$ for all $\alpha \in \Lambda(n, r)$, by (4). Assume that $\lambda \in \Lambda^+(n, r)$ and we have $\mathrm{inj}(L(\mu)) < \infty$ for all $\mu \in \Lambda^+(n, r)$ which are less than λ in the dominance order. The simple module $L(\lambda)$ occurs exactly once as a composition factor of $\bigwedge^\alpha E$, where α is the transpose of λ, and all other composition factors have highest weight smaller than λ in the dominance order. Thus we have submodules $N_1 < N_2$ of $\bigwedge^\alpha E$, with $N_2/N_1 \cong L(\lambda)$ and all weights of N_1 and $\bigwedge^\alpha E/N_2$ less than λ. By the inductive hypothesis we have $\mathrm{inj}(N_1), \mathrm{inj}(\bigwedge^\alpha E/N_2) < \infty$. It follows that $\mathrm{inj}(N_2) < \infty$ and therefore $\mathrm{inj}(L(\lambda)) = \mathrm{inj}(N_2/N_1) < \infty$.

Thus, by induction, we have $\mathrm{inj}(L(\lambda)) < \infty$ for all $\lambda \in \Lambda^+(n, r)$, i.e. $\mathrm{inj}(L) < \infty$ for all simple modules L and therefore $\mathrm{glob}(S) < \infty$.

Appendix: Quasihereditary Algebras

We give a self contained account of the theory of quasihereditary algebras and their associated tilting modules. References are given in the discussion in A5.

A1 Let k be a field and let S be a finite dimensional k-algebra. We assume that S is Schurian, in the (other) sense that $\mathrm{End}_S(L) = k$, for every simple S-module L. We fix a complete set of pairwise non-isomorphic simple S-modules $\{L(\lambda) \mid \lambda \in \Lambda^+\}$. For $\lambda \in \Lambda^+$, we fix a (minimal) projective cover $P(\lambda)$ and a (minimal) injective envelope $I(\lambda)$ of $L(\lambda)$.

We write $\mathrm{mod}(S)$ for the category of finite dimensional left S-modules and, for $X \in \mathrm{mod}(S)$, $\lambda \in \Lambda^+$, write $[X : L(\lambda)]$ for the multiplicity of $L(\lambda)$ as a composition factor of X.

Let π be a subset of Λ^+. We say that $V \in \mathrm{mod}(S)$ *belongs* to π if all composition factors of V belong to $\{L(\lambda) \mid \lambda \in \pi\}$. Among all submodules belonging to π, of an arbitrary $V \in \mathrm{mod}(S)$, there is a unique maximal one which we denote $O_\pi(V)$ (by analogy with the standard notation $O_\pi(G)$ for the largest normal π subgroup of G, for G a finite group and π a set of primes). Moreover, among all submodules U of V such that V/U belongs to π there is a unique minimal one, which we denote by $O^\pi(V)$ (also by analogy with standard notation in finite group theory). Note that if $\phi : V \to V'$ is a morphism in $\mathrm{mod}(S)$ then $\phi(O_\pi(V)) \leq O_\pi(V')$ and $\phi(O^\pi(V)) \leq O^\pi(V')$. Defining $O_\pi(\phi) : O_\pi(V) \to O_\pi(V')$ and $O^\pi(\phi) : O^\pi(V) \to O^\pi(V')$ to be the restrictions of ϕ, we have functors O_π and O^π from $\mathrm{mod}(S)$ to the category of k-spaces. It is easy to check that O_π is left exact and that O^π is right exact.

Let $x \in S$ and let $\phi : S \to S$ be right multiplication by x. Then we have $\phi(O^\pi(S)) \leq O^\pi(S)$, by functoriality, i.e. $O^\pi(S)x \leq O^\pi(S)$ so that $O^\pi(S)$ is an ideal of S.

(1)   *For $V \in \mathrm{mod}(S)$ we have $O^\pi(V) = O^\pi(S) \cdot V$. In particular we have $O^\pi(S) \cdot V = 0$ if V belongs to π.*

This trivially holds for $V = S$ and hence also for V a direct sum of copies of S. In general we write $V = F/T$, where $F \in \mathrm{mod}(S)$ is free and T is a submodule. By right exactness we have $O^\pi(F/T) = (O^\pi(F) + T)/T = (O^\pi(S) \cdot F + T)/T = O^\pi(S) \cdot (F/T)$.

Now if V belongs to π then $O^\pi(V) = 0$ so that $O^\pi(S) \cdot V = 0$.

We put $S(\pi) = S/O^\pi(S)$ and regard $O_\pi(V)$ and $V/O^\pi(V)$ as $S(\pi)$-modules, for $V \in \mathrm{mod}(S)$. Note that, for $\lambda \in \pi$, we have that $L(\lambda)$ is naturally an $S(\pi)$-module and indeed it is easy to check that:

(2) $\{L(\lambda) \mid \lambda \in \pi\}$ *is a complete set of pairwise non-isomorphic simple* $S(\pi)$-*modules,* $P(\lambda)/O^{\pi}(P(\lambda))$ *is an* $S(\pi)$-*module projective cover of* $L(\lambda)$, *and* $O_{\pi}(I(\lambda))$ *is an* $S(\pi)$-*module injective envelope of* $L(\lambda)$, *for* $\lambda \in \pi$.

As well as the "truncation functor" O_{π} we shall also consider a truncation functor defined by an idempotent. Let $\xi \in S$ be a non-zero idempotent and let S_{ξ} denote the algebra $\xi S \xi$. We have the Schur functor $f : \mathrm{mod}(S) \rightarrow \mathrm{mod}(S_{\xi})$. For V a finite dimensional left S-module, fV is the subspace ξV of V regarded as a left $\xi S \xi$-module, and for a morphism $\theta : V \rightarrow V'$ of left S-modules, $f\theta : fV \rightarrow fV'$ is the restriction of θ. Note that we have a natural k-space isomorphism $\mathrm{Hom}_S(S\xi, V) \rightarrow \xi V$, for $V \in \mathrm{mod}(S)$, and since $S\xi$ is a projective module (or arguing directly) we get:

(3) $f : \mathrm{mod}(S) \rightarrow \mathrm{mod}(S_{\xi})$ *is exact.*

Let $\Lambda_{\xi}^+ = \{\lambda \in \Lambda^+ \mid \xi L(\lambda) \neq 0\}$. For a finite dimensional left S-module V we regard the dual space $V^* = \mathrm{Hom}_k(V, k)$ as a left module for the opposite algebra S^{op} in the usual way. Note that the natural map $\mathrm{Hom}_S(U, V) \rightarrow \mathrm{Hom}_{S^{\mathrm{op}}}(V^*, U^*)$ is a k-space isomorphism, for $U, V \in \mathrm{mod}(S)$. In particular, for $X \in \mathrm{mod}(S)$, the algebra $\mathrm{End}_S(X)$ is local if and only if $\mathrm{End}_{S^{\mathrm{op}}}(X^*)$ is local so that X is indecomposable if and only if X^* is indecomposable. Similarly, we get that $X \in \mathrm{mod}(S)$ is projective (resp. injective) if and only if $X^* \in \mathrm{mod}(S^{\mathrm{op}})$ is injective (resp. projective).

Let $g : \mathrm{mod}(S^{\mathrm{op}}) \rightarrow \mathrm{mod}(S_{\xi}^{\mathrm{op}})$ be the Schur functor.

(4) *(i) For* $V \in \mathrm{mod}(S)$ *restriction* $gV^* \rightarrow (fV)^*$ *is an isomorphism of* S_{ξ}^{op}-*modules.*

(ii) Λ_{ξ}^+ *is the set of* $\lambda \in \Lambda^+$ *such that* $P(\lambda)$ *is a direct summand of* $S\xi$.

(iii) For $\lambda \in \Lambda_{\xi}^+$ *and* $V \in \mathrm{mod}(S)$, *the natural map* $\mathrm{Hom}_S(P(\lambda), V) \rightarrow \mathrm{Hom}_{S_{\xi}}(fP(\lambda), fV)$ *is an isomorphism.*

(iv) $\{fL(\lambda) \mid \lambda \in \Lambda_{\xi}^+\}$ *is a complete set of pairwise non-isomorphic irreducible* S_{ξ}-*modules.*

(v) $fP(\lambda)$ *is a projective cover of* $fL(\lambda)$, *for* $\lambda \in \Lambda_{\xi}^+$.

(vi) $fI(\lambda)$ *is an injective envelope of* $fL(\lambda)$, *for* $\lambda \in \Lambda_{\xi}^+$.

(vii) For $X \in \mathrm{mod}(S)$ *and* $\lambda \in \Lambda_{\xi}^+$, *we have* $[X : L(\lambda)] = [fX : fL(\lambda)]$.

Proof We leave (i) for the reader to check.

We take parts (ii) to (v) together. We note that if U is a simple left S-module then fU is either simple or zero. Suppose $0 \neq \xi u \in \xi U$. Then we have $S_{\xi} \xi u = \xi(S \xi u)$. Since U is simple, we have $S\xi u = U$ and hence $S_{\xi} \xi u = \xi U$. Thus fU, if non-zero, is generated by each non-zero element and hence is simple. Now let $0 = V_0 < V_1 < \cdots < V_n$ be a composition series of $S\xi$. Then we get a series $0 = fV_0 \leq fV_1 \leq \cdots \leq fV_n$ for the left regular module S_ξ. For $1 \le i \le n$, the section fV_i/fV_{i-1} is isomorphic to $f(V_i/V_{i-1})$ and hence is either 0 or simple. Thus every composition factor of S_ξ is isomorphic to fU, for some simple module S-module U. Since every simple left S_ξ-module is a composition factor of the left regular module every simple left S_ξ-module is isomorphic to fU, for some simple S-module U.

If $fL(\lambda) \ne 0$ then $\mathrm{Hom}_S(S\xi, L(\lambda)) \ne 0$ so that $P(\lambda)$ occurs as a direct summand of $S\xi$. Now suppose that $P(\lambda)$ is a direct summand of $S\xi$. We write $\xi = \xi_1 + \cdots + \xi_n$, as an orthogonal sum of primitive idempotents. Thus we have $S\xi = S\xi_1 \oplus \cdots \oplus S\xi_n$ and $P(\lambda) \cong S\xi_i$, for some $1 \le i \le n$. Let $V \in \mathrm{mod}(S)$ and consider the restriction map $\mathrm{Hom}_S(S\xi_i, V) \to \mathrm{Hom}_{S_\xi}(\xi S\xi_i, \xi V)$. Let $\theta \in \mathrm{Hom}_S(S\xi_i, V)$. Then we have $\theta(s\xi_i) = sv$ for all $s \in S$, where $v = \theta(\xi_i) = \xi_i \theta(\xi_i) \in \xi_i V$. If the restriction of θ to $S_\xi \xi_i$ is zero then $v = \theta(\xi_i) = 0$ and $\theta = 0$. Now suppose that $\eta \in \mathrm{Hom}_{S_\xi}(\xi S\xi_i, \xi V)$. Then we have $\eta(s\xi_i) = sv'$, for all $s \in S_\xi$, where $v' = \eta(\xi_i) \in \xi_i V$. Then η is the restriction to $\xi S\xi_i$ of the S-module homomorphism $\theta' : S\xi \to V$, defined by $\theta'(s\xi) = sv'$, for $s \in S$. Hence restriction $\mathrm{Hom}_S(P(\lambda), V) \to \mathrm{Hom}_{S_\xi}(fP(\lambda), fV)$ is an isomorphism. Since $fP(\lambda) \cong \xi S\xi_i = S_\xi \xi_i$ is a direct summand of the left regular module S_ξ, it is projective. Taking $V = P(\lambda)$ we have a k-algebra isomorphism $\mathrm{End}_S(P(\lambda)) \to \mathrm{End}_{S_\xi}(fP(\lambda))$ and, since $P(\lambda)$ is indecomposable, we obtain that $fP(\lambda)$ is indecomposable. Taking $V = L(\mu)$, for $\mu \in \Lambda^+$, we get $\mathrm{Hom}_{S_\xi}(fP(\lambda), fL(\mu)) \cong \mathrm{Hom}_S(P(\lambda), L(\mu))$, which is 0 for $\mu \ne \lambda$. This gives that $fP(\lambda)$ has head $fL(\lambda)$, in particular that $fL(\lambda) \ne 0$ and also that $fL(\lambda) \not\cong fL(\mu)$, for $\mu \ne \lambda$. Thus Λ_ξ^+ is precisely the set of $\lambda \in \Lambda^+$ such that $P(\lambda)$ is a direct summand of $S\xi$ and $fL(\lambda) \not\cong fL(\mu)$, $fP(\lambda) \not\cong fP(\lambda)$, for λ, μ distinct elements of Λ_ξ^+. This completes the proof of (ii)–(v).

Now let I be an injective indecomposable with socle L. Then $P = I^*$ is a projective S^{op}-module with head $H \cong L^*$. If $fI \ne 0$ then $gP \ne 0$, by (i), and has head $gH \cong (fL)^*$, by (v) and (i). Thus $fL \ne 0$, i.e. $L \cong L(\lambda)$ and $I \cong I(\lambda)$, for some $\lambda \in \Lambda_\xi^+$. Moreover $gP \cong (fI)^*$ is projective indecomposable and hence fI is injective indecomposable. Conversely, for $\lambda \in \Lambda_\xi^+$ we have $fL(\lambda) \ne 0$ and hence $fI(\lambda) \ne 0$ and so is the injective indecomposable S_ξ-module with socle $fL(\lambda)$. This proves (vi).

We get (vii) from (iv) and (3) above.

We now fix a partial ordering $\le$ on Λ^+. For $\lambda \in \Lambda^+$ we define $\pi(\lambda) = \{\mu \in \Lambda^+ \mid \mu < \lambda\}$. Let $M(\lambda)$ be the unique maximal submodule of $P(\lambda)$. We define $K(\lambda) = O^{\pi(\lambda)}(M(\lambda))$ and $\Delta(\lambda) = P(\lambda)/K(\lambda)$. Similarly we define $\nabla(\lambda) \le I(\lambda)$ by the formula $\nabla(\lambda)/L(\lambda) = O_{\pi(\lambda)}(I(\lambda)/L(\lambda))$. We call the modules $\Delta(\lambda)$, $\lambda \in \Lambda^+$, the *standard* modules and the modules $\nabla(\lambda)$, $\lambda \in \Lambda^+$, the *costandard* modules.

(5)   *We have* $\mathrm{End}_S(\Delta(\lambda)) = k$ *and* $\mathrm{End}_S(\nabla(\lambda)) = k$, *for all* $\lambda \in \Lambda^+$.

Proof Let $\theta \in \mathrm{End}_S(\Delta(\lambda))$. Now $\Delta(\lambda)$ has a unique maximal submodule

$M = M(\lambda)/K(\lambda)$ and θ induces a homomorphism $\bar{\theta}$, say, on the head $\Delta(\lambda)/M$. But we have $\Delta(\lambda)/M \cong L(\lambda)$ and $\mathrm{End}_S(L(\lambda)) = k$ so that $\bar{\theta}$ is scalar multiplication by c, say. Thus, putting $\phi = \theta - c.\mathrm{id}$ (where id is the identity map on $\Delta(\lambda)$) we have $\phi(\Delta(\lambda)) \leq M$. Thus $L(\lambda)$ is not a composition factor of $\mathrm{Im}(\phi)$ and hence is a composition factor of $\mathrm{Ker}(\phi)$. If $\mathrm{Ker}(\phi) \neq \Delta(\lambda)$ then $\mathrm{Ker}(\phi) \leq M$, which does not have $L(\lambda)$ as a composition factor. Hence $\mathrm{Ker}(\phi) = \Delta(\lambda)$ so $\phi = 0$ and θ is multiplication by c. This shows that $\mathrm{End}_S(\Delta(\lambda)) = k$, and one similarly shows that $\mathrm{End}_S(\nabla(\lambda)) = k$.

(6) *(i) For $\lambda, \mu \in \Lambda^+$, we have*

$$\mathrm{Hom}_S(\Delta(\lambda), \nabla(\mu)) = \begin{cases} k, & \text{if } \lambda = \mu; \\ 0, & \text{otherwise.} \end{cases}$$

(ii) Let $X \in \mathrm{mod}(S)$ and $\lambda \in \Lambda^+$. If $\mathrm{Ext}^1_S(\Delta(\lambda), X) \neq 0$ or $\mathrm{Ext}^1_S(X, \nabla(\lambda)) \neq 0$ then X has a composition factor $L(\mu)$ with $\mu \not< \lambda$.

Proof (i) Suppose $0 \neq \phi \in \mathrm{Hom}_S(\Delta(\lambda), \nabla(\mu))$. Since $\nabla(\mu)$ has simple socle $L(\mu)$, we have that $L(\mu)$ occurs as a composition factor of $\mathrm{Im}(\phi)$ and hence of $\Delta(\lambda)$. This gives $\lambda \geq \mu$. We have $\mathrm{Ker}(\phi) \neq \Delta(\lambda)$ and $\Delta(\lambda)$ has a unique maximal submodule $M = M(\lambda)/K(\lambda)$ and $L(\lambda) \cong \Delta(\lambda)/M$. Thus $L(\lambda)$ is a composition factor of $\Delta(\lambda)/\mathrm{Ker}(\phi)$ and hence of $\mathrm{Im}(\phi) \leq \nabla(\mu)$. This gives $\lambda \leq \mu$ and hence $\lambda = \mu$.

Let $\psi : \Delta(\lambda) \to \nabla(\lambda)$ be any homomorphism. Now M, and hence $\psi(M)$, does not have $L(\lambda)$ as a composition factor. However, $\nabla(\lambda)$ has simple socle $L(\lambda)$ so that every non-zero submodule of $\nabla(\lambda)$ has $L(\lambda)$ as a composition factor. Thus $M \leq \mathrm{Ker}(\psi)$ and ψ induces a homomorphism $\bar{\psi} : L(\lambda) \to \nabla(\lambda)$. Moreover, the image of $\bar{\psi}$ is contained in the socle $L(\lambda)$ of $\nabla(\lambda)$. In this way we obtain an endomorphism of $L(\lambda)$ and obtain an isomorphism $\mathrm{Hom}_S(\Delta(\lambda), \nabla(\lambda)) \to \mathrm{End}_S(L(\lambda)) = k$.
(ii) By the long exact sequence we may assume that $X = L(\mu)$ for some $\mu \in \Lambda^+$. Suppose that $\mathrm{Ext}^1_S(\Delta(\lambda), L(\mu)) \neq 0$. From the short exact sequence $0 \to K(\lambda) \to P(\lambda) \to \Delta(\lambda) \to 0$, we get an exact sequence

$$\mathrm{Hom}_S(K(\lambda), L(\mu)) \to \mathrm{Ext}^1_S(\Delta(\lambda), L(\mu)) \to 0$$

(since $P(\lambda)$ is projective). Thus we have $\mathrm{Hom}_S(K(\lambda), L(\mu)) \neq 0$ and hence there is a submodule K' of $K(\lambda)$ such that $K(\lambda)/K' \cong L(\mu)$. But now, if $\mu < \lambda$ then both $M(\lambda)/K(\lambda)$ and $K(\lambda)/K'$ belong to $\pi(\lambda)$ and therefore $M(\lambda)/K'$ belongs to $\pi(\lambda)$. Thus we have $K(\lambda) = O^{\pi(\lambda)}(M(\lambda)) \leq K' < K(\lambda)$, a contradiction. Hence we have $\mu \not< \lambda$. If $\mathrm{Ext}^1_S(L(\mu), \nabla(\lambda)) \neq 0$ we similarly obtain $\mu \not< \lambda$.

For $X \in \mathrm{mod}(S)$ we write $[X]$ for the class of X in $\mathrm{Grot}(S)$, the Grothendieck group of $\mathrm{mod}(S)$. Thus $\mathrm{Grot}(S)$ is free abelian on $\{[L(\lambda)] \mid \lambda \in$

$\Lambda^+\}$. Moreover, $L(\lambda)$ occurs with multiplicity 1 in $\Delta(\lambda)$ and other composition factors have the form $L(\mu)$, with $\mu < \lambda$. Thus we have

$$[\Delta(\lambda)] = [L(\lambda)] + \sum_{\mu < \lambda} a_\mu [L(\mu)]$$

for certain non-negative integers a_μ. Thus the sets of the elements $[\nabla(\lambda)]$, $\lambda \in \Lambda^+$, and $[L(\lambda)]$, $\lambda \in \Lambda^+$, are related by a unitriangular matrix and therefore the elements $[\nabla(\lambda)]$, $\lambda \in \Lambda^+$, form a $\mathbb{Z}$-basis of $\mathrm{Grot}(S)$. Similar remarks apply to the modules $\nabla(\lambda)$, $\lambda \in \Lambda^+$. To summarize, we have the following.

(7) *The Grothendieck group* $\mathrm{Grot}(S)$ *of* $\mathrm{mod}(S)$ *has* $\mathbb{Z}$-*bases:*
(i) $\{[L(\lambda)] \mid \lambda \in \Lambda^+\}$, *(ii)* $\{[\Delta(\lambda)] \mid \lambda \in \Lambda^+\}$, *and, (iii)* $\{[\nabla(\lambda)] \mid \lambda \in \Lambda^+\}$.

By exactness of $\mathrm{Hom}_S(P(\lambda), -)$ and $\mathrm{Hom}_S(-, I(\lambda))$ we have:

(8) $\dim \mathrm{Hom}_S(P(\lambda), X) = \dim \mathrm{Hom}_S(X, I(\lambda)) = [X : L(\lambda)]$, for $X \in \mathrm{mod}(S)$ *and* $\lambda \in \Lambda^+$.

For $X \in \mathrm{mod}(S)$, in addition to the composition multiplicities, we have the integers $(X : \Delta(\lambda))$ and $(X : \nabla(\lambda))$ (for $\lambda \in \Lambda^+$) defined by the equations

$$[X] = \sum_{\lambda \in \Lambda^+} (X : \Delta(\lambda))[\Delta(\lambda)]$$

and

$$[X] = \sum_{\lambda \in \Lambda^+} (X : \nabla(\lambda))[\nabla(\lambda)].$$

Note that the functions $X \mapsto (X : \Delta(\lambda))$ and $X \mapsto (X : \nabla(\lambda))$ are additive on short exact sequences of S-modules.

Let $X \in \mathrm{mod}(S)$. We call an S-module filtration $0 = X_0 \leq X_1 \leq \cdots \leq X_r = X$ of X a Δ-*filtration* (resp. ∇-*filtration*) if, for $1 \leq i \leq r$, the factor X_i/X_{i-1} is either 0 or isomorphic to $\Delta(\lambda)$ (resp. isomorphic to $\nabla(\lambda)$) for some $\lambda \in \Lambda^+$. We write $X \in \mathcal{F}(\Delta)$ (resp. $X \in \mathcal{F}(\nabla)$) to indicate that X is a finite dimensional left S-module which has some Δ-filtration (resp. ∇-filtration). Note that if $X \in \mathcal{F}(\Delta)$ (resp. $X \in \mathcal{F}(\nabla)$) then $(X : \Delta(\lambda))$ (resp. $(X : \nabla(\lambda))$) is the multiplicity of $\Delta(\lambda)$ (resp. $\nabla(\lambda)$) as a factor in any Δ-filtration (resp. ∇-filtration) of X.

A2 We are now ready to give the key definition.

Definition A2.1 *We say that* $\mathrm{mod}(S)$ *is a high weight category (with respect to the ordering* $\leq$*) if the following properties hold for all* $\lambda \in \Lambda^+$:
(i) $I(\lambda)/\nabla(\lambda) \in \mathcal{F}(\nabla)$;
(ii) whenever $(I(\lambda)/\nabla(\lambda) : \nabla(\mu)) \neq 0$, *for* $\mu \in \Lambda^+$, *we have* $\mu > \lambda$.

To emphasize the dependence on the ordering we will sometimes say that $(\mathrm{mod}(S), \leq)$ is a high weight category, or that $(\mathrm{mod}(S), \Lambda^+)$ is a high weight category. We assume from now on that $\mathrm{mod}(S)$ is a high weight category. We call the elements of Λ^+ *dominant weights*.

Some basic properties are summarized as follows.

Proposition A2.2 *(i) Let* $X \in \mathrm{mod}(S)$ *and* $\lambda \in \Lambda^+$. *If* $\mathrm{Ext}^1_S(\Delta(\lambda), X) \neq 0$
or
$\mathrm{Ext}^1_S(X, \nabla(\lambda)) \neq 0$ *then* X *has a composition factor* $L(\mu)$ *with* $\mu > \lambda$.
In particular if $\mathrm{Ext}^1_S(\Delta(\lambda), \Delta(\mu)) \neq 0$ *or* $\mathrm{Ext}^1_S(\nabla(\mu), \nabla(\lambda)) \neq 0$, *for some*
$\mu \in \Lambda^+$, *then we have* $\mu > \lambda$.
(ii) For $X \in \mathcal{F}(\Delta)$ *and* $Y \in \mathcal{F}(\nabla)$ *we have*

$$\dim \mathrm{Ext}^i_S(X, Y) = \begin{cases} \sum_{\nu \in \Lambda^+}(X : \Delta(\nu))(Y : \nabla(\nu)), & \text{if } i = 0; \\ 0, & \text{otherwise.} \end{cases}$$

In particular, for $\lambda, \mu \in \Lambda^+$, *we have*

$$\mathrm{Ext}^i_S(\Delta(\lambda), \nabla(\mu)) = \begin{cases} k, & \text{if } i = 0, \ \lambda = \mu; \\ 0, & \text{otherwise.} \end{cases}$$

We have

$$(X : \Delta(\lambda)) = \dim \mathrm{Hom}_S(X, \nabla(\lambda))$$

and

$$(Y : \nabla(\lambda)) = \dim \mathrm{Hom}_S(\Delta(\lambda), Y)$$

for $\lambda \in \Lambda^+$.
(iii) For $X \in \mathrm{mod}(S)$ *we have* $X \in \mathcal{F}(\Delta)$ *(resp.* $X \in \mathcal{F}(\nabla)$*) if and only if*

$$\mathrm{Ext}^1_S(X, \nabla(\lambda)) = 0$$

(resp. $\mathrm{Ext}^1_S(\Delta(\lambda), X) = 0$*) for all* $\lambda \in \Lambda^+$.
(iv) For $\lambda \in \Lambda^+$ *we have* $P(\lambda) \in \mathcal{F}(\Delta)$ *(resp.* $I(\lambda) \in \mathcal{F}(\nabla)$*) and*

$$(P(\lambda) : \Delta(\mu)) = [\nabla(\mu) : L(\lambda)]$$

(resp. $(I(\lambda) : \nabla(\mu)) = [\Delta(\mu) : L(\lambda)]$*) for* $\mu \in \Lambda^+$.
(v) Let $0 \to X' \to X \to X'' \to 0$ *be a short exact sequence in* $\mathrm{mod}(S)$. *If* $X', X \in \mathcal{F}(\nabla)$ *(resp.* $X, X'' \in \mathcal{F}(\Delta)$*) then* $X'' \in \mathcal{F}(\nabla)$ *(resp.* $X' \in \mathcal{F}(\Delta)$*).*

(vi) If $X \in \mathcal{F}(\Delta)$ *(resp.* $X \in \mathcal{F}(\nabla)$*) and* Y *is a direct summand of* X *then* $Y \in \mathcal{F}(\Delta)$ *(resp.* $Y \in \mathcal{F}(\nabla)$*).*

Proof (i) Assume that $\mathrm{Ext}^1_S(X, \nabla(\lambda)) \neq 0$. By the long exact sequence we can assume that $X = L(\mu)$ for some dominant weight μ. The short exact sequence $0 \to \nabla(\lambda) \to I(\lambda) \to I(\lambda)/\nabla(\lambda) \to 0$ gives rise to an exact sequence

$$\mathrm{Hom}_S(L(\mu), I(\lambda)/\nabla(\lambda)) \to \mathrm{Ext}^1_S(L(\mu), \nabla(\lambda)) \to 0$$

(since $I(\lambda)$ is injective). Thus we have $\mathrm{Hom}_S(L(\mu), I(\lambda)/\nabla(\lambda)) \neq 0$. Now $I(\lambda)/\nabla(\lambda)$ has a filtration with sections of the form $\nabla(\tau)$, with $\tau > \lambda$. By left exactness of $\mathrm{Hom}_S(L(\mu), -)$, we get $\mathrm{Hom}_S(L(\mu), \nabla(\tau)) \neq 0$ for some $\tau > \lambda$. But $\nabla(\tau)$ has simple socle $L(\tau)$ so we get $\mu = \tau > \lambda$.

We deal with the remaining part of (i) at the end of the proof of the proposition.

(ii) We suppose first that $X = \Delta(\lambda)$ and $Y = \nabla(\mu)$. The case $i = 0$ is A1(6)(i). Now consider the case $i = 1$. If $\mathrm{Ext}^1_S(\Delta(\lambda), \nabla(\mu)) \neq 0$ then we have $\lambda > \mu$, by what we proved of part (i), and $\lambda \not> \mu$, by A1(6)(ii), a contradiction. By induction on filtration length, and the long exact sequence, we get $\mathrm{Ext}^1_S(X, Y) = 0$ for all $X \in \mathcal{F}(\Delta)$, $Y \in \mathcal{F}(\nabla)$. Now suppose that $i > 1$ and we have proved that $\mathrm{Ext}^{i-1}_S(X, Y) = 0$ for all $X \in \mathcal{F}(\Delta)$, $Y \in \mathcal{F}(\nabla)$. Let $X \in \mathcal{F}(\Delta)$ and $\mu \in \Lambda^+$. From the short exact sequence $0 \to \nabla(\mu) \to I(\mu) \to I(\mu)/\nabla(\mu) \to 0$ we obtain

$$\mathrm{Ext}^{i-1}_S(X, I(\mu)/\nabla(\mu)) \cong \mathrm{Ext}^i_S(X, \nabla(\mu))$$

and, since $I(\mu)/\nabla(\mu) \in \mathcal{F}(\nabla)$, we get $\mathrm{Ext}^i_S(X, \nabla(\mu)) = 0$. By the long exact sequence (and induction on filtration length) we get $\mathrm{Ext}^i_S(X, Y) = 0$ for all $Y \in \mathcal{F}(\nabla)$. Hence we have $\mathrm{Ext}^i_S(X, Y) = 0$ for all $X \in \mathcal{F}(\Delta)$, $Y \in \mathcal{F}(\nabla)$ and $i \geq 1$ by induction.

The formula $\dim \mathrm{Hom}_S(X, Y) = \sum_{\nu \in \Lambda^+} (X : \Delta(\nu))(Y : \nabla(\nu))$ is valid for $X = \Delta(\lambda)$, $Y = \nabla(\mu)$ by A1(6)(i). For arbitrary $X \in \mathcal{F}(\Delta)$, $Y \in \mathcal{F}(\nabla)$, the formula follows by induction on filtration length (and the vanishing of $\mathrm{Ext}^1_S(X', Y')$ for $X' \in \mathcal{F}(\Delta)$, $Y' \in \mathcal{F}(\nabla)$).

(iii) If $X \in \mathcal{F}(\Delta)$ (resp. $X \in \mathcal{F}(\nabla)$) then $\mathrm{Ext}^1_S(X, \nabla(\lambda)) = 0$ (resp. $\mathrm{Ext}^1_S(\Delta(\lambda), X) = 0$) for all $\lambda \in \Lambda^+$ by (ii).

Now suppose that $X \in \mathrm{mod}(S)$ and $\mathrm{Ext}^1_S(\Delta(\lambda), X) = 0$ for all $\lambda \in \Lambda^+$. (The other case is similar.) We argue by induction on the dimension of X. There is nothing to prove if $X = 0$. We now assume $X \neq 0$ and that $X' \in \mathcal{F}(\nabla)$, whenever $X' \in \mathrm{mod}(S)$ with $\dim X' < \dim X$ satisfies the condition $\mathrm{Ext}^1_S(\Delta(\lambda), X') = 0$ for all $\lambda \in \Lambda^+$.

Let $\mu \in \Lambda^+$ be as small as possible such that $\mathrm{Hom}_S(L(\mu), X) \neq 0$. We claim that $\mathrm{Ext}^1_S(L(\nu), X) = 0$ for all $\nu < \mu$. For such an element ν we have a short exact sequence $0 \to N \to \Delta(\nu) \to L(\nu) \to 0$ and hence an exact sequence

$$\mathrm{Hom}_S(N, X) \to \mathrm{Ext}^1_S(L(\nu), X) \to \mathrm{Ext}^1_S(\Delta(\nu), X).$$

For a composition factor $L(\nu')$ of N we have $\nu' < \nu$ and therefore $\nu' < \mu$. Hence we have $\mathrm{Hom}_S(N, X) = 0$ by the choice of μ. But we also have $\mathrm{Ext}^1_S(\Delta(\nu), X) = 0$, by the hypothesis, and so we get $\mathrm{Ext}^1_S(L(\nu), X) = 0$, proving the claim.

We have a short exact sequence $0 \to L(\mu) \to \nabla(\mu) \to Q \to 0$ and hence an exact sequence

$$\mathrm{Hom}_S(\nabla(\mu), X) \to \mathrm{Hom}_S(L(\mu), X) \to \mathrm{Ext}^1_S(Q, X).$$

Composition factors of Q have the form $L(\nu)$, with $\nu < \mu$, so we have $\mathrm{Ext}^1_S(Q, X) = 0$, by the claim. Thus the map

$$\mathrm{Hom}_S(\nabla(\mu), X) \to \mathrm{Hom}_S(L(\mu), X)$$

is onto, and we have a homomorphism $\nabla(\mu) \to X$, whose restriction to $L(\mu)$ is non-zero. Since $\nabla(\mu)$ has simple socle $L(\mu)$, this homomorphism is injective. Thus we have a copy X_1, say, of $\nabla(\mu)$ in X. Now for $\lambda \in \Lambda^+$, by the long exact sequence, we have an exact sequence

$$\mathrm{Ext}^1_S(\Delta(\lambda), X) \to \mathrm{Ext}^1_S(\Delta(\lambda), X/X_1) \to \mathrm{Ext}^2_S(\Delta(\lambda), X_1).$$

We have $\mathrm{Ext}^1_S(\Delta(\lambda), X) = 0$ by hypothesis and

$$\mathrm{Ext}^2_S(\Delta(\lambda), X_1) \cong \mathrm{Ext}^2_S(\Delta(\lambda), \nabla(\mu)) = 0,$$

by (ii), and hence we have $\mathrm{Ext}^1_S(\Delta(\lambda), X/X_1) = 0$ for all $\lambda \in \Lambda^+$. By the inductive hypothesis we have $X/X_1 \in \mathcal{F}(\nabla)$. But $X_1 \cong \nabla(\mu)$ so we get $X \in \mathcal{F}(\nabla)$.

(iv) We have $\mathrm{Ext}^1_S(P(\lambda), \nabla(\mu)) = 0$, for all $\mu \in \Lambda^+$, since $P(\lambda)$ is projective. Hence $P(\lambda) \in \mathcal{F}(\Delta)$, by (iii). We have

$$(P(\lambda) : \Delta(\mu)) = \dim \mathrm{Hom}_S(P(\lambda), \nabla(\mu)) = [\nabla(\mu) : L(\lambda)]$$

by (ii) and A1(8). We have $I(\lambda) \in \mathcal{F}(\nabla)$ by hypothesis and

$$(I(\lambda) : \nabla(\mu)) = \dim \mathrm{Hom}_S(\Delta(\mu), I(\lambda)) = [\Delta(\mu) : L(\lambda)]$$

by (ii) and A1(8) again.

(v),(vi) These follow from (iii).

Finally, returning to (i), we now have, by (iv) and (v), that $K(\lambda)$ has a filtration with sections of the form $\Delta(\lambda')$, with $\lambda' > \lambda$. The dual of the argument given above at the start of the proof shows that if $\text{Ext}^1_S(\Delta(\lambda), X) \neq 0$ then X has a composition factor $L(\mu)$ with $\mu > \lambda$.

For $\lambda \in \Lambda^+$ we define $l(\lambda) = l$, where l is the length of a longest chain $\lambda_0 < \lambda_1 < \cdots < \lambda_l = \lambda$ in Λ^+. Define $l(\Lambda^+)$ to be the maximum of the lengths $l(\lambda)$, for $\lambda \in \Lambda^+$.

Proposition A2.3 *We have* $\text{Ext}^i_S(L(\lambda), L(\mu)) = 0$ *for* $\lambda, \mu \in \Lambda^+$ *and* $i > l(\lambda) + l(\mu)$. *Thus S has finite global dimension, bounded by $2l(\Lambda^+)$.*

Proof We argue by induction on $l(\lambda)+l(\mu)$. We assume that for all $\lambda', \mu' \in \Lambda^+$ with $l(\lambda') + l(\mu') < l(\lambda) + l(\mu)$ we have $\text{Ext}^i_S(L(\lambda'), L(\mu')) = 0$ for all $i > l(\lambda')+l(\mu')$. We have a short exact sequence $0 \to N \to \Delta(\lambda) \to L(\lambda) \to 0$ and, for every composition factor $L(\nu)$ of N, we have $\nu < \lambda$ and hence $l(\nu) < l(\lambda)$. For $i > l(\lambda) + l(\mu)$, we have the exact sequence

$$\text{Ext}^{i-1}_S(N, L(\mu)) \to \text{Ext}^i_S(L(\lambda), L(\mu)) \to \text{Ext}^i_S(\Delta(\lambda), L(\mu)).$$

Now $i - 1 > l(\nu) + l(\mu)$ for every composition factor $L(\nu)$ of N and hence, by the inductive hypothesis, we have $\text{Ext}^{i-1}_S(N, L(\mu)) = 0$. Thus it suffices to prove that $\text{Ext}^i_S(\Delta(\lambda), L(\mu)) = 0$. We have a short exact sequence $0 \to L(\mu) \to \nabla(\mu) \to Q \to 0$ and for every composition factor $L(\nu)$ of Q, we have $\nu < \mu$. We get an exact sequence

$$\text{Ext}^{i-1}_S(\Delta(\lambda), Q) \to \text{Ext}^i_S(\Delta(\lambda), L(\mu)) \to \text{Ext}^i_S(\Delta(\lambda), \nabla(\mu)).$$

Now, for every composition factor $L(\lambda')$ of $\Delta(\lambda)$ and every composition factor $L(\nu)$ of Q, we have $l(\lambda') + l(\nu) \leq l(\lambda) + l(\nu) < l(\lambda) + l(\mu)$ and therefore $i - 1 > l(\lambda') + l(\nu)$. Thus we have $\text{Ext}^{i-1}_S(\Delta(\lambda), Q) = 0$, by the inductive hypothesis. We also have $\text{Ext}^i_S(\Delta(\lambda), \nabla(\mu)) = 0$, by Proposition A2.2(ii), and therefore $\text{Ext}^i_S(\Delta(\lambda), L(\mu)) = 0$ and hence $\text{Ext}^i_S(L(\lambda), L(\mu)) = 0$, as required.

A3 We call a set π of dominant weights *saturated* if it has the property that $\lambda \in \pi$ whenever $\lambda < \mu$ and $\mu \in \pi$.

Lemma A3.1 *(i) Let $X \in \mathcal{F}(\nabla)$ (resp. $X \in \mathcal{F}(\Delta)$) and let μ be a minimal element of the set $\{\nu \in \Lambda^+ \mid (X : \nabla(\nu)) \neq 0\}$ (resp. $\{\nu \in \Lambda^+ \mid (X : \Delta(\nu)) \neq 0\}$). Then some submodule (resp. quotient module) of X is isomorphic to $\nabla(\mu)$ (resp. $\Delta(\mu)$).*
Let π be a saturated set of dominant weights.
(ii) If $X \in \mathcal{F}(\nabla)$ then $O_\pi(X) \in \mathcal{F}(\nabla)$ and, for $\lambda \in \Lambda^+$, we have

$$(O_\pi(X) : \nabla(\lambda)) = \begin{cases} (X : \nabla(\lambda)), & \text{if } \lambda \in \pi; \\ 0, & \text{otherwise.} \end{cases}$$

(iii) If $X \in \mathcal{F}(\Delta)$ then $O^\pi(X), X/O^\pi(X) \in \mathcal{F}(\Delta)$ and we have

$$(X/O^\pi(X) : \Delta(\lambda)) = \begin{cases} (X : \Delta(\lambda)), & \text{if } \lambda \in \pi; \\ 0, & \text{otherwise.} \end{cases}$$

Proof Part (i) follows from Proposition A2.2(i).

We now prove part (ii), and leave part (iii) to the reader. We argue by induction on dimension. If $X = 0$ there is nothing to prove. We leave the case in which $X \cong \nabla(\mu)$, for some $\mu \in \Lambda^+$, to the reader. Now assume $X \neq 0$ and the result holds for all $X' \in \mathcal{F}(\nabla)$ with $\dim X' < \dim X$. Let μ be a minimal element of the set $\{\nu \in \Lambda^+ \mid (X : \nabla(\nu)) \neq 0\}$. By (i), X contains a submodule X_1 isomorphic to $\nabla(\mu)$. Suppose first that $\mu \in \pi$. Then $X_1 \leq O_\pi(X)$ and it follows that $O_\pi(X)/X_1 \cong O_\pi(X/X_1)$. By the inductive hypothesis $O_\pi(X/X_1)$ has a ∇-filtration and

$$(O_\pi(X/X_1) : \nabla(\lambda)) = \begin{cases} (X/X_1 : \nabla(\lambda)), & \text{if } \lambda \in \Lambda^+; \\ 0, & \text{otherwise.} \end{cases}$$

Thus $O_\pi(X)/X_1 \in \mathcal{F}(\nabla)$ and therefore $O_\pi(X) \in \mathcal{F}(\nabla)$. For $\lambda \in \pi$ we get

$$\begin{aligned}(O_\pi(X) : \nabla(\lambda)) &= (O_\pi(X)/X_1 : \nabla(\lambda)) + (X_1 : \nabla(\lambda)) \\ &= (X/X_1 : \nabla(\lambda)) + (X_1 : \nabla(\lambda)) \\ &= (X : \nabla(\lambda))\end{aligned}$$

and for $\lambda \notin \pi$ we get

$$\begin{aligned}(O_\pi(X) : \nabla(\lambda)) &= (O_\pi(X)/X_1 : \nabla(\lambda)) + (X_1 : \nabla(\lambda)) \\ &= 0 + 0 = 0\end{aligned}$$

using the inductive hypothesis and the fact that $\lambda \neq \mu$.

Thus we may assume that no minimal element of the support of X belongs to π. But then $(X : \nabla(\lambda)) = 0$ for every $\lambda \in \pi$ so that X has a filtration in which we have $O_\pi(Y) = 0$ for each section Y. Hence $O_\pi(X) = 0$, and the result holds.

For a set of dominant weights π we now regard O_π as a functor from $\mathrm{mod}(S)$ to $\mathrm{mod}(S(\pi))$ (as in A1). We consider now the right derived functors $R^i O_\pi$. We regard an $S(\pi)$-module X also as an S-module (also denoted X) via the natural map $S \to S(\pi)$. Note that, for $V \in \mathrm{mod}(S(\pi))$ and $W \in \mathrm{mod}(S)$, the image of any S-module homomorphism $V \to W$ lies in $O_\pi(W)$. Thus we have $\mathrm{Hom}_S(V, W) = \mathrm{Hom}_{S(\pi)}(V, O_\pi(W))$ and hence we have the factorization $\mathrm{Hom}_S(V, -) = \mathrm{Hom}_{S(\pi)}(V, -) \circ O_\pi$. Note that $O_\pi(I(\lambda)) = 0$ if $\lambda \notin \pi$ and, by A1(2), $O_\pi(I(\lambda))$ is the injective envelope of $L(\lambda)$ in $\mathrm{mod}(S(\pi))$, if $\lambda \in \pi$. Thus $O_\pi : \mathrm{mod}(S) \to \mathrm{mod}(S(\pi))$ takes injectives to acyclics and we therefore have, for $W \in \text{mod}(S)$, a Grothendieck spectral sequence with E_2-page $\text{Ext}^i_{S(\pi)}(V, R^j O_\pi(W))$, converging to $\text{Ext}^*_S(V, W)$.

Proposition A3.2 *Let π be a saturated set of dominant weights.*
(i) For $W \in \mathcal{F}(\nabla)$ we have $R^i O_\pi(W) = 0$ for all $i > 0$.
(ii) For $X \in \text{mod}(S)$ belonging to π we have $R^i O_\pi(X)$ for all $i > 0$.

Proof (i) Consider first the case $i = 1$. For $\lambda \in \Lambda^+$ we have a short exact sequence $0 \to \nabla(\lambda) \to I(\lambda) \to Q(\lambda) \to 0$, where $Q(\lambda)$ has a ∇-filtration with sections of the form $\nabla(\mu)$, $\mu > \lambda$. Hence we get an exact sequence

$$0 \to O_\pi(\nabla(\lambda)) \to O_\pi(I(\lambda)) \to O_\pi(Q(\lambda)) \to RO_\pi(\nabla(\lambda)) \to 0.$$

Suppose that $\lambda \in \pi$. Then we have $O_\pi(I(\lambda)/\nabla(\lambda)) = O_\pi(I(\lambda))/\nabla(\lambda)$ so that $O_\pi(I(\lambda)) \to O_\pi(Q(\lambda))$ is surjective and $RO_\pi(\nabla(\lambda)) = 0$. Now suppose $\lambda \notin \pi$. Then $Q(\lambda)$ is filtered by modules $\nabla(\mu)$, with $\mu > \lambda$, and no such μ belongs to π. Hence $O_\pi(Q(\lambda)) = 0$ and again $RO_\pi(\nabla(\lambda)) = 0$. Hence in all cases $RO_\pi(\nabla(\lambda)) = 0$ and therefore, by the long exact sequence, we have $RO_\pi(X) = 0$ for all $X \in \mathcal{F}(\nabla)$. Now suppose, for some $i > 0$, we have shown that $R^i O_\pi(X) = 0$ for all $X \in \mathcal{F}(\nabla)$. We get $R^{i+1} O_\pi(\nabla(\lambda)) = R^i O_\pi(Q(\lambda)) = 0$ and hence $R^{i+1} O_\pi(X) = 0$, for all $X \in \mathcal{F}(\nabla)$, by the long exact sequence. This proves (i) by induction on i.
(ii) For $\lambda \in \pi$ we have a short exact sequence $0 \to L(\lambda) \to \nabla(\lambda) \to \nabla(\lambda)/L(\lambda) \to 0$ and hence, by (i), an exact sequence

$$0 \to L(\lambda) \to \nabla(\lambda) \to \nabla(\lambda)/L(\lambda) \to RO_\pi(L(\lambda)) \to 0$$

and isomorphisms $R^{i-1} O_\pi(\nabla(\lambda))/L(\lambda)) \to R^i O_\pi(L(\lambda))$, for $i > 1$. Thus we get
$RO_\pi(L(\lambda)) = 0$, and hence $RO_\pi(X) = 0$ for X belonging to π. Now suppose that $i > 1$ and that $R^{i-1} O_\pi(X)$ for all X belonging to π. Then, for $\lambda \in \pi$, we get $R^i O_\pi(L(\lambda)) \cong R^{i-1} O_\pi(\nabla(\lambda)/L(\lambda)) = 0$ and hence, $R^i O_\pi(X) = 0$ for all X belonging to π. Thus we have $R^i O_\pi(X) = 0$ for all $i \geq 1$ and X belonging to π, by induction.

We leave it to the reader to formulate and prove the corresponding results for O^π.

Proposition A3.3 *Let M, N be finite dimensional S-modules belonging to the saturated set π. Then, for all $i \geq 0$, we have*

$$\text{Ext}^i_{S(\pi)}(M, N) \cong \text{Ext}^i_S(M, N).$$

Proof By the discussion before Proposition A3.2, we have a Grothendieck spectral sequence with E_2-page $\text{Ext}^i_{S(\pi)}(M, R^j O_\pi(N))$ which converges to

$\text{Ext}_S^*(M, N)$. But $R^j O_\pi(N) = 0$ for $j > 0$, so the spectral sequence degenerates and we have $\text{Ext}_S^r(M, N) \cong \text{Ext}_{S(\pi)}^r(M, N)$ for all $r \geq 0$.

Let π be a saturated set of dominant weights. By A1(2) $\{L(\lambda) \mid \lambda \in \pi\}$ is a complete set of pairwise non-isomorphic irreducible $S(\pi)$-modules and, moreover, for $\lambda \in \pi$, the $S(\pi)$-module $I_\pi(\lambda) = O_\pi(I(\lambda))$ is the injective envelope of $L(\lambda)$. By Lemma A3.1(ii), $I_\pi(\lambda)/\nabla(\lambda)$ has a filtration with sections of the form $\nabla(\mu)$, with $\lambda < \mu \in \pi$. Let $\sigma = \{\nu \in \Lambda^+ \mid \nu < \lambda\}$. We have $O_\sigma(\nabla(\mu)) = 0$ for $\mu > \lambda$ (since $\nabla(\mu)$ has socle $L(\mu)$) and hence, by left exactness of O_σ, we have $O_\sigma(I_\pi(\lambda)/\nabla(\lambda)) = 0$. Hence, applying O_σ to the short exact sequence $0 \to \nabla(\lambda)/L(\lambda) \to I_\pi(\lambda)/L(\lambda) \to I_\pi(\lambda)/\nabla(\lambda) \to 0$, we get $O_\sigma(I_\pi(\lambda)/L(\lambda)) = O_\sigma(\nabla(\lambda)/L(\lambda)) = \nabla(\lambda)/L(\lambda)$. Hence the modules $\nabla(\lambda)$, $\lambda \in \pi$, are the costandard modules for $S(\pi)$. Now we have that $I_\pi(\lambda)/\nabla(\lambda)$ has a ∇-filtration, by Lemma A3.1(ii), and $I_\pi(\lambda)/\nabla(\lambda)$ has a filtration with sections $\nabla(\mu)$. Moreover, we have $(I_\pi(\lambda)/\nabla(\lambda) : \nabla(\mu)) = (I_\pi(\lambda) : \nabla(\mu)) - (\nabla(\lambda) : \nabla(\mu))$, which gives that we have $\lambda < \mu \in \pi$, whenever $(I_\pi(\lambda)/\nabla(\lambda) : \nabla(\mu)) \neq 0$. This shows that $\text{mod}(S(\pi))$ is a high weight category with costandard modules $\nabla(\lambda)$, $\lambda \in \pi$. A similar argument, using Lemma A3.1(iii), shows that the modules $\Delta(\lambda)$, $\lambda \in \pi$, are the standard modules. We collect these facts together for future reference.

Proposition A3.4 *For a saturated set of dominant weights π we have that $\text{mod}(S(\pi))$ is a high weight category with standard modules $\Delta(\lambda)$, $\lambda \in \pi$, and costandard modules $\nabla(\lambda)$, $\lambda \in \pi$, with respect to the labelling of the complete set of irreducible modules $\{L(\lambda) \mid \lambda \in \pi\}$ and the induced partial ordering on π.*

We drop, for the moment, our standing assumption that $\text{mod}(S)$ is a high weight category.

Lemma A3.5 (S, Λ^+) *is a high weight category if and only if the following properties hold for all $\lambda \in \Lambda^+$:*
(i) $K(\lambda) \in \mathcal{F}(\Delta)$;
(ii) *whenever* $(K(\lambda) : \Delta(\mu)) \neq 0$, *for $\mu \in \Lambda^+$, we have $\mu > \lambda$.*

Proof Suppose that (S, Λ^+) is a high weight category. Let $\lambda \in \Lambda^+$ and let $\pi = \{\mu \in \Lambda^+ \mid \mu \not> \lambda\}$. Note that $P(\lambda)/K(\lambda)$ belongs to π so that $O^\pi(P(\lambda)) \leq K(\lambda)$. For $\mu \in \pi$ we have $(P(\lambda)/O^\pi(P(\lambda)) : \Delta(\mu)) = (P(\lambda) : \Delta(\mu)) = [\nabla(\mu) : L(\lambda)]$, by Lemma A3.1(iii), so when this is non-zero we have $\mu \geq \lambda$, and hence $\mu = \lambda$. In this case we get $(P(\lambda)/O^\pi(P(\lambda)) : \Delta(\lambda)) = [\nabla(\lambda) : L(\lambda)] = 1$. Hence we have $\dim P(\lambda)/O^\pi(P(\lambda)) = \dim \Delta(\lambda) = \dim P(\lambda)/K(\lambda)$ so that $\dim O^\pi(P(\lambda)) = \dim K(\lambda)$ and hence $O^\pi(P(\lambda)) = K(\lambda)$. Thus we have $K(\lambda) \in \mathcal{F}(\Delta)$ and $(K(\lambda) : \Delta(\mu)) = 0$, for $\mu \in \pi$, by Lemma A3.1(iii), so that whenever $(K(\lambda) : \Delta(\mu)) \neq 0$, we have $\mu > \lambda$, as required.

Now suppose that (i) and (ii) hold for all $\lambda \in \Lambda^+$. These conditions are dual to the defining conditions for $\mathrm{mod}(S)$ to be a high weight category and imply that $\mathrm{mod}(S^{\mathrm{op}})$ is a high weight category with respect to the labelling $\{L(\lambda)^* \mid \lambda \in \Lambda^+\}$ of the irreducible S^{op}-modules, and moreover, the modules $(P(\lambda)/K(\lambda))^*$, $\lambda \in \Lambda^+$, are the costandard modules for S^{op}. Writing $P^*(\lambda) = I(\lambda)^*$ we then get that the maximal submodule $M^*(\lambda)$ of $P^*(\lambda)$ is $(I(\lambda)/L(\lambda))^*$ and that the submodule $K^*(\lambda)$, such that $M^*(\lambda)/K^*(\lambda)$ is the largest quotient of $M^*(\lambda)$ belonging to $\pi(\lambda)$, is $(I(\lambda)/\nabla(\lambda))^*$ and $\Delta^*(\lambda) = P^*(\lambda)/K^*(\lambda) \cong \nabla(\lambda)^*$. Now, by what we have proved so far, the conditions (i) and (ii) hold for the algebra S^{op}. Hence $(I(\lambda)/\nabla(\lambda))^*$ has a filtration with sections $\nabla(\mu)^*$, with $\mu > \lambda$, and it follows that $I(\lambda)/\nabla(\lambda)$ has a filtration with sections $\nabla(\mu)$, $\mu > \lambda$. Hence $\mathrm{mod}(S)$ is a high weight category, with respect to the given ordering $\leq$.

An ideal H of S is called a *hereditary* ideal if it satisfies the following conditions:
(i) H is projective as a left S-module;
(ii) $\mathrm{Hom}_S(H, S/H) = 0$;
(iii) $HNH = 0$, where N is the radical of S.

Definition A3.6 The algebra S is called quasihereditary if there exists a chain of ideals $S = H_0 > H_1 > \cdots > H_n = 0$ with H_i/H_{i+1} hereditary in S/H_{i+1}, for $0 \leq i < n$. Such a chain of ideals is called a hereditary chain.

Proposition A3.7 *(i) Suppose that $\mathrm{mod}(S)$ is a high weight category with respect to a given partial order $\leq$. Write out the elements of Λ^+ as $\lambda_1, \ldots, \lambda_n$ in such a way that $i < j$ whenever $\lambda_i < \lambda_j$ and define $\pi(i) = \{\lambda_1, \ldots, \lambda_i\}$, for $1 \leq i \leq n$. Then $S > O^{\pi(1)}(S) > \cdots > O^{\pi(n)}(S) = 0$ is a hereditary chain of ideals, and hence S is quasihereditary.*
(ii) Suppose given a hereditary chain $S = H_0 > H_1 > \cdots > H_n = 0$ in S. Let $\{L(\lambda) \mid \lambda \in \Lambda^+\}$ be a complete set of pairwise inequivalent irreducible S-modules. Let $\Lambda^+(i)$ be the set of $\lambda \in \Lambda^+$ such that $L(\lambda)$ occurs as a composition factor of S/H_i, for $1 \leq i \leq n$, and define a positive integer $r(\lambda) \leq n$, for $\lambda \in \Lambda^+$, by the condition $\lambda \in \Lambda^+(r(\lambda)) \backslash \Lambda^+(r(\lambda) - 1)$ (where $\Lambda^+(0)$ is the empty set). Let $\leq$ be the partial ordering such that $\lambda < \mu$ if and only if $r(\lambda) < r(\mu)$ (for $\lambda, \mu \in \Lambda^+$). Then $\mathrm{mod}(S)$ is a high weight category with respect to $\leq$.

Proof (i) We first show that if μ is any maximal element of Λ^+ and $\sigma = \Lambda^+ \backslash \{\mu\}$ then $O^\sigma(S)$ is a hereditary ideal of S. Since S is a direct sum of the modules $P(\lambda)$, $\lambda \in \Lambda^+$, and each $P(\lambda) \in \mathcal{F}(\Delta)$, the left regular module S has a Δ-filtration. By Lemma A3.1(iii), $O^\sigma(S)$ has a filtration with sections all of the form $\Delta(\mu)$. Hence, by Proposition A2.2(i), $O^\sigma(S)$ is a direct sum of copies of $\Delta(\mu)$. But $K(\mu) \leq P(\mu)$ has a filtration with sections of the form $\Delta(\nu)$, $\nu > \mu$, and μ is maximal in Λ^+ so that $K(\mu) = 0$, i.e.

$P(\mu) = \Delta(\mu)$ and $\Delta(\mu)$ is projective. Hence $O^\sigma(S)$ is projective. We have
dim $\text{Hom}_S(P(\mu), S/O^\sigma(S)) = [S/S^\sigma(S) : L(\mu)] = 0$, since all composition
factors of $S/O^\sigma(S)$ come from $\{L(\lambda) \mid \lambda \in \sigma\}$ and $\mu \notin \sigma$. Hence we have
$\text{Hom}_S(O^\sigma(S), S/O^\sigma(S)) = 0$.

We write $O^\sigma(S) = \Delta_1 \oplus \cdots \oplus \Delta_t$, with $\Delta_1, \ldots, \Delta_t$ isomorphic to $\Delta(\mu)$.
Now $\Delta(\mu)$ has unique maximal submodule $M(\mu)$, so that $N \cdot \Delta_i \cong M(\mu)$
belongs to σ, for $1 \leq i \leq t$. By A1(2) we therefore have $O^\sigma(S)N\Delta_i = 0$,
$1 \leq i \leq t$, and hence $O^\sigma(S)NO^\sigma(S) = 0$. Thus $O^\sigma(S)$ is a hereditary ideal
in S.

Now we must show that $O^{\pi(i)}(S)/O^{\pi(i+1)}(S)$ is a hereditary ideal in
$S/O^{\pi(i+1)}(S)$, for $0 \leq i < n$. Let $\pi = \pi(i+1)$, $\sigma = \pi(i)$ and define $\lambda \in \pi$
by $\{\lambda\} = \pi \backslash \sigma$. We must show that $O^\sigma(S)/O^\pi(S)$ is a hereditary ideal of
$S/O^\pi(S)$, i.e. that $O^\sigma(S(\pi))$ is a hereditary ideal of $S(\pi)$. But this is true
by the previous paragraph and Proposition A3.4.

(ii) Let $H = H_{n-1}$ and let $\pi = \Lambda^+(n-1)$ with the partial ordering induced from the specified partial ordering on Λ^+. We assume inductively that
$\text{mod}(S/H)$ is a high weight category for the partial ordering on π induced by
that on Λ^+. Since H is projective, we have $H \cong \bigoplus_{\lambda \in \Lambda^+} P(\lambda)^{(d_\lambda)}$, for non-
negative integers d_λ. If $d_\lambda > 0$ with $\lambda \in \pi$ then we have dim $\text{Hom}_S(H, S/H) \geq$
dim $\text{Hom}_S(P(\lambda), S/H) = [S/H : L(\lambda)] \neq 0$, a contradiction. Hence we have
$H \cong \bigoplus_{\mu \in \pi_c} P(\mu)^{(d_\mu)}$, where $\pi_c = \Lambda^+ \backslash \pi$, the complement of π. Hence S/H
is the largest quotient of S belonging to π, i.e. we have $H = O^\pi(S)$.

Now if $d_\lambda \neq 0$ then $M(\lambda) = NP(\lambda)$ embeds in $NO^\pi(S)$. But we have
$O^\pi(S)NO^\pi(S) = 0$ so that $O^\pi(S)M(\lambda) = 0$ and $M(\lambda)$ is an S/H-module
and so belongs to π, by A1(1). Thus all composition factors of $M(\lambda)$ have
the form $L(\mu)$, for $\mu < \lambda$, and so $K(\lambda) = 0$, i.e. $P(\lambda) = \Delta(\lambda)$. Moreover,
since each $L(\mu)$, with $\mu \in \pi_c$, occurs as a composition factor of H and hence
as a composition factor of $P(\lambda)$ for some $\lambda \in \pi_c$ with $d_\lambda > 0$, we must have
$d_\lambda > 0$ for all $\lambda \in \pi_c$. Thus we have $H \cong \bigoplus_{\lambda \in \pi_c} P(\lambda)^{(d_\lambda)}$, with $d_\lambda > 0$ for
all $\lambda \in \pi_c$. For $\lambda \in \pi_c$ we have shown $P(\lambda) = \Delta(\lambda)$ and $K(\lambda) = 0$ so that
$K(\lambda) \in \mathcal{F}(\Delta)$ and trivially $K(\lambda)$ has a filtration with sections $\Delta(\mu)$, $\mu > \lambda$.

For $\lambda \in \pi$ define $R(\lambda) = O^\pi(P(\lambda)) = HP(\lambda)$. Then $P_0(\lambda) = P(\lambda)/R(\lambda)$
is the projective cover of $L(\lambda)$, as an $S(\pi) = S/H$-module, by A1(2). We
define $M_0(\lambda)$ to be the unique maximal submodule of $P_0(\lambda)$ and define
$K_0(\lambda) \leq M_0(\lambda)$ by the condition that $M_0(\lambda)/K_0(\lambda)$ is the largest quotient
of $M_0(\lambda)$ with all composition factors of the form $L(\nu)$, with $\nu < \lambda$, i.e.
$K_0(\lambda) = O^{\pi(\lambda)}(M_0(\lambda))$. By hypothesis, $H = O^\pi(S)$ is projective and, since
$P(\lambda)$ is a direct summand of S, we get that $R(\lambda) = O^\pi(P(\lambda))$ is projective. Moreover, $R(\lambda)$ has no quotient belonging to π. Hence if $P(\mu)$ is a
summand of $R(\lambda)$ then $\mu \in \pi_c$ and hence $P(\mu) = \Delta(\mu)$. In particular, we
have $O^{\pi(\lambda)}(R(\lambda)) = 0$ and so, applying $O^{\pi(\lambda)}$ to the short exact sequence
$0 \to R(\lambda) \to M(\lambda) \to M_0(\lambda) \to 0$, we get $O^{\pi(\lambda)}(M(\lambda)) \cong O^{\pi(\lambda)}(M_0(\lambda))$ and
hence $P(\lambda) \to P_0(\lambda)$ induces an isomorphism $\Delta(\lambda) \to \Delta_0(\lambda)$. Now we have
a short exact sequence $0 \to R(\lambda) \to K(\lambda) \to K_0(\lambda) \to 0$. By the inductive

hypothesis, $K_0(\lambda)$ has a filtration with sections $\Delta(\mu)$, with $\lambda < \mu \in \pi$, and, as we have shown, $R(\lambda)$ has a filtration with sections $\Delta(\mu)$ with $\mu \in \pi_c$, and therefore $\mu > \lambda$. Hence $\mathrm{mod}(S)$ is a high weight category with respect to $\leq$, by Lemma A3.5.

From now on we shall use the expressions "S is quasihereditary" and "$\mathrm{mod}(S)$ is a high weight category" interchangeably. We return to our standing assumption that (S, Λ^+) is a quasihereditary algebra.

We now explore a form of truncation, defined by an idempotent, analogous to that brought about by the O_π functor. It is convenient to introduce at this point the concept of a theory of weights for a quasihereditary algebra.

Definition   A3.8   A *theory of weights* for (S, Λ^+) is an injective map $\theta :$ $\Lambda \to S$ which has the following properties:
(i) $\{\theta(\alpha)\}$ consists of pairwise orthogonal non-zero idempotents whose sum is 1;
(ii) for each $\alpha \in \Lambda$, writing $S\theta(\alpha) \cong \bigoplus_{\lambda \in \Lambda^+} P(\lambda)^{(d_\lambda)}$, the set $\{\lambda \in \Lambda^+ \mid d_\lambda \neq 0\}$ has a unique minimal element, which we write α^+, and $d_{\alpha^+} = 1$;
(iii) the map $\Lambda \to \Lambda^+$, given by $\alpha \mapsto \alpha^+$, is surjective.

Note that the quasihereditary algebra (S, Λ^+) has the following trivial theory of weights. Let $1 = \sum_{i=1}^n e_i$ be a primitive orthogonal decomposition of 1. We take $\Lambda = [1, n]$ and define $\theta : \Lambda \to S$ by $\theta(i) = e_i$, $i \in [1, n]$. Then θ is a theory of weights and, for $i \in [1, n]$, we have $i^+ = \lambda$, where λ is the dominant weight such that $Se_i \cong P(\lambda)$.

We assume from now on that (S, Λ^+) is a quasihereditary algebra with a given theory of weights θ. We call the elements of Λ *weights*. For $\Gamma \subseteq \Lambda$ we set $\Gamma^+ = \{\alpha^+ \mid \alpha \in \Gamma\}$. We write ξ_α for $\theta(\alpha)$, for $\alpha \in \Lambda$, and write ξ_Γ for $\sum_{\alpha \in \Gamma} \xi_\alpha$, for $\Gamma \subseteq \Lambda$. For $U \in \mathrm{mod}(S)$ we define the *weight space* $U^\alpha = \xi_\alpha U$ and note that $U^\alpha \cong \mathrm{Hom}_S(S\xi_\alpha, U)$. Thus we have

$$\dim U^\alpha = \sum_{\mu \in \Lambda^+} d_\mu \dim \mathrm{Hom}_S(P(\mu), U) = \sum_{\mu \in \Lambda^+} d_\mu [U : L(\mu)]$$

where $S\xi_\alpha \cong \bigoplus_{\mu \in \Lambda^+} P(\mu)^{(d_\mu)}$. From the property (ii) we get:

(1) $\dim L(\lambda)^\alpha = 1$ if $\lambda = \alpha^+$ and $L(\lambda)^\alpha \neq 0$ implies $\alpha^+ \leq \lambda$, for $\lambda \in \Lambda^+, \alpha \in \Lambda$.

From the defining properties of $\Delta(\lambda)$ and $\nabla(\lambda)$ we also get the following.

(2)   For $U = \Delta(\lambda)$ or $\nabla(\lambda)$ we have that $\dim U^\alpha = 1$ if $\alpha^+ = \lambda$ and that $U^\alpha \neq 0$ implies $\alpha^+ \leq \lambda$.

Lemma A3.9 *Let π be a saturated set of dominant weights. We have $O^\pi(S) = S\xi_\Gamma S$, where $\Gamma = \{\alpha \in \Lambda \mid \alpha^+ \notin \pi\}$.*

Proof Let $\alpha \in \Lambda$ and suppose that $O^\pi(S\xi_\alpha) \neq S\xi_\alpha$. Then we have $O^\pi(P(\lambda)) \neq P(\lambda)$ for some $\lambda \geq \alpha^+$. Since $P(\lambda)$ has unique simple quotient $L(\lambda)$ we get $\lambda \in \pi$ and hence, by saturation, $\alpha^+ \in \pi$. Hence we get $O^\pi(S\xi_\alpha) = S\xi_\alpha \leq O^\pi(S)$, for all $\alpha \in \Lambda$ with $\alpha^+ \notin \pi$, and therefore $S\xi_\Gamma \leq O^\pi(S)$. Since $O^\pi(S)$ is an ideal, we have $S\xi_\Gamma S \leq O^\pi(S)$. Suppose, for a contradiction, that $S\xi_\Gamma S \neq O^\pi(S)$. Then we have a maximal submodule M, say, of $O^\pi(S)$ containing $S\xi_\Gamma S$ with $O^\pi(S)/M \cong L(\lambda)$ and $\lambda \notin \pi$. Let $\alpha \in \Lambda$ with $\alpha^+ = \lambda$. Then we have $\xi_\alpha L(\lambda) = L(\lambda)^\alpha \neq 0$ and therefore $\xi_\Gamma L(\lambda) \neq 0$ and hence $\xi_\Gamma O^\pi(S) \nsubseteq S\xi_\Gamma S$, which is not true. Thus we have $O^\pi(S) = S\xi_\Gamma S$, as required.

Lemma A3.10 *Let π be a saturated set of dominant weights. The quasihereditary algebra $S = S(\pi)$ has the theory of weights $\bar{\theta} : \bar{\Lambda} \to \pi$, where $\bar{\Lambda} = \{\alpha \in \Lambda \mid \alpha^+ \in \pi\}$ and $\bar{\theta}(\alpha) = \theta(\alpha) + O^\pi(S)$, for $\alpha \in \bar{\Lambda}$.*

Proof We set $\bar{\xi}_\alpha = \xi_\alpha + O^\pi(S)$, for $\alpha \in \bar{\Lambda}$. Since $1 = \sum_{\alpha \in \Lambda} \xi_\alpha$ is an orthogonal decomposition in S and $\xi_\alpha \in O^\pi(S)$ for $\alpha \notin \bar{\Lambda}$, by Lemma A3.9, we have the orthogonal decomposition $1 = \sum_{\alpha \in \bar{\Lambda}} \bar{\xi}_\alpha$ in $S(\pi)$. Moreover, if $\alpha \in \bar{\Lambda}$ and $\bar{\xi}_\alpha = 0$ then we have $\xi_\alpha \in O^\pi(S) = \bigoplus_{\beta \in \Lambda} O^\pi(S\xi_\beta)$, giving $\xi_\alpha \in O^\pi(S\xi_\alpha)$ and hence $S\xi_\alpha \leq O^\pi(S\xi_\alpha)$ and therefore $S\xi_\alpha = O^\pi(S\xi_\alpha)$. From the defining property (ii), of a theory of weights, we get that $P(\lambda) = O^\pi(P(\lambda))$, where $\lambda = \alpha^+ \in \pi$. But $P(\lambda)$ has simple quotient $L(\lambda)$ so $O^\pi(P(\lambda)) \neq P(\lambda)$, a contradiction. Hence the image of $\bar{\theta}$ consists of non-zero idempotents. For $\alpha \in \bar{\Lambda}$ we have $S(\pi)\bar{\xi}_\alpha = (S\xi_\alpha + O^\pi(S))/O^\pi(S) \cong S\xi_\alpha/O^\pi(S\xi_\alpha)$. Writing $S\xi_\alpha \cong \bigoplus_{\lambda \in \Lambda^+} P(\lambda)^{(d_\lambda)}$ we get $S\xi_\alpha/O^\pi(S\xi_\alpha) \cong \bigoplus_{\lambda \in \pi}(P(\lambda)/O^\pi(P(\lambda))^{(d_\lambda)}$ which gives property (ii).

We now consider the form of truncation defined by idempotents corresponding to cosaturated subsets.

Proposition A3.11 *Let π be a set of dominant weights which is cosaturated (i.e. $\Lambda^+ \backslash \pi$ is saturated). Let Γ be a set of weights such that $\Gamma^+ = \pi$, let $\xi = \xi_\Gamma$ and let $f : \mathrm{mod}(S) \to \mathrm{mod}(S_\xi)$ be the Schur functor.*
(i) For $\lambda \in \Lambda^+$ we have $fL(\lambda) \neq 0$ (resp. $f\Delta(\lambda) \neq 0$, resp. $f\nabla(\lambda) \neq 0$) if and only if $\lambda \in \pi$.
(ii) $\{fL(\lambda) \mid \lambda \in \pi\}$ is a complete set of pairwise non-isomorphic irreducible S_ξ-modules and (S_ξ, π) is a quasihereditary algebra (where the ordering on π is induced from that on Λ^+) with standard modules $f\Delta(\lambda)$, $\lambda \in \pi$, and costandard modules $f\nabla(\lambda)$, $\lambda \in \pi$.

Proof (i) This follows from (2) above.

(ii) The first statement holds by A1(4)(iv). Let $\lambda \in \pi$. The surjection $P(\lambda) \to \Delta(\lambda)$ gives rise to a surjection $fP(\lambda) \to f\Delta(\lambda)$. Now $fP(\lambda)$ has simple head $fL(\lambda)$, by A1(4)(v), so that $f\Delta(\lambda)$ has simple head $fL(\lambda)$. Similarly $f\nabla(\lambda)$ has simple socle $fL(\lambda)$. By exactness of f, we have $fP(\lambda)/fM(\lambda) \cong fL(\lambda)$ so that $fM(\lambda)$ is the unique maximal submodule of $fP(\lambda)$. Let $K_\xi(\lambda)$ be the smallest submodule of $fM(\lambda)$ such that all composition factors of $fM(\lambda)/K_\xi(\lambda)$ come from $\{fL(\mu) \mid \mu < \lambda, \mu \in \pi\}$. Certainly all composition factors of $fM(\lambda)/fK(\lambda)$ have this form (by exactness) so we have $K_\xi(\lambda) \leq fK(\lambda)$. Now if $K_\xi(\lambda) \neq fK(\lambda)$ then $fK(\lambda)/K_\xi(\lambda)$, and hence $fK(\lambda)$, has a quotient $fL(\mu)$ with $\mu < \lambda$, $\mu \in \pi$. But $fK(\lambda)$ is filtered by modules of the form $f\Delta(\nu)$, with $\nu > \lambda$, so that some $f\Delta(\nu)$, with $\nu > \lambda$, would have to have a quotient $fL(\mu)$. But $f\Delta(\nu)$ has simple head $fL(\nu)$ so we would get $\mu = \nu > \lambda$, a contradiction. Hence we have $K_\xi(\lambda) = fK(\lambda)$ and, putting $\Delta_\xi(\lambda) = f\Delta(\lambda)$ we have, by exactness, that $K_\xi(\lambda)$ is filtered by the modules $\Delta_\xi(\mu)$, with $\mu > \lambda$. Hence $\mathrm{mod}(S_\xi)$ is quasihereditary with standard modules $\Delta_\xi(\lambda) = f\Delta(\lambda)$, $\lambda \in \pi$. Similarly we obtain that $f\nabla(\lambda)$, $\lambda \in \pi$ are the costandard modules.

We record the following for future use.

Lemma A3.12 *Assume the hypotheses and notation of Proposition A3.11. For $X \in \mathcal{F}(\Delta)$ and $Y \in \mathcal{F}(\nabla)$ the natural map*

$$\mathrm{Hom}_S(X, Y) \to \mathrm{Hom}_{S_\xi}(fX, fY)$$

is surjective.

Proof Suppose we have a short exact sequence $0 \to X' \to X \to X'' \to 0$ with $X', X'' \in \mathcal{F}(\Delta)$. Then we get the commutative diagram

$$
\begin{array}{ccccccccc}
0 & \to & \mathrm{Hom}_S(X'', Y) & \to & \mathrm{Hom}_S(X, Y) & \to & \mathrm{Hom}_S(X', Y) & \to & 0 \\
& & \downarrow & & \downarrow & & \downarrow & & \\
0 & \to & \mathrm{Hom}_{S_\xi}(fX'', fY) & \to & \mathrm{Hom}_{S_\xi}(fX, fY) & \to & \mathrm{Hom}_{S_\xi}(fX', fY) & \to & 0
\end{array}
$$

with rows exact (by Proposition A2.2(ii)). If the outer vertical maps are surjective then so is the middle one. Thus, by induction on the dimension of X, we are reduced to the case $X = \Delta(\lambda)$, for some $\lambda \in \Lambda^+$. A similar reduction allows us to assume that $Y = \nabla(\mu)$, for some $\mu \in \Lambda^+$. If either λ or μ does not belong to π then $fX = 0$ or $fY = 0$ so that $\mathrm{Hom}_{S_\xi}(fX, fY) = 0$ and the map is certainly surjective. Thus we may assume $\lambda, \mu \in \pi$. If $\lambda \neq \mu$ we get $\mathrm{Hom}_{S_\xi}(fX, fY) = 0$ by Proposition A2.2(ii). Thus we may assume that $X = \Delta(\lambda)$ and $Y = \nabla(\lambda)$. But now $\dim \mathrm{Hom}_{S_\xi}(fX, fY) = 1$, by Proposition A2.2(ii), and it suffices to prove that $\mathrm{Hom}_S(X, Y) \to \mathrm{Hom}_{S_\xi}(fX, fY)$ is not the zero map. Let $\theta : X \to Y$ be the composition $\Delta(\lambda) \to L(\lambda) \to \nabla(\lambda)$, where the first map is the canonical surjection and the second is inclusion. By exactness $f\theta$ is the composition $f\Delta(\lambda) \to fL(\lambda) \to f\nabla(\lambda)$, and the the first map is surjective and the second injective. Thus $f\theta \neq 0$ and $\mathrm{Hom}_S(X,Y) \to \mathrm{Hom}_{S_\xi}(fX, fY)$ is surjective.

For $\pi \subseteq \Lambda^+$ we write $V \in \mathcal{F}_\pi(\Delta)$ to indicate that $V \in \mathrm{mod}(S)$ has a filtration with sections belonging to $\{\Delta(\lambda) \mid \lambda \in \pi\}$.

We have the following relationship between the homological algebra of S and S_ξ.

Proposition A3.13 *Let π be a cosaturated subset of Λ^+, let $\Gamma \subseteq \Lambda$ be such that $\Gamma^+ = \pi$, and let $\xi = \xi_\Gamma$. For $X \in \mathcal{F}_\pi(\Delta)$, $Y \in \mathrm{mod}(S)$ and $i \geq 0$ we have $\mathrm{Ext}_S^i(X,Y) \cong \mathrm{Ext}_{S_\xi}^i(\xi X, \xi Y)$.*

Proof We consider first the case $i = 0$. For $U \in \mathcal{F}_\pi(\Delta)$ we define

$$\mathcal{S}(U) = \{\mu \in \Lambda^+ \mid \mu \geq \lambda \text{ for some } \lambda \in \Lambda^+ \text{ with } (U : \Delta(\lambda)) \neq 0\}$$

and define the *depth* of U to be the cardinality of $\mathcal{S}(U)$. We claim that, for given $X \in \mathcal{F}_\pi(\Delta)$, the restriction map $\mathrm{Hom}_S(X,Y) \to \mathrm{Hom}_{S_\xi}(\xi X, \xi Y)$ is an isomorphism for all $Y \in \mathrm{mod}(S)$.

We first prove injectivity. Suppose that $X = \Delta(\lambda)$, for some $\lambda \in \pi$. If $g \in \mathrm{Hom}_S(\Delta(\lambda), Y)$ maps to 0 then $g(\xi\Delta(\lambda)) = 0$. If $g \neq 0$ then the kernel of g lies in the unique maximal submodule M, say, of $\Delta(\lambda)$. Choosing $\alpha \in \Gamma$ such that $\alpha^+ = \lambda$, we have $g(\Delta(\lambda)^\alpha) = 0$, giving $\Delta(\lambda)^\alpha \leq M$. But $\Delta(\lambda)$ is generated by any weight space $\Delta(\lambda)^\alpha$ with $\alpha^+ = \lambda$ (by (1) and (2)) so this is impossible. Hence $g = 0$ and restriction $\mathrm{Hom}_S(\Delta(\lambda), Y) \to \mathrm{Hom}_{S_\xi}(\xi\Delta(\lambda), \xi Y)$ is injective. Now suppose that X is not isomorphic to $\Delta(\lambda)$, for any $\lambda \in \pi$. Then we have a short exact sequence $0 \to X' \to X \to X'' \to 0$, where $0 \neq X', X'' \in \mathcal{F}_\pi(\Delta)$. Thus we get a commutative diagram

$$
\begin{array}{ccccccc}
0 & \to & \mathrm{Hom}_S(X'', Y) & \to & \mathrm{Hom}_S(X, Y) & \to & \mathrm{Hom}_S(X', Y) \\
& & \downarrow & & \downarrow & & \downarrow \\
0 & \to & \mathrm{Hom}_{S_\xi}(\xi X'', \xi Y) & \to & \mathrm{Hom}_{S_\xi}(\xi X, \xi Y) & \to & \mathrm{Hom}_{S_\xi}(\xi X', \xi Y)
\end{array}
$$

with rows exact. Assuming inductively that the first and third vertical maps are injective, a diagram chase reveals that the second is too. Hence we get that $\mathrm{Hom}_S(X,Y) \to \mathrm{Hom}_{S_\xi}(\xi X, \xi Y)$ is injective for all $X \in \mathcal{F}_\pi(\Delta)$, $Y \in \mathrm{mod}(S)$, by induction on dimension.

Now suppose that $X = P(\lambda)$, for some $\lambda \in \pi$. The dimension of $\mathrm{Hom}_S(P(\lambda), X)$ is the composition multiplicity $[X : L(\lambda)]$ and the dimension of $\mathrm{Hom}_{S_\xi}(\xi P(\lambda), \xi Y)$ is the composition multiplicity $[\xi X : \xi L(\lambda)]$. However, we have $[X : L(\lambda)] = [\xi X : \xi L(\lambda)]$, by A1(4)(vii), so that

$$\dim \mathrm{Hom}_S(P(\lambda), Y) = \dim \mathrm{Hom}_{S_\xi}(\xi P(\lambda), \xi Y).$$

Since we know that the map $\mathrm{Hom}_S(P(\lambda), Y) \to \mathrm{Hom}_{S_\xi}(\xi P(\lambda), \xi Y)$ is injective, it must be an isomorphism. Hence the claim holds for projective modules in $\mathcal{F}_\pi(\Delta)$. Now let P be the projective cover of X and let J be the kernel of a surjection $P \to X$. Then $P \in \mathcal{F}_\pi(\Delta)$. Hence we also have $J \in \mathcal{F}_\pi(\Delta)$, by Proposition A2.2(v), and the additivity of $(- : \Delta(\lambda))$ on $\mathrm{mod}(S)$, for $\lambda \in \Lambda^+$ (see A1). We get a commutative diagram

$$
\begin{array}{ccccc}
0 & \to & \mathrm{Hom}_S(X,Y) & \to & \mathrm{Hom}_S(P,Y) & \to \mathrm{Hom}_S(J,Y) \\
& & \downarrow & & \downarrow & \downarrow \\
0 & \to & \mathrm{Hom}_{S_\xi}(\xi X, \xi Y) & \to & \mathrm{Hom}_{S_\xi}(\xi P, \xi Y) & \to \mathrm{Hom}_{S_\xi}(\xi J, \xi Y)
\end{array}
$$

with rows exact, vertical maps injective and middle vertical map an isomorphism. A diagram chase reveals that $\mathrm{Hom}_S(X,Y) \to \mathrm{Hom}_{S_\xi}(\xi X, \xi Y)$ is surjective.

Thus we have, for fixed $X \in \mathcal{F}_\pi(\Delta)$, an isomorphism of (left exact) functors $\mathrm{Hom}_S(X,-) \to \mathrm{Hom}_{S_\xi}(\xi X, \xi-)$. Taking derived functors we obtain an isomorphism

$$
\mathrm{Ext}_S^i(X,-) \to R^i \mathrm{Hom}_{S_\xi}(\xi X, \xi-)
$$

in each degree. Now $F = \mathrm{Hom}_{S_\xi}(\xi X, \xi-)$ is the composite $G \circ H$, where $H = \xi- : \mathrm{mod}(S) \to \mathrm{mod}(S_\xi)$ (the Schur functor) and $G = \mathrm{Hom}_{S_\xi}(\xi X, -) : \mathrm{mod}(S_\xi) \to \mathrm{mod}(k)$. If $I \in \mathrm{mod}(S)$ is injective then $I \in \mathcal{F}(\nabla)$ and hence $H(I) = \xi I$ is a S_ξ-module which has a costandard filtration. Since ξX has a standard filtration we have $\mathrm{Ext}_{S_\xi}^i(\xi X, \xi I) = 0$ for $i > 0$. Thus H takes injective modules to G-acyclic modules and thus we have a Grothendieck spectral sequence with E_2-page $R^i G \circ R^j H(Y)$ converging to $R^* F(Y)$. But H is exact, so the spectral sequence degenerates and we get $R^i F(Y) \cong R^i G(H(Y))$, in other words $R^i \mathrm{Hom}_{S_\xi}(\xi X, \xi Y) \cong \mathrm{Ext}_{S_\xi}^i(\xi X, \xi Y)$, for all $i \geq 0$. Thus we have $\mathrm{Ext}_S^i(X,Y) \cong \mathrm{Ext}_{S_\xi}^i(\xi X, \xi Y)$, for all $i \geq 0$.

A4 We now describe the theory of tilting modules for a quasihereditary algebra (S, Λ^+), due to Ringel. We call a finite dimensional S-module X a *tilting module* if $X \in \mathcal{F}(\Delta)$ and $X \in \mathcal{F}(\nabla)$, i.e. if X has both a Δ-filtration and a ∇-filtration.

For $X \in \mathrm{mod}(S)$ define the *defect set* of X to be the set of $\lambda \in \Lambda^+$ such that for some $\mu \in \Lambda^+$ we have $\lambda \leq \mu$ and $\mathrm{Ext}_S^1(\Delta(\mu), X) \neq 0$.

Lemma A4.1 *Let $X \in \mathcal{F}(\Delta)$. Then X embeds in some tilting module T. Moreover, if there is some $\lambda \in \Lambda^+$ such that $(X : \Delta(\lambda)) = 1$ and $\nu \leq \lambda$ whenever $(X : \Delta(\nu)) \neq 0$ then we can choose T so that we also have $(T : \Delta(\lambda)) = 1$ and $\nu \leq \lambda$ whenever $(T : \Delta(\nu)) \neq 0$.*

Proof Suppose not and let A be a defect set of smallest possible size among all counterexamples X. If A is empty then $X \in \mathcal{F}(\nabla)$, by Proposition A2.2(iii), so we may take $T = X$, a contradiction. Thus A is non-empty.

Let μ be a maximal element of A and let X be a counterexample with defect set A such that $\dim \operatorname{Ext}_S^1(\Delta(\mu), X)$ is as small as possible. Now we have a non-split extension $0 \to X \to \tilde{X} \to \Delta(\mu) \to 0$ and this gives rise to an exact sequence

$$0 \to \operatorname{Hom}_S(\Delta(\nu), X) \to \operatorname{Hom}_S(\Delta(\nu), \tilde{X}) \to \operatorname{Hom}_S(\Delta(\nu), \Delta(\mu))$$
$$\to \operatorname{Ext}_S^1(\Delta(\nu), X) \to \operatorname{Ext}_S^1(\Delta(\nu), \tilde{X}) \to \operatorname{Ext}_S^1(\Delta(\nu), \Delta(\mu))$$

for $\nu \in \Lambda^+$. Thus if $\operatorname{Ext}_S^1(\Delta(\nu), \tilde{X}) \neq 0$ then either $\operatorname{Ext}_S^1(\Delta(\nu), X) \neq 0$, in which case $\nu \in A$, or $\operatorname{Ext}_S^1(\Delta(\nu), \Delta(\mu)) \neq 0$, in which case we get $\nu < \mu$, by Proposition A2.2(i), and again $\nu \in A$. Hence the defect set of $\tilde{X}$ is contained in A. We now take $\nu = \mu$. Any homomorphism $\theta : \Delta(\mu) \to \tilde{X}$ has image in X, for otherwise θ would induce a non-zero map $\bar{\theta} : \Delta(\mu) \to \tilde{X}/X$ and since $\tilde{X}/X \cong \Delta(\mu)$ and $\operatorname{End}_S(\Delta(\mu)) = k$, this map would have to be an isomorphism. But then we would have $\tilde{X} = X \oplus \operatorname{Im}(\theta)$, contradicting the fact that $\tilde{X}$ is a non-split extension. Hence the map $\operatorname{Hom}_S(\Delta(\mu), X) \to \operatorname{Hom}_S(\Delta(\mu), \tilde{X})$ is an isomorphism and $\operatorname{Hom}_S(\Delta(\mu), \tilde{X}) \to \operatorname{Hom}_S(\Delta(\mu), \Delta(\mu))$ is the zero map. Thus we have an exact sequence

$$0 \to \operatorname{Hom}_S(\Delta(\mu), \Delta(\mu)) \to \operatorname{Ext}_S^1(\Delta(\mu), X) \to \operatorname{Ext}_S^1(\Delta(\mu), \tilde{X}) \to \operatorname{Ext}_S^1(\Delta(\mu), \Delta(\mu)).$$

But now $\operatorname{Hom}_S(\Delta(\mu), \Delta(\mu)) = k$ and $\operatorname{Ext}_S^1(\Delta(\mu), \Delta(\mu)) = 0$ so we get $\dim \operatorname{Ext}_S^1(\Delta(\mu), \tilde{X}) = \dim \operatorname{Ext}_S^1(\Delta(\mu), X) - 1$. Thus, by the choice of X, the module $\tilde{X}$ embeds in a tilting module. Hence X embeds in a tilting module.

Suppose $\lambda \in \Lambda^+$ is such that $(X : \Delta(\lambda)) = 1$ and $\nu \leq \lambda$ for all ν such that $(X : \Delta(\nu)) \neq 0$. Since $\operatorname{Ext}_S^1(\Delta(\mu), X) \neq 0$ we have $\operatorname{Ext}_S^1(\Delta(\mu), \Delta(\nu)) \neq 0$ for some $\nu \in \Lambda^+$ with $(X : \Delta(\nu)) \neq 0$. Thus we have $\mu < \nu$ and $\nu \leq \lambda$ and hence $\mu < \lambda$. Thus $\tilde{X}$ also has the property that $(\tilde{X} : \Delta(\lambda)) = 1$ and $\nu < \lambda$ whenever $(\tilde{X} : \Delta(\nu)) \neq 0$. Again, by the choice of X, the module $\tilde{X}$, and hence X, embeds in some tilting module T with the property that $(T : \Delta(\lambda)) = 1$ and $\nu \leq \lambda$ whenever $(T : \Delta(\nu)) \neq 0$.

We can now prove the classification of tilting modules.

Theorem A4.2 *(i) For each $\lambda \in \Lambda^+$ there exists a unique (up to isomorphism) indecomposable tilting module $T(\lambda)$ such that $[T(\lambda) : L(\lambda)] = 1$ and $\mu \leq \lambda$ whenever $[T(\lambda) : L(\mu)] \neq 0$.*
(ii) Every tilting module is a direct sum of the modules $T(\lambda)$, $\lambda \in \Lambda^+$.
(iii) Every indecomposable tilting module is absolutely indecomposable.

Proof For $\lambda \in \Lambda^+$ we take $X = \Delta(\lambda)$ in Lemma A4.1 and obtain a tilting module T_λ such that $(T_\lambda : \Delta(\lambda)) = 1$ and $\mu \leq \lambda$ whenever $(T : \Delta(\mu)) \neq 0$. Note that every indecomposable summand of T is also a tilting module, by Proposition A2.2(vi). Let $T(\lambda)$ be the indecomposable direct summand of T

such that $(T(\lambda) : \Delta(\lambda)) = 1$. Then $T(\lambda)$ is an indecomposable tilting module with the required properties.

For $\lambda \in \Lambda^+$ we put $\pi(\lambda) = \{\mu \in \Lambda^+ \mid \mu < \lambda\}$ and define $U(\lambda) = O_{\pi(\lambda)}(T(\lambda))$ and $V(\lambda) = O^{\pi(\lambda)}(T(\lambda))$. Then we have $T(\lambda)/U(\lambda) \cong \nabla(\lambda)$ and $V(\lambda) \cong \Delta(\lambda)$, by Lemma A3.1(ii),(iii). Let $\theta \in \mathrm{End}_S(T(\lambda))$. Then θ induces an endomorphism of $T(\lambda)/U(\lambda)$ and, since $\mathrm{End}_S(\nabla(\lambda)) = k$, the induced endomorphism is multiplication by a scalar c, say. Thus, putting $\phi = \theta - c.\mathrm{id}$ (where id is the identity map on $T(\lambda)$), we have $\phi(T(\lambda)) \le U(\lambda)$. By indecomposability and the Fitting lemma, we have that ϕ is nilpotent. Thus the ideal $I = \{\theta \in \mathrm{End}_S(T(\lambda)) \mid \theta(T(\lambda)) \le U(\lambda)\}$ of $\mathrm{End}_S(T(\lambda))$ is nilpotent and $\mathrm{End}_S(T(\lambda)) = k.\mathrm{id} \oplus I$. Hence I is the nilpotent radical of $\mathrm{End}_S(T(\lambda))$ and, since $\mathrm{End}_S(T(\lambda))/I = k$, the module $T(\lambda)$ is absolutely indecomposable.

Let T be a non-zero tilting module and let $\lambda \in \Lambda^+$ be maximal such that $(T : \nabla(\lambda)) \ne 0$. From Proposition A2.2(i), we get that T has a homomorphic image isomorphic to $\nabla(\lambda)$, and hence there exists an epimorphism $T \to T(\lambda)/U(\lambda) \cong \nabla(\lambda)$. Now we have an exact sequence

$$\mathrm{Hom}_S(T, T(\lambda)) \to \mathrm{Hom}_S(T, T(\lambda)/U(\lambda)) \to \mathrm{Ext}^1_S(T, U(\lambda)).$$

Moreover $\mathrm{Ext}^1_S(T, U(\lambda)) = 0$ (since $T \in \mathcal{F}(\Delta)$, $U(\lambda) \in \mathcal{F}(\nabla)$) and hence there exists a homomorphism $\phi : T \to T(\lambda)$ inducing a surjection $T \to T(\lambda) \to T(\lambda)/U(\lambda)$. Let $U = \mathrm{Ker}(\phi)$. By the same argument we have a homomorphism $\psi : T(\lambda) \to T$ inducing a surjection $\overline{\psi} : T(\lambda) \to T \to T/U \cong \nabla(\lambda)$. Now $L(\lambda)$ does not occur as a composition factor of $U(\lambda)$ and hence does not occur as a composition factor of $\overline{\psi}(U(\lambda))$. However, the module $\nabla(\lambda)$ has simple socle $L(\lambda)$ and so we must have $\overline{\psi}(U(\lambda)) = 0$. Since $\dim T(\lambda)/U(\lambda) = \dim T/U = \dim \nabla(\lambda)$, we must have $U(\lambda) = \mathrm{Ker}(\overline{\psi})$ and so $\overline{\psi}$ induces an isomorphism $\overline{\overline{\psi}} : T(\lambda)/U(\lambda) \to T/U$. Thus the composite $\phi \circ \psi : T(\lambda) \to T(\lambda)$ induces an isomorphism $T(\lambda)/U(\lambda) \to T(\lambda)/U(\lambda)$. Thus $\phi \circ \psi \in \mathrm{End}_S(T(\lambda))$ is not nilpotent. But the nilpotent radical of $\mathrm{End}_S(T(\lambda))$ has codimension 1, so $\phi \circ \psi$ is an isomorphism and hence $\phi : T \to T(\lambda)$ is a splitting. Thus $T(\lambda)$ is a direct summand of T. By induction on the dimension of T we obtain that T is a direct sum of copies of $T(\lambda)$, $\lambda \in \Lambda^+$. In particular an indecomposable tilting module is isomorphic to $T(\lambda)$, for some $\lambda \in \Lambda^+$, and so is absolutely indecomposable.

For $X \in \mathcal{F}(\nabla)$ we define the *support*, denoted $\mathrm{supp}(X)$, to be the set of $\lambda \in \Lambda^+$ such that we have $\lambda \le \mu$ for some $\mu \in \Lambda^+$ such that $(X : \nabla(\mu)) \ne 0$.

Lemma A4.3 *Let $0 \ne X \in \mathcal{F}(\nabla)$. Then there exists a short exact sequence $0 \to X' \to T \to X \to 0$ in $\mathrm{mod}(S)$ such that T is a tilting module and $X' \in \mathcal{F}(\nabla)$ has support strictly contained in the support of X.*

Proof We suppose, inductively, that the result holds for all non-zero modules in $\mathcal{F}(\nabla)$ which have support strictly contained in $\mathrm{supp}(X)$. Let λ be maximal such that $L(\lambda)$ is a composition factor of X and let $A = \text{supp}(X)\backslash\{\lambda\}$. Then by Proposition A2.2(i), we have a submodule Y, say, such that X/Y is a direct sum of copies of $\nabla(\lambda)$, say n of them, and $\lambda \notin \text{supp}(Y)$. Now $T(\lambda)$ has a submodule $U(\lambda)$, say, such that $T(\lambda)/U(\lambda) \cong \nabla(\lambda)$, and $U(\lambda) \in \mathcal{F}(\nabla)$. Putting $T_0 = T(\lambda)^{(n)}$ and $U_0 = U(\lambda)^{(n)}$, we have $T_0/U_0 \cong X/Y$. We choose an epimorphism $\pi : T_0 \to X/Y$ with kernel U_0. We have an exact sequence

$$\text{Hom}_S(T_0, X) \to \text{Hom}_S(T_0, X/Y) \to \text{Ext}^1_S(T_0, Y)$$

and $\text{Ext}^1_S(T_0, Y) = 0$ by Proposition A2.2(ii), so there is an S-homomorphism $\phi : T_0 \to X$ such that the composite $T_0 \to X \to X/Y$ is the surjective map π. Let $\psi : T_0 \oplus Y \to X$ be the sum of ϕ and the inclusion map $Y \to X$. Then ψ is surjective and the kernel consists of those pairs $(t_0, y) \in T_0 \oplus Y$ with $\phi(t_0) + y = 0$. Thus we have $\text{Ker}(\psi) = \{(t_0, -\phi(t_0)) \mid t_0 \in U_0\}$, which is isomorphic to U_0.

We have $\text{supp}(Y) \leq \text{supp}(X)\backslash\{\lambda\}$. Hence, by minimality, there is an epimorphism $\xi : T_1 \to Y$ such that T_1 is a tilting module and $\text{Ker}(\xi) \in \mathcal{F}(\nabla)$ has support in $\text{supp}(Y)$. Let $\zeta : T = T_0 \oplus T_1 \to T_0 \oplus Y$ be the sum of the identity map and ξ and let $\sigma : T \to X$ be the composite $\psi \circ \zeta$.

We must show that the kernel X' of σ belongs to $\mathcal{F}(\nabla)$ and has support contained in A. We have $X' = \text{Ker}(\sigma) = \text{Ker}(\psi \circ \zeta) = \zeta^{-1}(\text{Ker}(\psi))$. Thus ζ induces an isomorphism $\bar{\zeta} : X'/\text{Ker}(\zeta) \to \text{Ker}(\psi)$. Now both $\text{Ker}(\zeta)$ and $\text{Ker}(\psi)$ are in $\mathcal{F}(\nabla)$ and have support in A so that

$$\text{supp}(X') \subseteq \text{supp}(\text{Ker}(\zeta)) \bigcup \text{supp}(\text{Ker}(\psi)) \subseteq A,$$

as required.

Proposition A4.4 *For $X \in \text{mod}(S)$ we have $X \in \mathcal{F}(\nabla)$ if and only if X has a finite left resolution by tilting modules.*

Proof Suppose that X has a finite left resolution by tilting modules. In particular X has a resolution $0 \to X_r \to \cdots \to X_0 \to X \to 0$, for some $X_0, \ldots, X_r \in \mathcal{F}(\nabla)$, and this implies $X \in \mathcal{F}(\nabla)$ by Proposition A2.2(v) (and induction).

Now suppose $X \in \mathcal{F}(\nabla)$ and assume inductively that every module in $\mathcal{F}(\nabla)$ which has support strictly contained in the support of X has a finite resolution by tilting modules. By Lemma A4.3 we have a short exact sequence $0 \to X' \to T \to X \to 0$, where T is a tilting module and $X' \in \mathcal{F}(\nabla)$ has support strictly contained in the support of X. By the inductive hypothesis there is a resolution $0 \to T_r \to \cdots \to T_0 \to X' \to 0$, where each T_i is a tilting module. Combining these sequences, we obtain a resolution $0 \to T_r \to \cdots \to T_0 \to T \to X \to 0$, of the required form.

Remark Suppose that the global dimension of S is n and that we have a resolution $0 \to T_r \to \cdots \to T_1 \to T_0 \to X \to 0$ of $X \in \mathcal{F}(\nabla)$ by tilting modules, with $r > n$. Let I_j be the image of the map $T_j \to T_{j-1}$, for $r > j \geq 1$. Thus we have exact sequences $0 \to T_r \to T_{r-1} \to I_{r-1} \to 0$, and $0 \to I_j \to T_{j-1} \to I_{j-1} \to 0$, for $r - 1 \geq j \geq 2$, and $0 \to I_1 \to T_0 \to X \to 0$. Using Proposition A2.2(ii), we obtain

$$\mathrm{Ext}_S^1(I_{r-1}, T_r) = \mathrm{Ext}_S^2(I_{r-2}, T_r) = \cdots = \mathrm{Ext}_S^{r-1}(I_1, T_r) = \mathrm{Ext}_S^r(X, T_r) = 0.$$

Hence $T_r \to T_{r-1}$ is split. Thus I_{r-1} is a direct summand of a tilting module, and hence a tilting module. Hence we have a resolution by tilting modules $0 \to I_{r-1} \to T_{r-2} \to \cdots \to T_0 \to X \to 0$, of length $r - 1$. Continuing in this way, we obtain, for $X \in \mathcal{F}(\nabla)$, a tilting module resolution $0 \to T \to T_{n-1} \to \cdots \to T_0 \to X \to 0$, of length at most n.

We call the indecomposable tilting modules $T(\lambda)$, $\lambda \in \Lambda^+$, as above, the *partial tilting modules*. The partial tilting modules behave well under both forms of truncation.

Lemma A4.5 *Let $\lambda \in \Lambda^+$.*
(i) *Let π be a saturated subset of Λ^+. If $\lambda \in \pi$ then $O_\pi(T(\lambda)) = T(\lambda)$ and this is the partial tilting module for $S(\pi)$, labelled by λ.*
(ii) *Let π be a cosaturated subset of Λ^+. Assume that S has a theory of weights $\theta : \Lambda \to S$. Let Γ be a set of weights such that $\Gamma^+ = \pi$, let $\xi = \xi_\Gamma$ and let $f : \mathrm{mod}(S) \to \mathrm{mod}(S_\xi)$ be the Schur functor. If $\lambda \in \pi$ then $fT(\lambda)$ is the partial tilting module for S_ξ, labelled by λ. If $\lambda \notin \pi$ then $fT(\lambda) = 0$.*

Proof (i) Every composition factor of $T(\lambda)$ belongs to $\{L(\lambda) \mid \lambda \in \pi\}$, i.e. $T(\lambda)$ belongs to π and so $O_\pi(T(\lambda)) = T(\lambda)$. Since $T(\lambda)$, regarded as an $S(\pi)$-module, has a Δ-filtration and a ∇-filtration, it is a tilting module and since it is indecomposable $T(\lambda)$ is a partial tilting module for $S(\pi)$. Moreover, since $T(\lambda)$ has a ∇-filtration with $\nabla(\lambda)$ occurring once and other sections of the form $\nabla(\mu)$ with $\mu < \lambda$, we have that $T(\lambda)$ is the tilting module for $S(\pi)$ labelled by λ.
(ii) Let $\lambda \in \pi$. Then $f\nabla(\lambda) \neq 0$ and so, by exactness, we have $fT(\lambda) \neq 0$. Moreover, as an S_ξ-module, $fT(\lambda)$ has a ∇-filtration and a Δ-filtration, by PropositionA3.11(ii). Thus $fT(\lambda)$ is a tilting module. The natural map $\mathrm{End}_S(T(\lambda)) \to \mathrm{End}_{S_\xi}(fT(\lambda))$ is surjective by Lemma A3.12 and since $T(\lambda)$ is indecomposable, $fT(\lambda)$ is too. Thus $fT(\lambda)$ is a partial tilting module. Since $fT(\lambda)$ has a filtration by modules of the form $f\nabla(\mu)$, with $\mu \leq \lambda$, and $f\nabla(\lambda)$ occurring exactly once, $fT(\lambda)$ is the partial tilting module labelled by λ.

By a *full tilting module* we mean a tilting module T such that $T(\lambda)$ occurs as a component of T, for every $\lambda \in \Lambda^+$. Let T be a full tilting module. We define $S' = \mathrm{End}_S(T)^{\mathrm{op}}$ and call S' the *Ringel dual* of S. (Note that if T_0 is any full tilting module then $\mathrm{End}_S(T_0)$ is Morita equivalent, but not necessarily isomorphic, to $\text{End}_S(T)$ so that to be accurate we should perhaps call $\text{End}_S(T)$ a Ringel dual of S.) For $X \in \text{mod}(X)$ we define FX to be $\text{Hom}_S(T, X)$, regarded as an S'-module in the natural manner. For a morphism $\phi : X \to X'$ in $\text{mod}(S)$ we define $F\phi : FX \to FX'$ by $F\phi(\alpha) = \phi \circ \alpha$, for $\alpha \in \text{Hom}_S(T, X)$. In this way we have a left exact functor $F : \text{mod}(S) \to \text{mod}(S')$. We put $P'(\lambda) = FT(\lambda)$, for $\lambda \in \Lambda^+$, and note that $\{P'(\lambda) \mid \lambda \in \Lambda^+\}$ is a complete set of pairwise non-isomorphic projective indecomposable S'-modules. Thus defining $L'(\lambda)$ to be the head of $P'(\lambda)$, for $\lambda \in \Lambda^+$, we have that $\{L'(\lambda) \mid \lambda \in \Lambda^+\}$ is a complete set of pairwise non-isomorphic simple S'-modules. We further define $\Delta'(\lambda) = F\nabla(\lambda)$, $\lambda \in \Lambda^+$.

(1)   (i) *If* $0 \to X' \to X \to X'' \to 0$ *is a short exact sequence of finite dimensional S-modules with* $X' \in \mathcal{F}(\nabla)$ *then* $0 \to FX' \to FX \to FX'' \to 0$ *is exact.*
(ii) *For any tilting module T_0 and $X \in \text{mod}(S)$ the map* $\text{Hom}_S(T_0, X) \to \text{Hom}_{S'}(FT_0, FX)$ *is an isomorphism.*

Proof   (i) This follows from Proposition A2.2(ii).
(ii) Since T_0 is a direct summand of a direct sum of copies of T, it is enough to prove this with $T_0 = T$. If $\theta \in \text{Hom}_S(T, X)$ and $F(\theta) = 0$ then $\theta \circ \alpha = 0$ for all $\alpha \in \text{Hom}_S(T)$, in particular $\theta \circ \text{id} = 0$, where id is the identity map on T, and hence $\theta = 0$. Thus $\text{Hom}_S(T, X) \to \text{Hom}_{S'}(FT, FX)$ is injective. However, we have $\dim \text{Hom}_S(T, X) = \dim FX$ and $\dim \text{Hom}_{S'}(FT, FX) = \dim \text{Hom}_{S'}(S', FX) = \dim FX$ so that
$\text{Hom}_S(T, X) \to \text{Hom}_{S'}(FT, FX)$ is an isomorphism.

(2) S' *is Schurian, i.e. we have* $\text{End}_{S'}(L'(\lambda)) = k$ *for all* $\lambda \in \Lambda^+$.

Proof   For $\lambda \in \Lambda^+$ we have $\text{End}_S(T(\lambda)) \cong \text{End}_{S'}(P'(\lambda))$, by (1)(ii). Moreover, $T(\lambda)$ is absolutely indecomposable, by Theorem A4.2(iii). Hence $P'(\lambda)$ is absolutely indecomposable. Now the quotient map $P'(\lambda) \to L'(\lambda)$ gives rise to a surjective map $\text{End}_{S'}(P'(\lambda)) \to \text{End}_{S'}(L'(\lambda))$. But $\text{End}_{S'}(P'(\lambda))$ has a nilpotent ideal of codimension 1 and $\text{End}_{S'}(L'(\lambda))$ is a division ring. It follows that $\text{End}_{S'}(L'(\lambda)) = k$, and $L'(\lambda)$ is absolutely irreducible.

Lemma A4.6   *For* $\lambda, \mu \in \Lambda^+$ *we have* $[\Delta'(\lambda) : L'(\mu)] = (T(\mu) : \Delta(\lambda))$.

Proof   We have
$$[\Delta'(\lambda) : L'(\mu)] = \dim \text{Hom}_{S'}(P'(\mu), \Delta'(\lambda))$$
$$= \dim \text{Hom}_S(T(\mu), \nabla(\lambda)) = (T(\mu) : \Delta(\lambda))$$
by A1(8), and (1)(ii) above and Proposition A2.2(ii).

We denote the partial ordering on Λ^+ opposite to the given order $\leq$ by $\leq'$. We now prove that S' is a quasihereditary algebra with respect to the labelling of simples $\{L'(\lambda) \mid \lambda \in \Lambda^+\}$ and opposite order $\leq'$ on Λ^+.

Theorem A4.7 $(S', \leq')$ *is a quasihereditary algebra with standard modules* $\Delta'(\lambda)$, $\lambda \in \Lambda^+$.

Proof Applying F to a ∇-filtration of $T(\lambda)$ with final section $\nabla(\lambda)$, we have, by (1)(i), that $P'(\lambda)$ has filtration $0 = P'_0 < \cdots < P'_n = P$, say, with $P'_i/P'_{i-1} \cong \Delta'(\lambda_i)$ with $\lambda_n = \lambda$ and $\lambda_i >' \lambda$ (i.e. $\lambda_i < \lambda$) for $1 \leq i < n$. Note that all composition factors of $\Delta'(\lambda)$ come from the set $\{L'(\mu) \mid \mu <' \lambda\}$ and that $[\Delta'(\lambda) : L'(\lambda)] = 1$, by Lemma A4.5. Thus if $\Delta'(\lambda) = P'(\lambda)/K'(\lambda)$ and $L'(\lambda) = P'(\lambda)/M'(\lambda)$, then $M'(\lambda)/K'(\lambda)$ is the largest quotient of $M'(\lambda)$ all of whose composition factors come from $\{L'(\mu) \mid \mu <' \lambda\}$ and S' is quasihereditary, with respect to the ordering $\leq'$, as required.

For a homomorphism $\phi : S_1 \to S_2$, of finite dimensional algebras, and $V \in \mathrm{mod}(S_2)$ we write V^ϕ for the k-space V regarded as an S_1-module via the action $x * v = \phi(x)v$, for $x \in S_1$, $v \in V$. Now suppose given quasihereditary algebras (S_1, Λ_1^+), with simple modules $L_1(\lambda)$, $\lambda \in \Lambda^+$, and (S_2, Λ_2^+), with simple modules $L_2(\lambda)$, $\lambda \in \Lambda_2^+$. We say that an algebra isomorphism $\phi : S_1 \to S_2$ is an isomorphism of quasihereditary algebras if the bijection $\nu : \Lambda_1^+ \to \Lambda_2^+$, defined by $L_1(\lambda) \cong L_2(\nu(\lambda))^\phi$, for $\lambda \in \Lambda_1^+$, is order preserving.

For $\lambda \in \Lambda^+$ we write $T'(\lambda)$ for an indecomposable tilting module for S' with highest composition factor $L'(\lambda)$.

Proposition A4.8 (i) *Let* $X, Y \in \mathcal{F}(\nabla)$. *Then* F *induces isomorphisms* $\mathrm{Ext}_S^i(X, Y) \to \mathrm{Ext}_{S'}^i(FX, FY)$, *for all* $i \geq 0$.
(ii) *We have* $T'(\lambda) \cong FI(\lambda)$, *for* $\lambda \in \Lambda^+$.
(iii) *A suitable choice of full tilting modules for* S *and* S' *gives rise to an isomorphism* $S \to S''$ *of quasihereditary algebras.*

Proof (i) Let $I \in \mathrm{mod}(S)$ be injective. We have $FI \in \mathcal{F}(\Delta')$ by (1)(i). By Proposition A4.4, we have a finite resolution $0 \to T_r \to \cdots \to T_1 \to T_0 \to X \to 0$. Thus we get an exact sequence $0 \to FT_r \to \cdots \to FT_1 \to FT_0 \to FX \to 0$ and, since each FT_j is projective, this is a projective resolution of FX. Now we get complexes

$$0 \to \mathrm{Hom}_S(T_0, I) \to \mathrm{Hom}_S(T_1, I) \to \cdots \to \mathrm{Hom}_S(T_r, I) \to 0$$

and

$$0 \to \mathrm{Hom}_{S'}(FT_0, FI) \to \mathrm{Hom}_{S'}(FT_1, FI) \to \cdots \to \mathrm{Hom}_{S'}(FT_r, FI) \to 0.$$

Moreover, we have the canonical isomorphisms

$$\mathrm{Hom}_S(T_j, I) \to \mathrm{Hom}_{S'}(FI_j, FI),$$

for $0 \leq j \leq r$. Thus the complexes have the same homology. The top complex has homology $\mathrm{Hom}_S(X, I)$ in degree 0 and homology 0 in positive degree, since I is injective, and the homology of the bottom complex is $\mathrm{Ext}^*_{S'}(FX, FI)$. Hence we get that the natural map $\mathrm{Hom}_S(X, I) \to \mathrm{Hom}_{S'}(FX, FI)$ is an isomorphism and $\mathrm{Ext}^i_{S'}(FX, FI) = 0$, for $i > 0$.

Let $0 \to Y \to I \to Q \to 0$ be an exact sequence with I injective, and hence $Q \in \mathcal{F}(\nabla)$ by Proposition A2.2(iv),(v). We have a commutative diagram

$$
\begin{array}{ccccccc}
\mathrm{Hom}_S(X, Y) & \to & \mathrm{Hom}_S(X, I) & \to & \mathrm{Hom}_S(X, Q) & \to & \mathrm{Ext}^1_S(X, Y) \\
\downarrow & & \downarrow & & \downarrow & & \downarrow \\
\mathrm{Hom}_{S'}(FX, FY) & \to & \mathrm{Hom}_{S'}(FX, FI) & \to & \mathrm{Hom}_{S'}(FX, FQ) & \to & \mathrm{Ext}^1_{S'}(FX, FY)
\end{array}
$$

where, in each row, the sequence is exact, the first map is injective, and the final map is surjective. Moreover, the map $\mathrm{Hom}_S(X, I) \to \mathrm{Hom}_{S'}(FX, FI)$ is an isomorphism. It follows that $\mathrm{Hom}_S(X, Y) \to \mathrm{Hom}_{S'}(FX, FY)$ is injective, and hence also $\mathrm{Hom}_S(X, Q) \to \mathrm{Hom}_{S'}(FX, FQ)$ is injective. Now a diagram chase reveals that $\mathrm{Hom}_S(X, Y) \to \mathrm{Hom}_{S'}(FX, FY)$ is an epimorphism and hence an isomorphism. Hence $\mathrm{Hom}_S(X, Q) \to \mathrm{Hom}_{S'}(FX, FQ)$ is also an isomorphism. We now have that the first three vertical maps are isomorphisms, and hence so is the final map. Thus we have that for $i = 0, 1$ $\mathrm{Ext}^i_S(X, Y) \to \mathrm{Ext}^i_S(FX, FY)$ is an isomorphism. For $i \geq 1$ we have $\mathrm{Ext}^i_S(X, Y) \cong \mathrm{Ext}^{i-1}_S(X, Q)$ and $\mathrm{Ext}^i_{S'}(FX, FY) \cong \mathrm{Ext}^{i-1}_{S'}(FX, FQ)$ (since $\mathrm{Ext}^j_{S'}(FX, FI) = 0$, for $j \geq 2$, as proved above) so we get $\mathrm{Ext}^i_S(X, Y) \cong \mathrm{Ext}^i_{S'}(FX, FY)$ for all i by induction.

(ii) Taking $X = \nabla(\mu)$ in (i), with $\mu \in \Lambda^+$, we get $\mathrm{Ext}^1_{S'}(\Delta'(\mu), FI) = 0$, for I injective. Hence, by Proposition A2.2(iii), $FI \in \mathcal{F}(\nabla')$. Thus, FI is a tilting module, for I injective. Let $\lambda \in \Lambda^+$ and take $X = I = I(\lambda)$, in (i). We get an isomorphism $\mathrm{End}_S(I(\lambda)) \to \mathrm{End}_{S'}(FI(\lambda))$. Hence $FI(\lambda)$ is absolutely indecomposable. Moreover, $I(\lambda)$ has a filtration with sections $\nabla(\mu)$, with $\mu \geq \lambda$ and $\nabla(\lambda)$ occurring precisely once. Hence $FI(\lambda)$ has filtration with sections $\Delta'(\mu)$, with $\mu \leq' \lambda$ and $\Delta'(\lambda)$ occurring precisely once. Hence $FI(\lambda)$ is the partial tilting module $T'(\lambda)$ with highest composition factor $L'(\lambda)$.

(iii) We take $I = (S_S)^*$, the natural left dual of the right regular module. It follows from (ii) that I is a full tilting module for S'. Moreover, by (i), the map $\mathrm{End}_S((S_S)^*) \to \mathrm{End}_{S'}(F(S_S)^*)$ is an isomorphism, giving an isomorphism $\mathrm{End}_S((S_S)^*) \to (S'')^{\mathrm{op}}$. However, we have a natural isomorphism $S^{\mathrm{op}} \to \mathrm{End}_S((S_S)^*)$ and hence an isomorphism $S^{\mathrm{op}} \to (S'')^{\mathrm{op}}$, and hence an isomorphism $S \to S''$, which, as one can easily check, is an isomorphism of quasihereditary algebras.

Remark It follows from Proposition A4.8(i) that the category $\mathcal{F}(\nabla)$, of S-modules filtered by $\nabla(\lambda)$'s, is equivalent to the category $\mathcal{F}(\Delta')$, of S'-modules filtered by $\Delta'(\lambda)$'s. In view of the isomorphism $S \to S''$ we also get that that the category $\mathcal{F}(\Delta)$ of S-modules filtered by $\Delta(\lambda)$'s is equivalent to the category $\mathcal{F}(\nabla')$ of S'-modules filtered by $\nabla'(\lambda)$'s.

We leave it to the reader to check the following.

(3) *Suppose we have a finite set Ω and for each $\alpha \in \Omega$ a partial tilting module T_α such that, writing $T_\alpha \cong \bigoplus_{\lambda \in \Lambda^+} T(\lambda)^{(d_\lambda)}$, we have that the set $\{\lambda \in \Lambda^+ \mid d_\lambda \neq 0\}$ has a unique maximal element, which we denote α^+, and $d_{\alpha^+} = 1$. Suppose also that the map $\Omega \to \Lambda^+$, $\alpha \mapsto \alpha^+$, is surjective. Put $T = \bigoplus_{\alpha \in \Omega} T_\omega$. Then the Ringel dual algebra $S' = \mathrm{End}_S(T)$ has theory of weights $\eta : \Omega \to S'$ given by $\eta(\alpha) = \xi'_\alpha$, where $\xi'_\alpha \in S'$ is projection onto T_α, for $\alpha \in \Omega$.*

We shall call a system of tilting modules $\{T_\alpha \mid \alpha \in \Omega\}$ satisfying the hypotheses of (3) a *weighted system* of tilting modules. Note we always have the trivial weighted system $\Omega = \Lambda^+$, $T_\lambda = T(\lambda)$, for $\lambda \in \Lambda^+$.

We conclude by showing that the two forms of truncation are interchanged by Ringel's dual construction. Recall that, for $X, Y \in \mathrm{mod}(S)$ with Y indecomposable, we are writing $(X \mid Y)$ for the multiplicity of Y as a direct summand of X.

Proposition A4.9 *Let $\theta : \Lambda \to \Lambda^+$ be a theory of weights for (S, Λ^+). Let T be a full tilting module. Let π be a cosaturated subset of Λ^+ and let Γ be a subset of Λ such that $\Gamma^+ = \pi$. Restriction $\mathrm{End}_S(T) \to \mathrm{End}_{\xi_\Gamma S \xi_\Gamma}(\xi_\Gamma T)$ induces an isomorphism $S'(\pi) \to \mathrm{End}_{\xi_\Gamma S \xi_\Gamma}(\xi_\Gamma T)^{\mathrm{op}}$.*

Proof Put $\xi = \xi_\Gamma$. We first check that the dimension of $S'(\pi)$ agrees with that of $\mathrm{End}_{S_\xi}(\xi T)$. From Proposition A2.2(ii) for $X \in \mathcal{F}(\Delta)$, $Y \in \mathcal{F}(\nabla)$ we have $\dim \mathrm{Hom}_S(X, Y) = \sum_{\lambda \in \Lambda^+}(X : \Delta(\lambda))(Y : \nabla(\lambda))$. Applying this to the S_ξ-module ξT we get

$$\dim \mathrm{End}_{S_\xi}(\xi T) = \sum_{\lambda \in \pi}(\xi T : \xi \Delta(\lambda))(\xi T : \xi \nabla(\lambda)).$$

We have $(T \mid T(\lambda)) = (S' \mid P'(\lambda)) = \dim L'(\lambda)$. Thus we have

$$\dim S'(\pi) = \dim S'/O^\pi(S') = \sum_{\lambda \in \pi} \dim L'(\lambda). \dim P'(\lambda)/O^\pi(P'(\lambda))$$

$$= \sum_{\mu, \lambda \in \pi} \dim L'(\lambda).(P'(\lambda) : \Delta'(\mu)). \dim \Delta'(\mu)$$

$$= \sum_{\lambda, \mu \in \pi} \dim L'(\lambda).(T(\lambda) : \nabla(\mu)). \dim \mathrm{Hom}_S(T, \nabla(\mu))$$

$$= \sum_{\lambda, \mu \in \pi} (T \mid T(\lambda)).(T(\lambda) : \nabla(\mu)).(T : \Delta(\mu))$$

$$= \sum_{\mu \in \pi}(T : \nabla(\mu))(T : \Delta(\mu)) = \sum_{\mu \in \pi}(\xi T : \xi \nabla(\mu))(\xi T : \xi \Delta(\mu))$$

$$= \dim \mathrm{End}_{S_\xi}(\xi T).$$

Thus the dimensions agree. Moreover, the restriction map $\Phi : \mathrm{End}_S(T) \to \mathrm{End}_{S_\xi}(\xi T)$ is surjective, by Lemma A3.12. Let $T = T_1 \oplus \cdots \oplus T_m$ be a decomposition of T with indecomposable (non-zero) summands. We regard $\{T_i \mid 1 \leq i \leq m\}$ as a weighted system, in the trivial way, i.e. we put $\Omega = [1, m]$ so that, for $i \in [1, m]$, we have $i^+ = \lambda$ where $\lambda \in \Lambda^+$ is such that $T_i \cong T(\lambda)$. We let $\xi'_i \in \mathrm{End}_S(T)$ be projection onto T_i, for $i \in [1, m]$. Thus by (3) and Lemma A3.9, the ideal $O^\pi(S')$ is generated by the ξ'_i such that T_i is not isomorphic to $T(\lambda)$, for any $\lambda \in \pi$. Let $i \in [1, m]$. If $\Phi(\xi'_i) \neq 0$ then ξ'_i is not zero on ξT and hence $\xi T_i \neq 0$. This gives $\xi T_{i^+} \neq 0$ and hence $i^+ \in \pi$. Thus Φ is zero on the generators of $O^\pi(S')$ described above. But now, $S'(\pi) = S'/O^\pi(S')$, so that Φ induces a surjective map $\bar{\Phi} : S'(\pi) \to \mathrm{End}_{S_\xi}(\xi T)^{\mathrm{op}}$ and by dimensions this is an isomorphism.

A5   The definitions of quasihereditary algebra and high weight category first appeared in the paper by Cline, Parshall and Scott, [11]. These ideas provide a common framework for discussing representation theory in various situations, including the category $\mathcal{O}$ of Bernstein, Gel'fand and Gel'fand and the rational representation theory of reductive algebraic groups in positive characteristic. In the category $\mathcal{O}$ the notion of a Δ-filtration corresponds to a filtration by Verma modules and in the category of rational representations the notion of a ∇-filtration corresponds to that of a good filtration. We should also mention Jantzen's treatment of the basic properties of rational modules with a Weyl filtration (corresponding to a Δ-filtration in the general theory) in [59]. Many of the results and arguments were around in these contexts prior to [11] and much of the treatment of the fundamental properties that we give here is based on our own work on good filtrations in the algebraic group context. In particular we make consistent use of the O_π functors, introduced in [26].

Our notation and terminology is chosen to emphasize the analogy with weight theory for Lie algebras and algebraic groups. Thus we write Λ^+ for our partially ordered set, calling the elements of Λ^+ the dominant weights, and write Λ for the set of weights (as defined in A3), adopting the notation of [51] from the representation theory of the general linear group. Moreover we call a set of dominant weights which is downward closed under the partial order a saturated set, as in Lie theory, [54; 13.4], rather than an ideal. (This also avoids possible confusion with the algebraic structure of the quasihereditary algebra S.)

The algebra S_ξ is discussed in [3] and [51; Chapter 6]. We now make some remarks on parts of Proposition A2.2. The vanishing of

$$\mathrm{Ext}^i_S(\Delta(\lambda), \nabla(\mu)) = 0$$

for all $i > 0$, of part (ii), was proved in [11] (see the proof of [11; Theorem 3.11]). In the context of algebraic groups, it was proved by Cline, Parshall,

Scott and van der Kallen, [13; (3.3) Corollary], as a corollary of Kempf's vanishing theorem. From our perspective, [13; (3.3) Corollary] is seen as the origin and kernel of the theory of quasihereditary algebras. Part (iii) in the context of rational modules is the criterion, due to the author, for a module to have a good filtration appearing in [26; Corollary 1.3], from which the proof is taken. Part (iv) is often known as Brauer–Humphreys reciprocity because of the obvious analogy with the well known Brauer reciprocity for finite groups, [7; p. 257] and Humphreys reciprocity for Chevalley Lie algebras, [53; 4.4,4.5], though neither of these can be derived as a special case of Proposition A2.2(iv). However, the formula $((P(\lambda) : \Delta(\mu)) = [\nabla(\mu) : L(\lambda)]$ is the well known Bernstein–Gel'fand–Gel'fand reciprocity in the category $\mathcal{O}$, see e.g. [60; 2.24,(1)]. The formula $(I(\lambda) : \nabla(\mu)) = [\Delta(\mu) : L(\lambda)]$ in the category of rational modules was proved in [26; Theorem 2.6]. It was proved for "generalized Schur algebras" (which include the ordinary Schur algebras $S(n, r)$, see [32; (1.3)]) in [31; (2.2h)]. This proved in particular that $\mathrm{mod}(S(n, r))$ has the defining properties of a highest weight category (though somewhat before the phrase had been coined). For algebras Morita equivalent to the generalized Schur algebras see [10].

Proposition A2.3 is proved for generalized Schur algebras in [31; (2.2e)], from which we have taken the proof given here. For Lemma A3.1 (and its proof) in the context of rational representations see [29; (12.1.6)] (also [26; Remark (2)]). For Proposition A3.2 (in the rational module context) see [31; (2.1b),(2.1c)] and Proposition A3.3 (in the context of rational modules and generalized Schur algebras) see [31; (2.2d)]. The equivalence of the notions of high weight category and quasihereditary algebra is shown in [11; Theorem 3.6] (by arguments different from those given here). Lemma A3.9 is taken from [33; (3.3)]. A3 Proposition A3.11 and Lemma A3.12, due to Erdmann, are taken from [45; 1.6] and [45; 1.7] (in turn modelled on the algebraic group case [33; 1.5]). This covers the case when ξ is a sum of pairwise inequivalent idempotents: the notion of theory of weights is introduced in A3 to make a slight generalization to a context directly applicable to Schur algebras.

Section A4 is largely concerned with the theory of tilting modules associated with a quasihereditary algebra, as introduced by Ringel, [72]. The corresponding notion was introduced in the category $\mathcal{O}$ by Collingwood and Irving, [14], and in the rational module context by the author, [30], where a set of tilting modules was called a "special resolving system" (though it should be pointed out that we had at that time existence only for GL_n in general and under certain characteristic assumptions for other reductive groups, and we had no uniqueness statement). Most of A4 is based on Ringel's paper, though we have expressed the arguments so as to be independent of earlier work of Happel and of Auslander and Reiten. For Lemma A4.3 and Proposition A4.4, see [30; Section 1, Theorem and Lemma 2]. For Lemma A4.5 in the basic context see [45; 1.7], and see [33; 1.5] for the algebraic group setting.

Finally, we mention the excellent survey article by Dlab and Ringel, [**22**], which has similar aims to those of our appendix. The general set-up we have adopted is similar to theirs. In particular, for A1(6) see [**22**; Lemma 1.2(c) and Lemma 1.3], for Proposition A2.2(i) see [**22**; Lemma 1.3], for Proposition A2.2(ii),(iii) see [**22**; Theorem 1 and Lemma 2.4], and for Proposition A2.2(iv) see [**22**; Lemma 2.4 and Lemma 2.5].

References

1. K. Akin, "Extensions of symmetric tensors by alternating tensors", *J. Algebra* **121** (1989), 358–363

2. G. E. Andrews, "The Theory of Partitions", *Encyclopedia of Math. Appl.*, Vol. 2, Addison–Wesley, 1976

3. M. Auslander, "Representations of Artin algebras I", *Comm. Algebra* **1**, (1974), 177–268

4. D. Benson, *Representations and Cohomology I: Basic representation theory of finite groups and associative algebras*, Cambridge Studies in Advanced Mathematics **30**, Cambridge University Press, Cambridge 1991

5. A. Borel, "Properties and linear representations of Chevalley groups", in A. Borel (ed.) *Seminar on Algebraic Groups and Related Finite Groups*, Lecture Notes in Mathematics **131**, pp. 1–55, Springer-Verlag, Berlin/Heidelberg/ New York 1970

6. A. Borel, *Linear Algebraic Groups*, 2nd edn, Graduate Texts in Mathematics **126**, Springer, New York/Berlin/Heidelberg/London/Paris/Tokyo/Hong Kong/Barcelona, 1991

7. R. Brauer, "On modular and p-adic representations of algebras", *Proc. Nat. Acad. Sci. U.S.A.*, **25** (1939), 252–258

8. R. W. Carter and G. Lusztig, "On the modular representations of the general linear and symmetric groups", *Math. Zeit.* **136** (1974), 193–242

9. G. Cliff, "A tensor product theorem for quantum linear groups at even roots of unity", *J. Alg.* **165** (1994), 566–575

10. E. Cline, B. Parshall and L. L. Scott, "Algebraic stratification in representation categories", *J. Algebra* **117** (1988), 504–521

11. E. Cline, B. Parshall and L. L. Scott, "Finite dimensional algebras and highest weight categories", *J. reine angew. Math.* **391** (1988), 85–99

12. E. Cline, B. Parshall and L. L. Scott, "Stratifying endomorphism algebras", preprint

13. E. Cline, B. Parshall, L. L. Scott and W. van der Kallen, "Rational and

generic cohomology", *Invent. Math.* **39** (1977), 143–163

14. D. H. Collingwood and R. Irving, "A decomposition theorem for certain self dual modules in the category O", *Duke Math. J.* **58** (1989), 89–102

15. A. G. Cox, "On some applications of infinitesimal methods to quantum groups and related algebras", Ph.D. Thesis, University of London, 1997

16. C. W. Curtis and I. Reiner, *Representation Theory of Finite Groups and Associative Algebras*, Wiley Interscience, New York 1962

17. C. De Concini, D. Eisenbud and C. Procesi, "Young diagrams and determinantal varieties", *Invent. Math.* **56** (1980),129–165

18. R. Dipper and S. Donkin, "Quantum GL_n", *Proc. Lond. Math. Soc.* (3) **63** (1991), 165–211

19. R. Dipper and J. Du, "Trivial and alternating source modules of Hecke algebras of type A", *Proc. Lond. Math. Soc.* **66** (1993), 479–506

20. R. Dipper and G. D. James, "Representations of the Hecke algebras of general linear groups", *Proc. Lond. Math. Soc.* (3) **52** (1986), 20–52

21. R. Dipper and G. D. James, "The q-Schur algebra", *Proc. Lond. Math. Soc.* (3), **59** (1989), 23–50

22. V. Dlab and C. M. Ringel, "The module theoretic approach to quasi-hereditary algebras", in H. Tachikawa and S. Brenner (eds), *Representations of Algebras and Related Topics*, pp. 200–224, London Math. Soc. Lecture Note Series **168**, Cambridge University Press 1992

23. S. Donkin, "Problems in the representation theory of algebraic groups", Ph.D. Thesis, University of Warwick 1977

24. S. Donkin, "Hopf complements and injective comodules", *Proc. Lond. Math. Soc.*(3) **40** (1980), 298–319

25. S. Donkin, "On a question of Verma", *J. Lond. Math. Soc.* (2) **21** (1980), 445–455

26. S. Donkin, "A filtration for rational modules", *Math. Zeit.* **177** (1981), 1–8

27. S. Donkin, "A note on decomposition numbers for reductive algebraic groups", *J. Algebra* **80** (1983), 226–234

28. S. Donkin, "A note on decomposition numbers of general linear groups and symmetric groups", *Math. Proc. Camb. Phil. Soc.* **97** (1985), 473–488

29. S. Donkin, *Rational Representations of Algebraic Groups : Tensor Products and Filtrations*, Lecture Notes in Mathematics **1140**, Springer-Verlag, Berlin/Heidelberg/New York 1985

30. S. Donkin, "Finite resolutions of modules for reductive algebraic groups", *J. Algebra* **101**, (1986), 473–488

31. S. Donkin, "On Schur algebras and related algebras I", *J. Algebra* **104** (1986), 310–328

32. S. Donkin, "On Schur algebras and related algebras II", *J. Algebra* **111** (1987), 354–364

33. S. Donkin, "On tilting modules for algebraic groups", *Math. Zeit.* **212** (1993), 39–60

34. S. Donkin, "On Schur algebras and related algebras III: Integral representations", *Math. Proc. Camb. Phil. Soc.* **116** (1994), 37–55

35. S. Donkin, "On Schur algebras and related algebras IV. The blocks of the Schur algebras", *J. Algebra* **168** (1994), 400–429

36. S. Donkin, "Standard homological properties for quantum GL_n", *J. Algebra* **181** (1996), 235–266

37. S. Donkin, "On projective modules for algebraic groups", *J. Lond. Math. Soc.* **54** (1996), 75–88

38. S. Donkin, "The restriction of the regular module for a quantum group", in G.I. Lehrer (ed.), *Algebraic Groups and Lie Groups – A volume of papers in honour of the late R.W. Richardson*, Australian Math. Soc. Lecture Series 9, pp. 183–188, Cambridge University Press 1997

39. S. Donkin and K. Erdmann, "Tilting modules, symmetric functions and the module structure of the free Lie algebra", *J. Algebra* **203** (1998), 69–90

40. S. Donkin and I. Reiten, "On Schur algebras and related algebras V: Some quasi-hereditary algebras of finite type", *J. Pure App. Algebra* **97** (1994), 117–134

41. L. Dornhoff, *Group Representation Theory, Vol. B*, Marcel Dekker, Inc., New York 1972

42. S. R. Doty and D .K. Nakano, "Semisimple Schur algebras", preprint

43. P. Doubillet, G.-C. Rota and J. Stein, "Foundations of Combinatorics IX: Combinatorial methods in Invariant Theory", *Studies in Appl. Math.* **53** (1974), 185–216

44. J. Du, B. Parshall and L. Scott, "Quantum Weyl reciprocity and tilting modules", preprint 1997

45. K. Erdmann, "Symmetric groups and quasi-hereditary algebras", in V. Dlab and L.L. Scott (eds), *Finite Dimensional Algebras and Related Topics*, pp. 123–161, Kluwer, Dordrecht/Boston/London 1994

46. K. Erdmann, "Ext^1 for Weyl modules of $SL_2(K)$", *Math. Zeit.* **218** (1995), 447–459

47. K. Erdmann, "Decomposition numbers for symmetric groups and composition factors of Weyl modules", *J. Algebra* **180** (1996), 316–320

48. K. Erdmann, "On the kernel of the decomposition map for symmetric groups", preprint 1996

49. E. Friedlander, "A canonical filtration for certain rational modules", *Math. Zeit.* **188** (1984/5), 433–438

50. J. A. Green, "Locally finite representations", *J. Algebra* **41** (1976), 131–171

51. J. A. Green, *Polynomial Representations of GL_n*, Lecture Notes in Mathematics **830**, Springer, Berlin/Heidelberg/New York 1980

52. J. A. Green, "Schur algebras and general linear groups", in C. M. Campbell and E. F. Robertson (eds), *Groups, St. Andrews 1989 Vol. 1* London Mathematical Society Lecture Notes Series **159**, pp. 155–210, Cambridge University Press, Cambridge 1991

53. J. E. Humphreys, "Modular representations of classical Lie algebras and semisimple groups", *J. Algebra* **19** (1971), 51–79

54. J. E. Humphreys, *Lie Algebras and Representation Theory*, Graduate Text in Mathematics **9**, Springer 1972, Berlin/Heidelberg/New York

55. G. D. James, "On the decomposition matrices of the symmetric groups, III", *J. Algebra* **71** (1981), 115–122

56. G. D. James, *Representations of General Linear Groups, London Mathematical Society Lecture Note Series* **94**, Cambridge University Press 1984

57. G. D. James, "The decomposition matrices of $GL_n(q)$ for $n \le 10$", *Proc. Lond. Math. Soc.* **60** (1990), 225–265

58. J. C. Jantzen, "Über das Dekompositionsverhalten gewisser modularer Darstellungen halbeinfacher Gruppen und ihrer Lie-Algebren", *J. Algebra* **49** (1977), 441–469

59. J. C. Jantzen, "Über Darstellungen höherer Frobenius-Kerne halbeinfacher algebraischer Gruppen", *Math. Zeit.* **164** (1979), 271–292

60. J. C. Jantzen, *Moduln mit einem höchsten Gewicht*, Lecture Notes in Mathematics **750**, Springer, Berlin/Heidelberg/New York 1979

61. J. C. Jantzen, *Representations of Algebraic Groups*, Pure and Applied Mathematics **131**, Academic Press 1987

62. D. Krob and J.Y. Thibon, "Noncommutative symmetric functions IV, Quantum linear groups and Hecke algebras at $q = 0$", preprint 1996

63. I. G. Macdonald, *Symmetric Functions and Hall Polynomials*, Oxford University Press, 1979

64. S. MacLane, *Homology*, Springer, 1963

65. S. Martin, *Schur Algebras and Representation Theory*, Cambridge Tracts in Mathematics **112**, Cambridge University Press 1993

66. S. Martin, "Filtrations for q-Young modules", *Math. Proc. Camb. Phil. Soc.* **115** (1994), 397–406

67. D. G. Mead, "Determinantal ideals, identities and the Wronskian", *Pacific J. Math.* **42** (1972), 165–175

68. W. D. Nichols and M. B. Zoeller, "A Hopf algebra freeness theorem", *Amer. J. Math.* **111** (1989), 381–385

69. P. N. Norton, "0-Hecke algebras", *J. Australian Math. Soc. Ser. A.* **27** (1979), 337–357

70. J. Nuttall, "Modular symmetric functions and Doty's conjecture", Ph. D. Thesis, University of London 1997

71. B. Parshall and J. Wang, *Quantum linear groups* , Memoirs of the A.M.S, 439, (1991)

72. C. M. Ringel, "The category of modules with good filtrations over a quasi-hereditary algebra has almost split sequences", *Math. Zeit.* **208** (1991),209–225

73. L. Solomon, "A decomposition of the group algebra of a finite Coxeter group", *J. Algebra* **9** (1968), 220–239

74. R. Steinberg, *Conjugacy Classes in Algebraic Groups*, Lecture Notes in Mathematics **366**, Springer, Berlin/Heidelberg/New York 1970

75. L. Thams, "The subcomodule structure of quantum symmetric powers", *Bull. Australian Math. Soc.* **50** (1994), 29–39

76. B. Totaro, "Projective resolutions of representations of GL(n)", *J. reine ang. Math.* **482** (1997), 1–13

Index of notation

	basis of $A(n,r)$, 0.14
ξ_α	ξ_{ii}, where $i \in I(n,r)$ has content α, 0.14
$\leq$	the natural (dominance) partial order, 0.15
$X^+(n)$	the set of dominant weights, 0.15
$L(\lambda)$	the simple GL_n-module of high weight λ, 0.15
Ind_H^G	the induction functor from H-modules to G-modules, 0.16, 0.20
$B(n)$	the (Borel) subgroup of $G(n)$ consisting of lower triangular matrices, 0.16
k_λ	the 1-dimensional $B(n)$-module labelled by $\lambda \in X(n)$, 0.16, and 1-dimensional $B_q(n)$-module labelled by $\lambda \in X(n)$, 0.21
$\nabla(\lambda)$	$\mathrm{Ind}_{B(n)}^{G(n)} k_\lambda$, 0.16
$\bigwedge^\alpha E$	$\bigwedge^{\alpha_1} E \otimes \bigwedge^{\alpha_2} E \otimes \cdots$ (for $\alpha = (\alpha_1, \alpha_2, \ldots)$)), 0.16
$\Lambda^+(n,r)$	$\Lambda(n,r) \cap X^+(n)$, 0.16
$[\nabla(\lambda) : L(\mu)]$	the multiplicity of $L(\mu)$ as a composition factor of $\nabla(\lambda)$, 0.17
$\Lambda^+(n,r)_{\mathrm{col}}$	set of column regular partitions, 0.18
$\Lambda^+(n,r)_{\mathrm{row}}$	set of row regular partitions, 0.18
ω	$(1,1,\ldots,1) \in \Lambda(n,r)$, 0.18
e	ξ_ω, 0.18
f	the Schur functor, 0.18 (and 2.1)
$\mathrm{Hec}(r)$	the Hecke algebra of $\mathrm{Sym}(r)$, 0.19 (and $eS(n,r)e$ in 2.1)
$l(w)$	the length of $w \in \mathrm{Sym}(r)$, 0.19
T_w	basis element of the Hecke algebra, 0.19
$A_q(n)$	a q-deformation of $A(n)$, 0.20
$A_q(n,r)$	the degree r component of $A_q(n)$, 0.20
$S_q(n,r)$	the Schur q-algebra, i.e. the dual algebra of the coalgebra $A_q(n,r)$, 0.20
d_q	the quantum determinant, 0.20
$G_q(n)$	the quantum general linear group, 0.20
$B_q(n)$	the (Borel) subgroup of $G_q(n)$ with defining ideal generated by c_{ij}, $i < j$, 0.21
$\nabla_q(\lambda)$	$\mathrm{Ind}_{B_q(n)}^{G_q(n)} k_\lambda$, 0.21
$L_q(\lambda)$	the socle of $\nabla_q(\lambda)$, 0.21

Chapter 1

$\mathrm{cf}(E)$	the coefficient space of a comodule E, 1.1
E^*, V^*	dual comodules, 1.1
$\mathrm{ad}(\phi)$	the adjoint of the linear map ϕ, 1.1
$[a,b]$	$\{a, a+1, \ldots, b\}$, 1.2
$\mathrm{Sym}(X)$	the group of permutations of a set X, 1.2

Chapter 2

$S(E), S(V)$	the (quantum) symmetric algebras over E and V, 2.1
$S^r(E), S^r(V)$	the rth (quantum) symmetric powers of E and V, 2.1
$\bar{e}_i$	the image of e_i under the natural map $E^{\otimes r} \to S^r E$, 2.1
$\bar{v}_i$	the image of v_i under the natural map $V^{\otimes r} \to S^r V$, 2.1
$S^\alpha E$	$S^{\alpha_1} E \otimes S^{\alpha_2} E \otimes \cdots$, for $\alpha = (\alpha_1, \alpha_2, \ldots)$, 2.1
$S^\alpha V$	$S^{\alpha_1} V \otimes S^{\alpha_2} V \otimes \cdots$, for $\alpha = (\alpha_1, \alpha_2, \ldots)$, 2.1
$i \in \alpha$	$i \in I(n,r)$ has content $\alpha \in \Lambda(n,r)$, 2.1
$i \sim j$	$i, j \in I(n,r)$ have the same content, 2.1
i^α	$(1, \ldots, 1, 2 \ldots, 2, 3, \ldots)$, for $\alpha \in \Lambda(n,r)$, 2.1
j^α	$(\ldots, 3, 2, \ldots, 2, 1, \ldots, 1)$, for $\alpha \in \Lambda(n,r)$, 2.1
$I^+(\alpha)$	the set of $(i_1, \ldots, i_r) \in I(n,r)$ such that $i_1 \le \cdots \le i_{\alpha_1}$, $i_{\alpha_1+1} \le \cdots \le i_{\alpha_1+\alpha_2}$, $\ldots$, for $\alpha = (\alpha_1, \ldots, \alpha_r) \in \Lambda(n,r)$, 2.1
$I^-(\alpha)$	the set of $(i_1, \ldots, i_r) \in I(n,r)$ such that $i_1 \ge \cdots \ge i_{\alpha_1}$, $i_{\alpha_1+1} \ge \cdots \ge i_{\alpha_1+\alpha_2}, \ldots$, for $\alpha = (\alpha_1, \ldots, \alpha_r) \in \Lambda(n,r)$, 2.1
$A(n,r)^\alpha$	the k-span of the elements c_{ij} with $j \in \alpha$, 2.1
$^\alpha A(n,r)$	the k-span of the elements c_{ij} with $i \in \alpha$, 2.1
$^\alpha A(n,r)^\beta$	the k-span of the elements c_{ij} with $i \in \alpha, j \in \beta$, 2.1
ξ_{ij}	an element of a certain basis of $S(n,r)$, 2.1
ξ'_{ij}	an element of a certain basis of $S(n,r)$, 2.1
ξ_α	$\xi_{i^\alpha i^\alpha}$, for $\alpha \in \Lambda(n,r)$, 2.1
ω	$(1, \ldots, 1) \in \Lambda(n,r)$, 2.1
u	$(1, 2, \ldots, r)$, 2.1
v	$(r, \ldots, 2, 1)$, 2.1
e	ξ_ω, 2.1
b_σ	$\xi_{u,u\sigma}$, for $\sigma \in \mathrm{Sym}(r)$, 2.1
b'_σ	$\xi'_{v\sigma,v}$, for $\sigma \in \mathrm{Sym}(r)$, 2.1
T_σ	$b_{\sigma^{-1}}$, for $\sigma \in \mathrm{Sym}(r)$, 2.1
$\mathrm{Hec}(r), H(r), H$	$eS(n,r)e$, 2.1 (and the abstract Hecke algebra in 0.19)
f	the Schur functor, 2.1 (and 0.18)
$X \in \mathcal{F}(\nabla)$	$X \in \mathrm{mod}(G)$ admits a good filtration (∇-filtration), 2.1
w_0	the longest element of $\mathrm{Sym}(n)$, 2.1
λ^*	$-w_0\lambda$, for $\lambda \in X(n)$, 2.1
$\Delta(\lambda)$	$\nabla(\lambda^*)^*$, for $\lambda \in X^+(n)$, 2.1
ν	the trivial representation of the Hecke algebra, $\nu(T_w) = q^{l(w)}$, 2.1
ε	the sign representation of the Hecke algebra, $\varepsilon(T_w) = \mathrm{sgn}(w)$, 2.1
k	1-dimensional trivial module for a Hecke algebra, 2.1
k_s	1-dimensional sign module for a Hecke algebra, 2.1
$J(\alpha)$	the subset of $[1, r]$ defined by $\alpha \in \Lambda(n,r)$, 2.1
$\mathrm{Sym}(\alpha)$	the subgroup of $\mathrm{Sym}(r)$ defined by $\alpha \in \Lambda(n,r)$, 2.1
$H(\alpha)$	the subalgebra of $H(r)$ defined by $\alpha \in \Lambda(n,r)$, 2.1

$x(\alpha)$	$\sum_{w \in \mathrm{Sym}(\alpha)} T_w$, 2.1
$y(\alpha)$	$\sum_{w \in \mathrm{Sym}(\alpha)} (-q)^{N-l(w)} T_w$, for $\alpha \in \Lambda(n,r)$,
	where $N = \binom{n}{2}$, 2.1
ϵ_i	$(0, \ldots, 0, 1, 0, \ldots, 0)$, 2.2
$X_0^+(n)$	the set of $\lambda = (\lambda_1, \ldots, \lambda_m, 0, \ldots, 0) \in X(n)$ with
	$\lambda_1, \ldots, \lambda_m \neq 0$ (for some m), 2.2
$\lambda \subseteq \mu$	$J(\lambda) \subseteq J(\mu)$, 2.2

Chapter 3

G_1	infinitesimal subgroup of G, 3.1
B_1, B_1^+, T_1	infinitesimal subgroups $B \cap G_1$, $B^+ \cap G_1$, $T \cap G_1$
	of B, B^+, T, 3.1
$\hat{G}_1$	Jantzen subgroup of G, 3.1
$\hat{B}_1, \hat{B}_1^+$	Jantzen subgroups $\hat{G}_1 \cap B$, $\hat{G}_1 \cap B_1^+$ of B, B^+, 3.1
X_1	the set of $\lambda = (\lambda_1, \ldots, \lambda_n) \in X$ such that
	$0 \leq \lambda_1 - \lambda_2, \ldots, \lambda_{n-1} - \lambda_n, \lambda_n < l$, 3.1
$\|H\|$	the order of a finite quantum group H, 3.1
$[H : J]$	the index $\|H\|/\|J\|$ of a finite quantum subgroup in a
	finite quantum group H, 3.1
$I_1(\lambda)$	the injective hull of the B_1^+-module k_λ, 3.1
$\hat{I}_1(\lambda)$	the injective hull of the $\hat{B}_1^+$-module k_λ, 3.1
$\chi(\lambda)$	Weyl character and Schur symmetric function, 3.1
$\nabla_1(\lambda)$	$\mathrm{Ind}_{B_1}^{G_1} k_\lambda$, 3.1
$\hat{\nabla}_1(\lambda)$	$\mathrm{Ind}_{\hat{B}_1}^{\hat{G}_1} k_\lambda$, 3.1
$L_1(\lambda)$	the (simple) socle of the G_1-module $\nabla_1(\lambda)$, 3.1
$\hat{L}_1(\lambda)$	the (simple) socle of the $\hat{G}_1$-module $\hat{\nabla}_1(\lambda)$, 3.1
$w \cdot \lambda$	the "dot" action, i.e. $w \cdot \lambda = w(\lambda + \rho) - \rho$
	for $w \in \mathrm{Sym}(n), \lambda \in X$, 3.1
$\bar{G}$	ordinary GL_n, regarded as a k-group, 3.2
$F : G \to \bar{G}$	the (quantum) Frobenius morphism, 3.2
$Q_1(\lambda)$	the injective envelope of the G_1-module $L_1(\lambda)$,
	for $\lambda \in X$, 3.2
$\hat{Q}_1(\lambda)$	the injective envelope of the $\hat{G}_1$-module $\hat{L}_1(\lambda)$,
	for $\lambda \in X$, 3.2
$V \in \mathcal{F}(\hat{\nabla}_1)$	V is a $\hat{G}_1$-module which admits a filtration with sections
	of the form $\hat{\nabla}_1(\lambda)$, $\lambda \in X$, 3.2
$T(\lambda)$	indecomposable tilting module with highest weight λ, 3.3
$t(\lambda)$	$(l-1)\delta + l\lambda$, where $\delta = (n-1, \ldots, 1, 0)$, 3.3
$\bar{F} : \bar{G} \to \bar{G}$	the ordinary Frobenius morphism, 3.4

Chapter 4

$\bigwedge(E \otimes V)$	exterior algebra on $E \otimes V$, 4.1
J	a certain antiautomorphism of $H(r)$ and $S(n,r)$, 4.1
U°	the contravariant dual of an $H(r)$-module or $S(n,r)$-module U, 4.1
G_Σ	a subgroup of the quantum general linear group defined by a set Σ of simple roots, 4.2
$L_\Sigma(\lambda)$	simple G_Σ-module of highest weight λ, 4.2
$A(\pi)$	a generalized Schur coalgebra, 4.2
$S(\pi)$	a generalized Schur algebra, 4.2
$e_1, e_2, \ldots$	elementary symmetric functions, 4.3
$h_1, h_2, \ldots$	complete symmetric functions, 4.3
s_λ	Schur symmetric function, 4.3
$i(\lambda)$	the image of λ under a certain bijection $\Lambda^+(n,r)_{\text{row}} \to \Lambda^+(n,r)_{\text{col}}$, 4.3
$Y(\lambda)$	the Young module labelled by λ, 4.4
$Y_s(\lambda)$	the signed Young module labelled by λ, 4.4
$\text{Sp}(\lambda)$	the Specht module labelled by λ, 4.4
$D(\lambda)$	irreducible $H(r)$-module labelled by λ, 4.4
$\#$	a certain antiautomorphism of $H(r)$, 4.4
$A(\nu)$	$k[M(\nu)]$, for ν a composition of n, 4.6
$A(\nu, \rho)$	ρth component of the graded algebra $A(\nu)$, 4.6
$S(\nu, \rho)$	algebra dual of the coalgebra $A(\nu, \rho)$, 4.6
$D^\lambda E$	divided powers module, 4.8
$\text{inj}(X)$	the injective dimension of a module X, 4.8
$\text{proj}(X)$	the projective dimension of a module X, 4.8
$\text{glob}(S)$	the global dimension of an algebra S, 4.8

Appendix

$L(\lambda)$	simple S-module labelled by $\lambda \in \Lambda^+$, A1
$P(\lambda)$	projective cover of $L(\lambda)$, A1
$I(\lambda)$	injective envelope of $L(\lambda)$, A1
$\text{mod}(S)$	category of finite dimensional left S-modules, A1
$[X : L(\lambda)]$	multiplicity of $L(\lambda)$ as a composition factor of $X \in \text{mod}(S)$, A1
π	a subset of Λ^+, A1
$O_\pi(V)$	the largest submodule of V belonging to π, A1
$O^\pi(V)$	the smallest submodule of V such that $V/O^\pi(V)$ belongs to π, A1

Subject Index

adjoint, 1.1
affine variety, 0.2
antipode, 0.7
antistandard tableau, 1.2

belongs, 0.13, A1
bialgebra, 0.7
bicomodule, 1.1
bideterminant, 1.3
brute calculation, 4.1

character, 0.12
coalgebra, 0.7
coefficient functions, 0.4
coefficient space, 0.4
column regular, 0.18
comorphism, 0.20
content, 0.14
contravariant dual, 4.1
coordinate algebra, 0.2
costandard module, A1

Δ-filtration, A1
decomposition matrix, 0.21
decomposition number, 0.21
defect set, A4
descent, Preface
dominance order, 0.15
dominant, 0.15

exterior algebra, 1.2

full tilting module, A4

Gaussian polynomial, 4.8
global dimension, 4.8, A2
good epimorphism, 2.1
good filtration, 2.1
good monomorphism, 2.1
Grothendieck vanishing, 0.21

Hecke algebra, 0.19
hereditary ideal, A3
high weight category, A2
homogeneous module, 0.13
Hopf algebra, 0.7

induction, 0.16
inflation, 0.13

Jantzen subgroups, 3.1

k-form, 0.8
k-group, 0.8
k-variety, 0.8
Kempf's vanishing theorem, 0.21
Koszul resolution, 4.8

(l,p)-expansion, 3.4
Levi subalgebra, 4.6
Levi subgroups, 4.2
linear algebraic group, 0.3

∇-filtration, A1

order (of a quantum group), 3.1

Printed in the United States
By Bookmasters